Global Developments in Literacy Research for Science Education

Kok-Sing Tang · Kristina Danielsson
Editors

Global Developments in Literacy Research for Science Education

 Springer

Editors
Kok-Sing Tang
Curtin University
Perth
Australia

Kristina Danielsson
Linnaeus University
Växjö
Sweden

ISBN 978-3-319-69196-1   ISBN 978-3-319-69197-8 (eBook)
https://doi.org/10.1007/978-3-319-69197-8

Library of Congress Control Number: 2018930334

© Springer International Publishing AG 2018
This work is subject to copyright. All rights are reserved by the Publisher, whether the whole or part of the material is concerned, specifically the rights of translation, reprinting, reuse of illustrations, recitation, broadcasting, reproduction on microfilms or in any other physical way, and transmission or information storage and retrieval, electronic adaptation, computer software, or by similar or dissimilar methodology now known or hereafter developed.
The use of general descriptive names, registered names, trademarks, service marks, etc. in this publication does not imply, even in the absence of a specific statement, that such names are exempt from the relevant protective laws and regulations and therefore free for general use.
The publisher, the authors and the editors are safe to assume that the advice and information in this book are believed to be true and accurate at the date of publication. Neither the publisher nor the authors or the editors give a warranty, express or implied, with respect to the material contained herein or for any errors or omissions that may have been made. The publisher remains neutral with regard to jurisdictional claims in published maps and institutional affiliations.

Printed on acid-free paper

This Springer imprint is published by Springer Nature
The registered company is Springer International Publishing AG
The registered company address is: Gewerbestrasse 11, 6330 Cham, Switzerland

Foreword

Everywhere I turn, I see calls for more attention to science learning or education. Whether focused on science, engineering, design thinking, computational thinking, or coding, the calls for attention to science and science literacy are relentless. In many ways, this volume answers that call, providing researchers with richly layered, systematically studied, and robustly analyzed collections of data about teaching and learning of science, concepts, practices, and skills. But this volume does more. This volume draws our attention to the call that we need to make, a call for the study of the linguistic and humanistic dimensions of science teaching, learning, and practice.

As I read this book, I found myself asking, "What is science education for the 21st century?" What should be our foci, our principles, our raison d'être as scholars who care about science teaching and learning? In the USA the overwhelming emphasis in science-focused conversations is on developing technical skill, whether the natural scientific and mathematical concepts necessary to carry out research and development; the componential knowledge and skill necessary to build tools and instruments for design, data collection, engineering, and development; or the literacy skill to develop, track, and communicate designs and claims. In this technocratic approach to science teaching and learning, literacy skill in particular is conceptualized as facility with words and, at times, with other representations.

Such technical language skills matter, to be sure, but too often the human dimension of science literacy teaching and learning is left unacknowledged and unaddressed. That is, how do science practitioners think about their work? What is the passion that drives them to engage tirelessly in a seemingly mundane task all for the sake of producing knowledge, designing a thing, or solving a problem? Why do they use language as they do? Why do words and ways with words and other representations matter so much in communicating one's findings? And how often do educators provide students with opportunities to experience and grasp those purposes and passions? In a recent observation of engineers in an optical sensors lab, a colleague watched an engineer cheering on particles as he set up an experiment (Giroux & Moje, in press). Although many in science education would decry the anthropomorphizing of inanimate objects, claiming that such thinking diminishes the "science," what one sees when actually studying members of science disciplines and professions in practice is a strongly personal and impassioned stance toward their work. They are in it because they love it. That passion produces powerful thinking; it produces the necessary commitment to carry out the same task repeatedly until one produces results; that passion leads to asking new questions and seeking new knowledge or innovations. That passion turns what some would refer to as jargon into absolutely necessary technical language, language that signals membership in a specialized community of practice; language that matters because the work matters.

Science educators need to learn to teach that passion, to help learners see in different ways, and to give access to the purposes, the goals, and the dreams of people who do this work. When I was carrying out my own dissertation work many years ago, a young woman called Heather told me that she liked chemistry class because it was easy. It all made sense. The answers were all available. She went on to say that although it was easy, it wasn't really interesting because it wasn't, in her words, "about life and death" (Moje, 1996). At the time she shared this observation with me, her class was in the midst of a unit on nuclear chemistry. This young woman did not see the role that knowledge of chemistry could play in matters of life or death. She was bright, able, and curious. She found the subject matter "easy." But she had no passion or commitment to studying in the field because she saw no value, no purpose, no meaning to the work. So many years later, the field of science education is not, I would argue, much further along. But that is where this volume enters the picture.

Kok-Sing Tang and Kristina Danielsson have crafted a wonderful volume, one filled with reports of careful, systematic, and rigorous research on science literacy teaching and learning that, even as they advance an interest in the technical language skills needed for science learning, also place meaning, purpose, and engagement at the heart of building that technical skill. What Tang, Danielsson, and colleagues recognize is that science language research—and, therefore, science language learning, are human activities. From studies of science language learning in intercultural contexts to the study of figurative language learning and the practice of teaching writing in the midst of real scientific inquiry, these chapters assume that to learn oral and written languages of science, learners must have meaningful inquiry, richly articulated purposes, and powerful opportunities to practice what they are learning. They have to know why they need to use language in particular ways. They have to know why one representation might be more meaningful or useful than another. And they have to know why others might not be convinced by their arguments unless they offer ample and systematically collected evidence in an accessible manner. To engage learners in deep science learning and the appropriate use of scientific language and representation, one must understand who learners are and help them see why science matters. Tang, Danielsson, and their colleagues have contributed to this effort with a volume replete with careful, learner-focused, science literacy research.

The why of science is at the heart of science language and literacy learning. This volume makes that clear while also offering critical information for how to teach and engage children and youth in the most powerful ways. In that respect, this volume is a must-read for all those interested in understanding and advancing science and science literacy as meaningful, purposeful, human activity.

School of Education, University of Michigan Elizabeth Birr Moje, Dean
Ann Arbor, Michigan, USA

References

Giroux, C., & Moje, E. B. (in press). Learning from the professions: examining how, why, and when engineers read and write. *Theory into Practice*.

Moje, E. B. (1996). "I teach students, not subjects": teacher-student relationships as contexts for secondary literacy. *Reading Research Quarterly, 31*, 172–195.

Acknowledgements

We would like to thank the following reviewers who have given their time and expertise to review early drafts of the chapters and provide useful comments and suggestions for improvement.

Reviewers:

1. John Airey
   Uppsala University, Stockholm University, & Linnaeus University
2. Edenia Maria Ribeiro do Amaral
   Federal Rural University of Pernambuco
3. Monica Axelsson
   Stockholm University
4. Ana Lucia Gomes Cavalcanti Neto
   Secretariat of Education, Escada, Pernambuco, Brazil
5. Sibel Erduran
   University of Oxford & National Taiwan Normal University
6. Mariona Espinet
   Universitat Autònoma de Barcelona
7. Brian Hand
   University of Iowa
8. Qiuping He
   The Hong Kong Polytechnic University
9. Caroline Ho
   Singapore Ministry of Education
10. Britt Jakobson
    Stockholm University
11. Fredrik Jeppsson
    Linköping University
12. Rebecca Jesson
    University of Auckland
13. Erik Knain
    University of Oslo
14. Victor Lim-Fei
    Singapore Ministry of Education

15. Angel M. Y. Lin
    The University of Hong Kong
16. Inger Lindberg
    Stockholm University
17. Yu Liu
    Sichuan International Studies University
18. Yuen Yi Lo
    The University of Hong Kong
19. Ragnhild Löfgren
    Linköping University
20. Eva Maagerø
    University College of Southeast Norway
21. Silvija Markic
    University of Education Ludwigsburg
22. Felicia Moore Mensah
    Columbia University
23. Eduardo Fleury Mortimer
    Federal University of Minas Gerais
24. Audrey Msimanga
    University of the Witwatersrand
25. Marianne Ødegaard
    University of Oslo
26. Vaughan Prain
    Deakin University
27. Natasha Anne Rappa
    Murdoch University
28. Jon Smidt
    Norwegian University of Science and Technology
29. Aik Ling Tan
    Nanyang Technological University
30. Chew Lee Teo
    Singapore Ministry of Education
31. Russell Tytler
    Deakin University
32. Len Unsworth
    Australian Catholic University
33. Hwei Ming Wong
    Nanyang Technological University
34. Jason S. Wu
    Columbia University

In addition, the compilation and editing of this volume required the support and efforts of a number of people. We are thankful to Gde Buana Sandila Putra for his editorial assistance in facilitating the compilation of the volume. We also thank Natasha Anne Rappa for her assistance in proofreading the manuscript. Last but not least, we like to thank the anonymous reviewers engaged by Springer for their constructive feedback in improving and ensuring the quality of this volume.

Contents

1   The Expanding Development of Literacy Research in Science Education Around the World 1
Kok-Sing Tang and Kristina Danielsson

Part 1 National Curriculum and Initiatives 13

2   The Implementation of Scientific Literacy as Basic Skills in Norway After the School Reform of 2006. 15
Erik Knain and Marianne Ødegaard

3   But I'm Not an English Teacher!: Disciplinary Literacy in Australian Science Classrooms 29
Chris Davison and Sue Ollerhead

4   Meeting Disciplinary Literacy Demands in Content Learning: The Singapore Perspective 45
Caroline Ho, Natasha Anne Rappa and Kok-Sing Tang

Part 2 Content and Language Integrated Learning (CLIL) in Science 61

5   Learning Language and Intercultural Understanding in Science Classes in Germany. 63
Silvija Markic

6   Supporting English-as-a-Foreign-Language (EFL) Learners' Science Literacy Development in CLIL: A Genre-Based Approach   79
Yuen Yi Lo, Angel M. Y. Lin and Tracy C. L. Cheung

7   Language, Literacy and Science Learning for English Language Learners: Teacher Meta Talk Vignettes from a South African Science Classroom. 97
Audrey Msimanga and Sibel Erduran

8 The Content-Language Tension for English Language Learners in
   Two Secondary Science Classrooms 113
   Jason S. Wu, Felicia Moore Mensah and Kok-Sing Tang

Part 3 Science Classroom Literacy Practices 131

9 A Case Study of Literacy Teaching in Six Middle- and High-School
   Science Classes in New Zealand 133
   Aaron Wilson and Rebecca Jesson

10 Analyzing Discursive Interactions in Science Classrooms to
   Characterize Teaching Strategies Adopted by Teachers in Lessons
   on Environmental Themes 149
   Ana Lucia Gomes Cavalcanti Neto, Edenia Maria Ribeiro do Amaral
   and Eduardo Fleury Mortimer

11 Measuring Time. Multilingual Elementary School Students'
   Meaning-Making in Physics 167
   Britt Jakobson, Kristina Danielsson, Monica Axelsson and
   Jenny Uddling

12 Meaning-Making in a Secondary Science Classroom: A Systemic
   Functional Multimodal Discourse Analysis 183
   Qiuping He and Gail Forey

Part 4 Science Disciplinary Literacy Challenges 203

13 Literacy Challenges in Chemistry: A Multimodal Analysis of
   Symbolic Formulas 205
   Yu Liu

14 Gains and Losses: Metaphors in Chemistry Classrooms 219
   Kristina Danielsson, Ragnhild Löfgren and Alma Jahic Pettersson

15 Image Design for Enhancing Science Learning:
   Helping Students Build Taxonomic Meanings
   with Salient Tree Structure Images 237
   Yun-Ping Ge, Len Unsworth, Kuo-Hua Wang and Huey-Por Chang

Part 5 Disciplinary Literacy and Science Inquiry 259

16 Inquiry-Based Science and Literacy: Improving a Teaching Model
   Through Practice-Based Classroom Research 261
   Marianne Ødegaard

17 Infusing Literacy into an Inquiry Instructional Model to Support
   Students' Construction of Scientific Explanations 281
   Kok-Sing Tang and Gde Buana Sandila Putra

| 18 | Representation Construction as a Core Science Disciplinary Literacy 301
Russell Tytler, Vaughan Prain and Peter Hubber |

Part 6 Science Teacher Development 319

| 19 | Science and Language Experience Narratives of Pre-Service Primary Teachers Learning to Teach Science in Multilingual Contexts 321
Mariona Espinet, Laura Valdés-Sanchez and Maria Isabel Hernández |

| 20 | Examining Teachers' Shifting Epistemic Orientations in Improving Students' Scientific Literacy Through Adoption of the Science Writing Heuristic Approach 339
Brian Hand, Soonhye Park and Jee Kyung Suh |

| 21 | Developing Students' Disciplinary Literacy? The Case of University Physics 357
John Airey and Johanna Larsson |

Part 7 Commentary 377

| 22 | Commentary on the Expanding Development of Literacy Research in Science Education 379
Larry D. Yore |

Index 399

List of Contributors

John Airey Department of Physics and Astronomy, Uppsala University, Uppsala, Sweden
Department of Mathematics and Science Education, Stockholm University, Stockholm, Sweden
Department of Languages, Linnaeus University, Kalmar/Växjö, Sweden (john.airey @physics.uu.se)

Edenia Maria Ribeiro do Amaral Federal Rural University of Pernambuco, Recife, Brazil (edeniamramaral@gmail.com)

Monica Axelsson Department of Language Education, Stockholm University, Stockholm, Sweden (monica.axelsson@isd.su.se)

Ana Lucia Gomes Cavalcanti Neto Secretariat of Education, Escada, Pernambuco, Brazil (analuneto@gmail.com)

Huey-Por Chang Open University of Kaohsiung, Hualien, Taiwan (president @ouk.edu.tw)

Tracy C. L. Cheung Faculty of Education, The University of Hong Kong, Pokfulam, Hong Kong (tracyccl@hku.hk)

Kristina Danielsson Department of Swedish, Linnaeus University, Växjö, Sweden (kristina.danielsson@lnu.se)

Chris Davison School of Education, University of New South Wales, Sydney, Australia (c.davison@unsw.edu.au)

Sibel Erduran Department of Education, University of Oxford, Oxford, United Kingdom
National Taiwan Normal University, Taipei, Taiwan (sibel.erduran@education.ox. ac.uk)

Mariona Espinet Departament de Didàctica de la Matemàtica i de les Ciències Experimentals, Universitat Autònoma de Barcelona, Catalonia, Spain (mariona.espinet@uab.cat)

Gail Forey The Hong Kong Polytechnic University, Hung Hom, Hong Kong (gail.forey@polyu.edu.hk)

Yun-Ping Ge Department of Education and Human Potentials Development, National Dong-Hwa University, Hualien, Taiwan (yunpingge@yahoo.com.tw)

Brian Hand University of Iowa, Iowa City, Iowa, USA (brian-hand@uiowa.edu)

Qiuping He Department of English, The Hong Kong Polytechnic University, Hung Hom, Hong Kong (ares.he@connect.polyu.hk)

Maria Isabel Hernández Departament de Didàctica de la Matemàtica i de les Ciències Experimentals, Universitat Autònoma de Barcelona, Catalonia, Spain (mariaisabel.hernandez@uab.cat)

Caroline Ho English Language Institute of Singapore, Ministry of Education, Singapore (caroline_ho@moe.gov.sg)

Peter Hubber Deakin University, Victoria, Australia (peter.hubber@deakin.edu.au)

Britt Jakobson Department of Mathematics and Science Education, Stockholm University, Stockholm, Sweden (britt.jakobson@mnd.su.se)

Rebecca Jesson Woolf Fisher Research Centre, School of Curriculum and Pedagogy, Faculty of Education and Social Work, University of Auckland, Auckland, New Zealand (r.jesson@auckland.ac.nz)

Erik Knain Department of Teacher Education and School Research, University of Oslo, Oslo, Norway (erik.knain@ils.uio.no)

Johanna Larsson Department of Physics and Astronomy, Uppsala University, Uppsala, Sweden (johanna.larsson@physics.uu.se)

Angel M. Y. Lin Faculty of Education, The University of Hong Kong, Pokfulam, Hong Kong (yeonmia@gmail.com)

Yu Liu College of International Education, Sichuan International Studies University, Chongqing, China (liuyunus@gmail.com)

Yuen Yi Lo Faculty of Education, The University of Hong Kong, Pokfulam, Hong Kong (yuenyilo@hku.hk)

Ragnhild Löfgren Department of Social and Welfare Studies, Linköping University, Linköping, Sweden (ragnhild.lofgren@liu.se)

Silvija Markic Institute for Science and Technology – Chemistry Department, University of Education Ludwigsburg, Ludwigsburg, Germany (markic@ph-ludwigsburg.de)

List of Contributors

Felicia Moore Mensah Department of Mathematics, Science & Technology, Teachers College, Columbia University, New York, USA (moorefe@tc.columbia.edu)

Elizabeth Birr Moje School of Education, University of Michigan, Ann Arbor, Michigan, USA (moje@umich.edu)

Eduardo Fleury Mortimer Federal University of Minas Gerais, Belo Horizonte, Brazil (efmortimer@gmail.com)

Audrey Msimanga University of the Witwatersrand, Johannesburg, South Africa (audrey.msimanga@wits.ac.za)

Marianne Ødegaard Department of Teacher Education and School Research, University of Oslo, Oslo, Norway (marianne.odegaard@ils.uio.no)

Sue Ollerhead School of Education, University of New South Wales, Sydney, Australia (s.ollerhead@unsw.edu.au)

Soonhye Park North Carolina State University, Raleigh, North Carolina, USA (spark26@ncsu.edu)

Alma Jahic Pettersson Department of Social and Welfare Studies, Linköping University, Linköping, Sweden (alma.pettersson@liu.se)

Vaughan Prain Deakin University, Victoria, Australia (vaughan.prain@deakin.edu.au)

Gde Buana Sandila Putra National Institute of Education, Nanyang Technological University, Singapore, Singapore (buana.sandila@gmail.com)

Natasha Anne Rappa School of Education, Murdoch University, Perth, Western Australia, Australia (natashaannetang@gmail.com)

Jee Kyung Suh University of Alabama, Tuscaloosa, Alabama, USA (jksuh@ua.edu)

Kok-Sing Tang School of Education, Curtin University, Perth, Australia (kok-sing.tang@curtin.edu.au)

Russell Tytler Deakin University, Victoria, Australia (russell.tytler@deakin.edu.au)

Jenny Uddling Department of Language Education, Stockholm University, Stockholm, Sweden (jenny.uddling@isd.su.se)

Len Unsworth Learning Sciences Institute, Australian Catholic University, Sydney, Australia (len.unsworth@acu.edu.au)

Laura Valdés-Sanchez Departament de Didàctica de la Matemàtica i de les Ciències Experimentals, Universitat Autònoma de Barcelona, Catalonia, Spain (lauravaldessanchez@gmail.com)

Kuo-Hu Wang National Changhua University of Education, Changhua, Taiwan (sukhua@cc.ncue.edu.tw)

Aaron Wilson Woolf Fisher Research Centre, School of Curriculum and Pedagogy, Faculty of Education and Social Work, University of Auckland, Auckland, New Zealand (aj.wilson@auckland.ac.nz)

Jason S. Wu Department of Mathematics, Science & Technology, Teachers College, Columbia University, New York, USA (jasonwu@columbia.edu)

Larry D. Yore Distinguished Professor Emeritus, University of Victoria, Victoria, BC, Canada (lyore@uvic.ca)

Chapter 1
The Expanding Development of Literacy Research in Science Education Around the World

Kok-Sing Tang and Kristina Danielsson

Abstract This introductory chapter summarizes the research background that motivates this book volume and the broad conceptualizations of literacy adopted by the various contributors within the context of science education. It also provides an overview of the six sections in this book, namely (i) national curriculum and initiatives, (ii) content and language integrated learning (CLIL), (iii) classroom literacy practices, (iv) disciplinary literacy challenges, (v) disciplinary literacy and science inquiry, and (vi) teacher development, and summarizes the contributions within each section.

Keywords Literacy · content area literacy · disciplinary literacy · scientific literacy · multimodality · national curriculum · content and language integrated learning (CLIL) · classroom practices · literacy challenges · science inquiry · teacher development

1.1 Background

Literacy has been a major research area in science education for several decades. Researchers exploring the connection between literacy and science learning generally come from the language arts and science education communities, and they bring with them a diverse range of theoretical perspectives and methodological orientations. For some time, research in this area has tended to be situated in and originate from North America, culminating in two prominent conferences that brought scholars from literacy and science education together with a common interest in promoting literacy in science. The first conference – Crossing Borders: Connecting Science and Literacy,

K.-S. Tang (✉)
School of Education, Curtin University, Perth, Australia
e-mail: kok-sing.tang@curtin.edu.au

K. Danielsson
Department of Swedish, Linnaeus University, Växjö, Sweden
e-mail: kristina.danielsson@lnu.se

© Springer International Publishing AG 2018
K.-S. Tang, K. Danielsson (eds.), *Global Developments in Literacy Research for Science Education*, https://doi.org/10.1007/978-3-319-69197-8_1

was held at the University of Maryland in 2001 (Saul, 2004) while the second conference – Ontological, Epistemological, Linguistic, and Pedagogical Considerations of Language and Science Literacy, was held at the University of Victoria in 2002 (Hand et al., 2003; Yore et al., 2004; Yore & Treagust, 2006). This "border crossing" conversation continued with the "Literacy for science in the Common Core ELA Standards and the Next Generation Science Standards" workshop in Washington, DC, in December 2013 (National Research Council, 2014).

The development of literacy research in science education has continued to expand and gained increasing attention across the globe. More countries are currently examining the role of literacy in their national curriculum and have undertaken research to integrate various literacy practices into the teaching and learning of science (e.g., see Chaps. 2–4). This edited volume aims to highlight this growing development around the world, in addition to seeking new ideas and perspectives that emerge when researchers in different parts of the world address literacy-related issues in science classrooms within their respective linguistic, cultural, and political contexts. Specifically, this volume showcases recent developments in literacy-science research from countries and territories such as Australia, Brazil, China, Finland, Germany, Hong Kong, New Zealand, Norway, Singapore, South Africa, Spain, Sweden, Taiwan, and the USA.

The majority of the chapters in this volume were based on selected studies presented at several international conferences, notably the 7th International Conference on Multimodality (ICOM) in Hong Kong in June 2014, the 3rd International Science Education Conference (ISEC) in Singapore in November 2014, and the 11th European Science Education Research Association Conference (ESERA) in Helsinki in September 2015. It was evident that a significant number of conference presentations had a strong literacy focus in descriptive or intervention research conducted within science classrooms. This reflects the growing trend of literacy research in science education around the world, and consequently, the relevance and timeliness of this collected volume.

1.2 Evolving Views of Literacy

Research focusing on literacy in science dates back many decades under the terms "content area literacy" or "literacy across the curriculum" (Shanahan & Shanahan, 2008). Researchers in this area argue that the importance of reading and writing is not confined to the language classrooms, but should extend to all content areas. To support content area literacy, a range of reading and writing strategies for making sense of text have been developed and advocated to be used in the science classrooms. Well-known strategies include reciprocal teaching, CORI (Concept-Oriented Reading Instruction) and SQ3R (Survey, Question, Read, Recite, Review). These strategies are widely used in the USA by reading experts during remedial programs for struggling learners with the intention of improving their reading skills. While research in this area sowed the ideas of infusing literacy teaching into content subjects, the research mostly emphasized the application of generalizable reading and writing skills in all subject matter classrooms, rather than focused on a specific skill or practice that is characteristic of a discipline. As Shanahan and Shanahan (2008) argue, these strategies tend to be more effective for primary school students and less beneficial at the secondary and tertiary level. This partly explains why content area teachers in secondary and tertiary schools have largely been resistant to content area literacy despite the efforts of their literacy counterparts (Moje, 2008).

In the 1990s, content area literacy saw a gradual turn toward the language aspects of the discipline and most notably, the linguistic and discursive features of science. Researchers during this turn have begun to study classroom discourse as a "kind of applied linguistics" consisting of interactional moves and exchanges (Cazden, 1988). They have also developed analytical frameworks to characterize classroom talk, such as the triadic Initiate-Response-Evaluate exchange pattern (Mehan, 1979) or the Initiate-Response-Follow-up sequence (Wells, 1993), and the distinction between authoritative and dialogic talk (Mortimer & Scott, 2003). In science education, Lemke's (1990) *Talking Science* was generally attributed as the landmark study that foregrounds the role of language in science classroom discourse (Kelly, 2007). Based on in-depth analysis of talk and interaction among the teachers and students in science classrooms, Lemke (1990) concludes that learning science is largely learning the language of science.

In written texts, the theory of systemic functional linguistics (SFL) was widely used to characterize the literacy of science according to the unique linguistic features of scientific texts. For instance, Halliday and Martin (1993) describe several characteristics of science register such as interlocking definitions, technical taxonomies, lexical density, and nominalization. Other researchers have also analyzed the recurring patterns of science genres, such as information report, experimental procedure, explanation, and argument (Unsworth, 1998; Veel, 1997) that students typically go through and need to learn in science lessons. These studies subsequently led to a specific literacy approach known as "genre pedagogy" that explicitly teaches students to unpack the register and genre of written texts in science (Hyland, 2007; Unsworth, 2001).

At the turn of this century, developments in multimodality began to shift and expand the scope of scientific language to representations such as images, graphs, symbols, and gestures that are ubiquitously used in science classrooms (Jewitt, 2008). For example, Kress, Jewitt, Ogborn, and Tsatsarelis (2001) document the complex ensemble of multiple modes of representation orchestrated by the science teacher as a way of shaping scientific knowledge through a process of sign-making. In an analysis of the literacy demands of the science curriculum, Lemke (1998) also demonstrates that scientific knowledge is made through joint meaning-making across multiple semiotic modes. In particular, he stresses that the "concepts" of science are "semiotic hybrids of verbal, mathematical, visual-graphical, and actional-operational modes" (Lemke, 1998, p. 88). Subsequently, many other studies have examined how science teachers and students used multimodal representations to construct an understanding of different science concepts at various grade levels (e.g., Danielsson, 2016; Márquez, Izquierdo, & Espinet, 2006; Prain, Tytler, & Peterson, 2009; Tang, Delgado, & Moje, 2014). A common conclusion from these studies is the lack of emphasis on addressing the literacy demands of the multimodal practices in science.

In more recent years, there has been a growing "practice turn" toward getting students to learn the disciplinary practices of science (Erduran, 2015; Ford & Forman, 2006). This emphasis on practice arose mainly from the beliefs that students should be engaged in the activities of scientists, rather than passively learn the products of their practices in the form of scientific knowledge or genres. At the same time, there has been a shift toward viewing literacy as not just the conceptual or linguistic tools to support content or language learning in science, but also a form of social practice (Gee, 1992) to support the epistemic processes specific to a discourse community (Moje, 2008). Thus, the focus is on using literacy to engage students in the practices of science so that they develop a deeper understanding of how knowledge in science is formed as well as an appreciation of how the discipline develops its unique ways of knowing. In terms of classroom teaching, the emphasis has been on disciplinary literacy where students are supported in their use of scientific texts and language as part of scientific inquiry process (Moje, Collazo, Carrillo, & Marx, 2001). The literacy instruction includes teaching students how to use language more effectively to: (a) construct scientific explanations, (b) engage in evidence-based arguments, and (c) obtaining, evaluating, and communicating multimodal information. These three practices are among the eight core practices identified in the Next Generation Science Standards (NGSS) – a recent science education standards document in the USA (National Research Council, 2012).

The authors in this volume adopt a broad conceptualization of literacy that reflects the various linguistic, multimodal, and practice turns of literacy described earlier. The authors also come from a mix of disciplinary backgrounds such as science education, language education, applied linguistics, and the learning sciences. Although the range of theoretical perspectives and research methods varies, most authors raise similar research questions concerning the role of literacy in science teaching and learning, such as:

1. How do students, of various cultural and linguistic backgrounds, learn science through the various languages and representations of science?
2. How do teachers use various literacy activities and instructions to engage their students in constructing scientific meanings and practices?
3. What literacy approaches or pedagogies are suited to support diverse groups of students in science learning, broadly conceived in terms of gaining scientific knowledge, languages, practices, and dispositions?

In using the term "literacy" in science education, the authors are mindful that this is different from the notion of "scientific literacy," which is generally accepted as the educational vision of producing future citizens able to participate in or make informed decisions on science-related societal and political issues (Roberts, 2007). Instead, we are more concerned with the "fundamental" sense of literacy – as ways of using language, in a broad sense, including multimodal forms of representing scientific ideas – that will help students develop the "derived sense" of scientific literacy associated with the conceptual knowledge of and dispositions toward science (Norris & Phillips, 2003). Nevertheless, it is evident that scientific literacy presupposes a certain level of literacy in its fundamental sense (Hodson, 2008). Access to scientific knowledge and communication, participation in science-related activities, and debate about socio-scientific issues cannot be carried out other than through the language and representations of science. In other words, the focus on literacy in this book is more related to the means and how of achieving scientific literacy, rather than postulating or debating about the visions and problems of scientific literacy.

1.3 Overview of Chapters

This volume, consisting of 22 chapters, is organized around six thematic sections: (1) National Curriculum and Initiatives, (2) Content and Language Integrated Learning (CLIL), (3) Classroom Literacy Practices, (4) Disciplinary Literacy Challenges, (5) Disciplinary Literacy and Science Inquiry, and (6) Teacher Development.

The first section opens the volume with a broad overview of the changing national curriculum landscape toward literacy with examples from three countries: Norway, Australia, and Singapore. With each country representing a unique geopolitical and historical context, the three chapters in this section provide a snapshot of curriculum development in different parts of the world and, at the same time, they highlight a common global trend in terms of foregrounding the role of literacy in the content areas. In Chap. 2, Knain and Ødegaard describe how the curriculum reform, which began in Norway in 2006, increased the focus on basic literacy skills across all subject areas. They also describe the challenges and issues faced by Norwegian teachers and school leaders in developing reading, writing, and oral communication competencies as part of their disciplinary curriculum. In Australia, as explained by Davison and Ollerhead (Chap. 3), the importance of literacy in every subject has a long history given Australia's multicultural society and its national priority in supporting learners of English as a second language. With the new Australian Curriculum, there has been even greater focus on teachers integrating subject-specific language and literacy requirements into their content teaching. In Singapore, as Ho, Rappa, and Tang (Chap. 4) report, the emphasis on disciplinary literacy has a different origin. Under the banner of "effective communication" spearheaded by the Ministry of Education, there is a growing awareness that good teaching should include the skillful use of subject-specific language to help students better process and understand the subject.

Besides providing the curriculum background of each respective country, each chapter also focuses on different aspects of research and developmental work undertaken in the country. Knain and Ødegaard report on various research projects in Norway from both language and science education research that have been carried out to support the national research agenda on literacy (one such project is further elaborated in Chap. 16). Davison and Ollerhead focus on preservice teacher education in Australia as this is a crucial element that determines success in integrating subject-specific literacy in the classrooms. In Singapore, Ho et al. elaborate on the systemic support provided by the Ministry of Education in collaboration with the National Institute of Education, and how this support has been a key factor for change in the educational system.

The second section features research studies that focus on content and language integrated learning (CLIL) which deal with classrooms where students are learning a majority language at the same time as they learn science. CLIL refers to an instructional approach that combines content and language teaching (Markic, Chap. 5) and the practice of teaching content using a second or foreign language (Lo, Lin, & Cheung, Chap. 6). Given the diverse linguistic background of students in Germany, Hong Kong, South Africa, Singapore, and the USA, the chapters in this section collectively provide a complex picture of what it means to help second language learners who are learning a new medium of instruction in the country as well as the unfamiliar language of science, in what is effectively a "multilingual science lesson." First, Markic explains in Chap. 5 the situation of migrant children in Germany and the problem of getting science teachers to address the teaching of German as a foreign language in their science lessons. She presents a participatory action research involving science teachers and German as a second language (GSL) teachers, and reports some success in helping linguistically disadvantaged students. Chapter 6 by Lo et al. follows up with a similar study in Hong Kong in terms of the rationale and approach of having science and English language teachers work together. An interesting aspect in their study was the teachers' use of the Sydney School's genre-based pedagogy (Rose & Martin, 2012), followed by their illustrations of how the teachers weaved English language teaching into the process of science knowledge construction.

In Msimanga and Erduran's Chap. 7, the focus shifts to South Africa where the problem is further compounded by most students learning English as their third or subsequent language and being taught by science teachers who themselves are not proficient in English. This unique situation led Msimanga and Erduran to investigate how teachers who were nonnative speakers of English consciously or unconsciously used meta-talk as a discursive tool to engage students in discussing scientific ideas. The attention to meta-talk, or metadiscourse, reflects the growing awareness of this discursive resource that can potentially be used to engage students in the learning of science (Tang, 2017). The last chapter in this section by Wu, Mensah, and Tang (Chap. 8) provides a contrasting view to the argument of CLIL. While a key benefit of CLIL is to use content learning to support second or additional language development, Wu et al. question the role of the students' native language in learning science. With case studies from English language learners (ELLs) in New York and Singapore, this chapter illustrates the content-language tension faced by ELLs and aptly reminds us that language use in science classrooms is not merely shaped by cognitive or linguistic considerations, but is also largely ideologically and politically contested.

The third section explores the role of literacy in science classroom discourse. Wilson and Jesson (Chap. 9) provide an overview of the nature of literacy teaching in several science classrooms in Auckland where the teachers were identified by their school leaders as effective at developing students' literacy in science. Given the vision for literacy in science expressed in the New Zealand Curriculum (similar to Australia as reported in Chap. 3), what is telling from their findings was the limited range of literacy teaching confined to vocabulary instruction, repeated practice tasks, and the use of short teacher-designed texts. Wilson and Jesson thus advocate the need for a broader focus on more disciplinary-appropriate ways of reading, writing, and talking. In Chap. 10, Cavalcanti Neto, Amaral, and Mortimer analyze the discursive interactions in three biology classrooms on the topic of environment. By unpacking teaching strategies (e.g., oral exposure, questioning, reading) in terms of discursive interactions, they highlight how different discursive patterns (e.g., dialogic, authoritative) can facilitate or hinder the development of scientific literacy, both in the fundamental sense (Norris & Phillips, 2003) and Vision II of scientific literacy (Roberts, 2007). Chapters 11 and 12 shift the attention of literacy toward its multimodal aspect. Jakobson, Danielsson, Axelsson, and Uddling in Chap. 11 investigate the role of multimodal classroom interaction in students' science meaning-making in a grade 5 multilingual classroom in Sweden. They illustrate in microanalytic detail how the concept of measuring time is a multimodal assemblage of spoken and written language, gestures, physical objects, models, and metaphors. In a similar approach, He and Forey (Chap. 12) study how a range of semiotic resources was used to construct a sequential explanation of digestion in a grade 9 Australian science classroom, and further elucidate the "multiplying meaning" principle first proposed by Lemke (1998).

The fourth section addresses the question of what makes the literacy practices of science so challenging for many students. Liu tackles this question in Chap. 13 by examining the language of symbolic formulae in secondary school chemistry textbooks. His analysis reveals that a chemical formula often incorporates two different types of meanings to represent both the quantitative and qualitative composition of a compound. Making this distinction can help educators understand the literacy challenges posed by symbolic formulation in chemistry texts. In Chap. 14, Danielsson, Löfgren, and Pettersson examine another aspect of literacy challenge in the disciplinary discourse of chemistry – the use of metaphors. From their analysis of chemistry lessons in Sweden and Finland focusing on the atomic structure and ion formation, they found that the teachers used an abundance of scientific, everyday, and anthropomorphic metaphors to foreground different properties of the atom. With each metaphor having different affordances, or "gains and losses," in understanding chemistry, they argue for the need to have more discussions around metaphors in the classrooms. In Chap. 15, Ge, Unsworth, Wang, & Chang turn to the role of image in students' reading comprehension of science texts in Taiwan. Combining perspectives from cognitive and semiotic theories, they designed and conducted an experiment to test the hypothesis that image designs with salient tree structure can lead to better reading comprehension of the concept of biological classification system. Their findings offer empirical evidence to support the principle of sound image design in science teaching and instructional design.

The fifth section focuses on the role of disciplinary literacy in science inquiry and practices, and features three design research projects aimed at integrating literacy and inquiry. These projects start with the guiding principle that literacy in the classroom should mirror the disciplinary practices of scientists (Pearson, Moje, & Greenleaf, 2010). The first project (Ødegaard, Chap. 16) is the Budding Science and Literacy research implemented as part of the Norwegian curriculum emphasis on disciplinary literacy (Chap. 2). Inspired by the *Seeds of Science/Roots of Reading* (Barber et al., 2007), a teaching model was developed and enacted by six primary science teachers in Norway. From various data sources, Ødegaard argues that literacy activities embedded in science inquiry provide crucial support for students' meaning-making in science. A similar project by Tang and Putra (Chap. 17) was carried out in four secondary physics and chemistry classrooms in Singapore, in alignment with the country's new emphasis on subject-specific literacy (Chap. 4). They adapted the 5E Inquiry Model (Bybee et al., 2006) and infused literacy strategies designed to support students in constructing scientific explanations. Focusing on the teachers' enactment of the pedagogical model, Tang and Putra illustrate how literacy instruction can enable inquiry-based science. In Chap. 18, Tytler, Prain, and Hubber focus on another core aspect of science disciplinary literacy, which is the construction of multimodal representations. They developed a "representation construction" approach to inquiry in Australia to engage students in experimenting, generating, and refining representations to explain the material world. Through this inquiry approach, Tytler et al. argue that students can achieve a meta-representational understanding of how texts and knowledge are produced in scientific work.

The last section in this volume explores the important issues and tensions faced in preparing science teachers to integrate disciplinary literacy in their teaching. In the context of training a group of preservice primary teachers in Spain to become CLIL teachers, Espinet, Valdés-Sanchez, and Hernández (Chap. 19) examine the student teachers' beliefs and expectations through their science and language experience narratives. They found that the student teachers' science experiences were more related to school contexts and associated with negative experiences, whereas their language experiences were connected to a variety of out-of-school contexts and were more positive. Their result points to the need to address the isolation of science from the social and personal life of not only young children, but importantly of our preservice teachers as well. It also raises the question of how to use the teachers' positive language experiences as contexts for anchoring their science experiences. In Chap. 20, Hand, Park, and Suh focus attention on in-service science teachers' epistemic orientations toward the role of language. Through a 3-year professional development program in Midwestern USA, they tracked changes in the teachers' epistemic orientations and pedagogical practices as they learned and implemented the Science Writing Heuristic (SWH) approach. Their preliminary findings underscore the importance of teachers understanding the critical role of language as an epistemic tool, and the potential impact of such epistemic orientations on improving student learning. Lastly, Airey and Larsson (Chap. 21) explore the disciplinary literacy goals of undergraduate physics lecturers in Sweden and South Africa, and argue that each discipline emphasizes varying communicative practices for three different settings: the academy, the workplace, and society. These interdisciplinary differences pose significant challenges for physics trainee teachers who have to navigate across the disciplines of physics and education during their preservice teacher education.

1.4 Concluding Remarks

In sum, the studies featured in these chapters provide a landscape of the research within the literacy-science nexus stretching across the globe. This niche area of research has seen "cross-border" dialogue between researchers from literacy and science education predominantly from North America for the last few decades, and now increasingly involves researchers from around the world. In this regard, this volume not only contributes notable literature to the expanding conversation, but also represents a shift of this "cross-border" dialogue from the initial disciplinary "borders" between literacy and science to the present international borders across national boundaries. We hope that this volume will serve as a catalyst for more studies that will contribute to the continuing global conversation in the intersection of literacy and science education research.

References

Barber, J., Pearson, P., Cervetti, G., Bravo, M., Hiebert, E., Baker, J., et al. (2007). *An integrated science and literacy unit. Seeds of science. Roots of reading*. Nashville: Delta Education.

Bybee, R. W., Taylor, J. A., Gardner, A., Van Scotter, P., Powell, J. C., Westbrook, A., et al. (2006). *The BSCS 5E instructional model: Origins and effectiveness*. Colorado Springs: BSCS.

Cazden, C. B. (1988). *Classroom discourse: The language of teaching and learning* 1st ed. Portsmouth: Heinemann.

Danielsson, K. (2016). Modes and meaning in the classroom – the role of different semiotic resources to convey meaning in science classrooms. *Linguistics and Education, 35*, 88–99. https://doi.org/10.1016/j.linged.2016.07.005

Erduran, S. (2015). Introduction to the focus on … scientific practices. *Science Education, 99*(6), 1023–1025. https://doi.org/10.1002/sce.21192

Ford, M. J., & Forman, E. A. (2006). Redefining disciplinary learning in classroom contexts. *Review of Research in Education, 30*, 1.

Gee, J. P. (1992). *The social mind: Language, ideology, and social practice*. New York: Bergin & Garvey.

Halliday, M. A. K., & Martin, J. R. (1993). *Writing science: Literacy and discursive power*. Pittsburgh: University of Pittsburgh Press.

Hand, B., Alvermann, D. E., Gee, J., Guzzetti, B. J., Norris, S. P., Phillips, L. M., et al. (2003). Message from the "Island group": What is literacy in science literacy? *Journal of Research in Science Teaching, 40*(7), 607–615.

Hodson, D. (2008). *Towards scientific literacy*. Rotterdam: Sense Publishers.
Hyland, K. (2007). Genre pedagogy: Language, literacy and L2 writing instruction. *Journal of Second Language Writing, 16*, 148–164.
Jewitt, C. (2008). Multimodality and literacy in school classrooms. *Review of Research in Education, 32*, 241–267.
Kelly, G. J. (2007). Discourse in science classrooms. In S. K. Abell & N. G. Lederman (Eds.), *Handbook of research on science education* (pp. 443–469). Mahwah: Lawrence Erlbaum Associates.
Kress, G., Jewitt, C., Ogborn, J., & Tsatsarelis, C. (2001). *Multimodal teaching and learning: The rhetorics of the science classroom*: London: Continuum.
Lemke, J. L. (1990). *Talking science: Language, learning and values*: Norwood: Ablex.
Lemke, J. L. (1998). Multiplying meaning: Visual and verbal semiotics in scientific text. In J. Martin & R. Veel (Eds.), *Reading science* (pp. 87–113). London: Routledge.
Márquez, C., Izquierdo, M., & Espinet, M. (2006). Multimodal science teachers' discourse in modeling the water cycle. *Science Education, 90*(2), 202–226.
Mehan, H. (1979). *Learning lessons: Social organization in the classroom*. Cambridge: Harvard University Press.
Moje, E. B. (2008). Foregrounding the disciplines in secondary literacy teaching and learning: A call for change. *Journal of Adolescent & Adult Literacy, 52*(2), 96–107.
Moje, E. B., Collazo, T., Carrillo, R., & Marx, R. W. (2001). "Maestro, what is' quality'?": Language, literacy, and discourse in project-based science. *Journal of Research in Science Teaching, 38*(4), 469–496.
Mortimer, E. F., & Scott, P. (2003). *Meaning making in secondary science classrooms*. Buckingham: Open University Press.
National Research Council. (2012). *A framework for K-12 science education: Practices, crosscutting concepts, and core ideas*. Washington: The National Academies Press.
National Research Council. (2014). *Literacy for science: Exploring the intersection of the next generation science standards and common core for ELA standards, a workshop summary*. Washington: The National Academies Press.
Norris, S. P., & Phillips, L. M. (2003). How literacy in its fundamental sense is central to scientific literacy. *Science Education, 87*(2), 224–240.
Pearson, P. D., Moje, E., & Greenleaf, C. (2010). Literacy and science: Each in the service of the other. *Science, 328*(5977), 459–463. https://doi.org/10.1126/science.1182595
Prain, V., Tytler, R., & Peterson, S. (2009). Multiple representation in learning about evaporation. *International Journal of Science Education, 31*(6), 787–808. https://doi.org/10.1080/09500690701824249
Roberts, D. (2007). Scientific literacy/science literacy. In S. K. Abell & N. G. Lederman (Eds.), *Handbook of research on science education* (pp. 729–780). Mahwah: Lawrence Erlbaum.
Rose, D. & Martin, J.R. (2012). Learning to Write, Reading to Learn: Genre, knowledge and pedagogy in the Sydney School. London: Equinox.
Saul, W. (2004). *Crossing borders in literacy and science instruction: Perspectives on theory and practice*. Newark: International Reading Association; Arlington: NSTA Press.
Shanahan, T., & Shanahan, C. (2008). Teaching disciplinary literacy to adolescents: Rethinking content-area literacy. *Harvard Educational Review, 78*(1), 40–59.
Tang, K. S. (2017). Analyzing teachers' use of metadiscourse: The missing element in classroom discourse analysis. *Science Education, 101*(4), 548–583. https://doi.org/10.1002/sce.21275
Tang, K. S., Delgado, C., & Moje, E. B. (2014). An integrative framework for the analysis of multiple and multimodal representations for meaning-making in science education. *Science Education, 98*(2), 305–326. https://doi.org/10.1002/sce.21099
Unsworth, L. (1998). "Sound" explanations in school science: A functional linguistic perspective on effective apprenticing texts. *Linguistics and Education, 9*(2), 199–226.
Unsworth, L. (2001). *Teaching multiliteracies across the curriculum: Changing contexts of text and image in classroom practice*. Buckingham: Open University Press.

Veel, R. (1997). Learning how to mean-scientifically speaking: Apprenticeship into scientific discourse in the secondary school. In C. Frances & J. Martin (Eds.), *Genre and institutions: Social processes in the workplace and school* (pp. 161–195). London: Cassell.

Wells, G. (1993). Reevaluating the IRF sequence: A proposal for the articulation of theories of activity and discourse for the analysis of teaching and learning in the classroom. *Linguistics and Education, 5*, 1–37.

Yore, L. D., Hand, B., Goldman, S. R., Hildebrand, G. M., Osborne, J. F., Treagust, D. F., et al. (2004). New directions in language and science education research. *Reading Research Quarterly, 39*(3), 347–352.

Yore, L. D., & Treagust, D. F. (2006). Current realities and future possibilities: Language and science literacy – empowering research and informing instruction. *International Journal of Science Education, 28*(2), 291–314.

Part 1
National Curriculum and Initiatives

Chapter 2
The Implementation of Scientific Literacy as Basic Skills in Norway After the School Reform of 2006

Erik Knain and Marianne Ødegaard

Abstract In 2006, Norway implemented a new curriculum, which introduced basic literacy skills in every school subject. In this new curriculum, the basic skills are considered fundamental to learning in every school subject. As a consequence, since 2006, Norwegian teachers and school leaders have been grappling with how to develop and teach writing, reading, and oral communication as an integrated part of disciplinary education. Several initiatives have sought to develop literacy in school science since the introduction of the curriculum. However, teachers and school leaders are generally somewhat hesitant to explicitly address basic skills in the classroom. Some communities of teachers who discuss the purposes and qualities of writing have nonetheless been successfully established in Norway. We will discuss two approaches for the implementation of the curriculum reform. The first is based on first language (L1) literacy research, considering literacy as a general competence, however with a strong emphasis on writing in the disciplines. The second is from the science education community, with its emphasis on scientific literacy. Some of the major projects in Norway have focused on the intersection between the basic literacy skills and the process dimension of science in the main area "The Budding Researcher" in the new curriculum. We conclude that these two approaches each have their own strengths for developing scientific literacy and the basic skills introduced in the curriculum. Together, they offer opportunities for considering basic skills as both generic and discipline-specific, and for developing a metalanguage for discussing and reflecting on the teaching and learning of school science.

Keywords Curriculum reform · basic skills · scientific literacy · writing in the disciplines · inquiry-based science teaching

---

E. Knain (✉) · M. Ødegaard
Department of Teacher Education and School Research, University of Oslo, Oslo, Norway
e-mail: erik.knain@ils.uio.no

M. Ødegaard
e-mail: marianne.odegaard@ils.uio.no

2.1 Introduction: Basic Skills in a Curriculum Reform

The Knowledge Promotion Reform was introduced in autumn 2006 in Norway. The National Curriculum for Knowledge Promotion in Primary and Secondary Education and Training (LK06) comprises the following areas: the Core Curriculum, the Quality Framework, Subject Curricula, Distribution of Teaching Hours per Subject, and Individual Assessment. The reform covers primary, lower secondary, and upper secondary education and training. Another part of the reform was the introduction of national tests of students' abilities in reading, mathematics, and English to determine whether students' skills were consistent with the curriculum goals. The tests were intended to serve as a platform for professional development at the school and school owner levels (municipality or county councils) (Ministry of Education and Research, 2006). The published results include the municipality level as the smallest unit. The use of these tests is debated in Norway, with critics claiming that the tests lead to variations of "teaching to the test," rather than their intended goal of fostering the local development of teaching practices.

This chapter will discuss the nature of these basic skills and the efforts to implement them in the wake of the Knowledge Promotion Reform. Although the main focus of this chapter is school science, it is necessary to consider both the skills and the efforts to implement them in school teaching from a broader perspective. One of the key characteristics of the basic skills is that they were described at a general level as key competencies across all school subjects. Also, they were specifically described in each school subject with a clear understanding that they need to be adapted to and taught in accordance with the nature of the specific school subjects. The basic skills include the following: numeracy, the ability to read, the ability to express oneself orally, the ability to express oneself in writing, and the ability to use digital tools. The following citations from LK06 cover their main features:

> *Oral skills* in natural science means listening, speaking, and conversing to describe, share, and develop knowledge with content about natural science related to observations and experiences. This involves using natural science concepts to communicate knowledge and to formulate questions, arguments, and explanations. […] This involves an increasing use of natural science concepts to express understanding, to form opinions, and to participate in academic discussions.
>
> *Being able to express oneself in writing in natural science* means using text genres from the natural sciences to formulate questions and hypotheses, write plans and formulate explanations, compare and reflect on information, and use sources in a purposeful manner. This also involves describing one's observations and experiences, comparing information, arguing one's viewpoints, and reporting from field work, experiments, and processes related to technological development. […] The development of writing proficiency in the subject of natural science begins with using simple forms of expression before gradually using more precise natural science and scientific concepts, symbols, graphic presentation, and argumentation.
>
> *Being able to read in natural science* means understanding and using natural science and scientific concepts, symbols, diagrams, and arguments through goal-oriented work with natural science texts. This involves being able to identify, interpret, and use information from composite texts in books, newspapers, operating manuals, rules, brochures, and

digital sources. Reading in natural science includes critical assessment of how information is presented and used in arguments, e.g., by being able to distinguish between data, assumptions, assertions, hypotheses, and conclusions.

Numeracy in natural science means gathering, processing, and presenting figures and numbers. *Digital skills in natural science* means using digital tools to explore, record, calculate, visualize, document, and publish data from one's own and others' studies, experiments, and fieldwork (The Norwegian Directorate for Education and Training, 2016).

The notion of basic competencies was introduced by the official government report "I første rekke" ("First in line") (Official Norwegian Reports (NOU), 2003). LK06 includes two important references to the basic skills, that is, to literacy and to international efforts to define key competencies. The latter initiatives include the European Union's Asia-Europe Meeting (ASEM) framework and The Organization for Economic Co-operation and Development (OECD) project Definition and Selection of Competencies (DeSeCo) (Rychen & Salganik, 2003). The DeSeCo project developed a theoretical framing of competencies that aligns closely with "literacy." This framework understands "competence" as individual competencies that are found in social interactions and depend on cultural conditions. DeSeCo also understands competence as simultaneously situated and transferrable across situations. Interestingly, the term "literacy" is only used in the term "digital literacy," in the listing of basic skills in the "First in line" government report. The notion of liberal education (*allmenndannelse*) seems to be a stronger reference point in this official government report. In the later white paper "Culture for Learning" (Ministry of Education and Research, 2004), literacy is connected to the basic skills, as the following excerpt illustrates:

> The basic skills that are described here are necessary preconditions for learning and development in schools, the workplace, and civic society. They are independent of school subjects, but the school subjects are suitable for the development of such skills to varying degrees. These basic skills resemble the English concept of "literacy," which extends beyond only being able to read. It also encompasses "reading, writing, and numeracy" (English in original), and includes a variety of skills, such as the ability "to identify, to understand, to interpret, to create, and to communicate." (Ministry of Education and Research, 2004, p. 33) (author's translation).

The above excerpt alludes to literacy as a complete set of functional skills that are both cross-curricular and subject specific. The Subject Curricula in LK06 is organized according to main subject areas, though the basic skills cut across these main areas. Examples of subject areas include "Biological Diversity," "Body and Health," and "Phenomenon and Substances." Within each subject area, several specific competence goals are introduced by a verb that indicates the basic skills that are emphasized, that is, to describe, to explain, to argue, to discuss, to mention. For instance, the following goals are introduced for the tenth grade (our translation; key verbs are underlined):

- <u>Describe</u> the universe and different theories for how it has developed.
- <u>Explain</u> how we can produce electric energy from renewable and nonrenewable energy sources, and <u>discuss</u> the environmental effects of different ways of producing energy (The Norwegian Directorate for Education and Training, 2016).

To address the process dimension of science, the concept of "the budding researcher" was introduced as one of the main subject areas. The functional aspect of the basic skills in science can be identified if we consider the following description in LK06 of the budding researcher:

The Budding Researcher

> Teaching in natural science presents natural science as both a product that shows the knowledge we have acquired thus far in history and as processes that deal with how knowledge of natural science is developed and established. These processes involve the formulation of hypotheses, experimentation, systematic observations, discussions, critical assessment, argumentation, grounds for conclusion and presentation. The budding researcher shall uphold these dimensions while learning in the subject and integrate them into the other main subject areas (LK06).

This main area includes the basic skills cited above, that is, the formulation of hypotheses, discussions, critical assessment, argumentation, grounds for conclusion, and presentation.

The relationship between The Budding Researcher and the basic skills will be investigated later in this chapter, as it was paramount to substantive research and development initiatives in Norway, building on the idea that the concept of the budding researcher opened a training ground for basic skills, whereas basic skills became tools needed in inquiry.

2.2 Implementation and Development

The basic skills outlined in the curriculum fit into a "writing across the curriculum" (WAC) movement, which merged with the genre school into the Writing in the Disciplines pedagogy (WiD), "where disciplinary discourses are investigated and where students are helped to learn about the format and style guidelines for professional genres" (Hertzberg & Roe, 2015, p. 3). This development in the field of writing pedagogy is one main entry point to an understanding of basic skills. Further, within the field of science education, scientific literacy has been developing for decades—initially in a derived sense, but later also in a fundamental sense (Norris & Phillips, 2003). The fundamental sense is based on the essential role of text in science and involves reading, writing, and being fluent in the discourse patterns and communication systems of science. The derived sense of scientific literacy stems from the fundamental sense and involves being knowledgeable and educated in science and being able to take a critical stance on information. There has been a strong focus on scientific literacy in the Norwegian educational context as a functional and textual competence (Knain, 2015; Sørvik, Blikstad-Balas, & Ødegaard, 2015). Efforts to implement instruction that leads to scientific literacy driven by the science education community in Norway are the second main entry point in our attempt to answer the question: In light of the developments taking place after the introduction of the new curriculum, what happened in schools?

2.2.1 Assessments of the Curriculum Reform

In a review of assessments conducted after the implementation of the new curriculum, Rødnes and Gilje (2016) concluded that basic skills had not been systematically emphasized by teachers as intended. However, Sivesind (2012) concluded that there is evidence of Norwegian teachers' increased capability to explicitly focus on basic skills when teaching students. There is also significant variation between teachers and among school leaders regarding what "basic skills" actually means. Based on case studies conducted in ten schools, the researcher concluded that oral literacy was particularly emphasized in the early years; furthermore, while there is tentative progress in the development of teaching basic skills, there is great variation among schools regarding the degree of such progress (Hertzberg, 2010). While teachers are positive about teaching basic skills, they are often uncertain how to incorporate them in the classroom. To some, "basic skills" means an emphasis on fundamental skills that could be addressed in the first years of school. Paradoxically, the notion that basic skills are already used all the time in classrooms seems to bolster the lack of explicit focus on such skills in teachers' practices. Such thinking might go as follows: If the basic skills are already used, why focus on them? Teachers may also consider these skills the prime responsibility of the L1 teachers. Indeed, L1 teachers seem to be willing to take on this responsibility (Matre & Solheim, 2015).

These results underline one of the tenets of the DeSeCo project: for key competencies (i.e., the basic skills) to be taught and assessed, they need to be theoretically understood. A lack of an explicit description of what progression in basic skills actually involves was addressed in a revision of the curriculum in 2012.

2.2.2 Revision of Curriculum

Based on the finding that the practices of basic skills are not implemented as intended in classrooms, the curricula in science and four other subjects were revised to clarify the "basic skills." Since basic skills in science are considered a significant part of both learning science and scientific processes, basic skills were elucidated by integrating them in the science competence goals for all levels of science with progression, for example register, describe (age 9); systematize, revise (age 12); identify scientific argumentation and judge quality of sources (age 15) (Mork, 2013). This was done with the support of the Framework for Basic Skills (Norwegian Directorate for Education and Training, 2012), which is a document describing the functions of basic skills at different levels in compulsory and secondary education. Generic grids were developed for the basic skills that described their progression through the levels. For instance, in the main area of "Technology and Design" for fourth grade, the original learning goal (LK06)—that is, to "plan, build and test simple models of construction and document the process from idea to finished product"—was revised to include basic skills more explicitly. The new version includes a specification regarding how to document the inquiry process by combining different means of expression: "[…] and document the process from idea to finished product with text and illustrations." In order to insure progression in writing as a basic skill, one learning goal for seventh grade in "Biological Diversity" states the following: "plan and carry out an investigation in a natural environment, register observations and systematize the results." (The underlining indicates what was added after the curriculum revision.) The progression consists of using relevant terminology ("observations") and means of expression adapted to the subject of science ("systematize the results") (Norwegian Directorate for Education and Training, 2012).

In the revision, it was specified that the main area The Budding Researcher should also be integrated in all the main subject areas, as illustrated in the examples from "Technology and Design" and "Biological Diversity," where the scientific knowledge building process becomes more visible. The main area of "The Budding Researcher" is also more explicit in the revision in terms of considering basic skills, as it includes the following goals (for tenth grade): "write descriptive and argumentative texts with references to relevant sources, assess the quality of your own and others' text and revise the texts" (writing as a basic skill); and "identify scientific arguments, facts and claims in texts and graphics from newspapers, brochures and other media and assess the content critically" (reading as a basic skill). In the revised version of the curriculum, greater emphasis is placed on teaching "the budding researcher" and on how scientific knowledge is created (Mork, 2013), which thus reinforces the emphasis on basic skills.

2.2.3 Teacher Training Courses

In order for teachers to be equipped to meet the challenges of the new curriculum, several teacher training courses were developed by the National Centre of Science Education. One of these training courses led to the Budding Science and Literacy Project (presented in Chap. 16 in this book). Based on the results from the research project, additional courses were developed, which were funded by the Norwegian Directorate for Education and Training and the National Centre of Science Education. The focus of the courses was literacy, inquiry, and conceptual understanding in science. Utilizing a national network of science educators, these courses were conducted nationwide, engaging over 2000 teachers (the majority of whom were teaching science in elementary school). In parallel to these courses, the National Centres of Reading and Writing developed teacher training courses for reading and writing as basic skills across subjects.

2.2.4 Research Projects Involving Writing Across Different School Subjects

The project WRITE (Writing as a Basic Skill and Challenge), which was conducted between 2006 and 2011, was funded by Norwegian the Research Council and undertook a study of writing practices in L1, natural science, social science, religion and ethics education, and mathematics in different grades in K-12 education. The publications from the project emphasized how science teachers need to develop a vocabulary for making literacy skills explicit in their teaching, and also how a more explicit metalanguage for specific aspects of students' text is developed and can be used in discussing these texts. The functional approach to writing in the project is evident in the notion of "purpose" and "use" in the teaching of writing. In their investigations of writing practices across school subjects, the WRITE-researchers found a tension among L1 teachers in that they felt a dual responsibility to not only develop students' literacy in the mother tongue, but also to support the development of literacy in other school subjects. Based on the experiences from the project, the researchers presented "Ten theses about the teaching of writing," which include the following topics: (a) discussing the purpose of the writing; (b) developing disciplinary spaces for talking about language; (c) discussing assessment criteria for content, form, and use, depending on the purpose of the writing; (d) working with genres in all disciplines; and (e) offering support during the planning and writing processes (Smidt, Solheim, & Aasen, 2011, p. 13).

This interest in how writing is part of functional literacy practices as shared social norms in social groups is also found in the NORM project (Developing National Standards for the Assessment of Writing). The background for this project includes the recognition from earlier projects that there was a need to develop shared norms among teachers for assessing texts. The NORM project's ambition was, firstly, to develop explicit expectations (i.e., norms) describing different levels of achievement on a set of textual aspects. Using think-aloud interviews, in which teachers talked about the qualities in texts written by students, enabled the descriptions of the norms to be developed bottom-up, based on teachers' experiences and discussions. Secondly, this framework functioned as a set of tools in an intervention phase where researchers cooperated with school leaders and teachers in a professional development course, in which the teachers assessed and reflected upon students' texts using the developed framework. This implementation has been successful in that the experimental groups increased their skills to the same degree as if they had received an extra year of teaching (Berge & Skar, 2015). In an example on the use of the framework (Solheim & Matre, 2014), a student's text on the characteristics of iron is assessed according to the framework. In addition to the analysis of the linguistic features (author presence, conjunction, nominalization, passive or active voice, spelling, cohesion), the features that more specifically address the knowledge domain (what is happening is well described, good explanations, relevant knowledge) are assessed. This encounter between a language for describing text *as text* and what the text signifies in terms of learning and practices of science is significant in the implementation of the basic skills in school science.

In the Nadderud project (named after the upper secondary school that participated in the project), writing researchers from the University of Oslo cooperated with a group of social science, natural science, English, religion and ethics, and L1 teachers. The group met regularly during a 4-year period to discuss texts written by students. The focus of this project is akin to the WRITE and NORM projects, as it focuses on the significance of making norms on writing explicit and accessible for reflection for the teachers in the school subjects. The discussion addressed the following questions: "What role does this text have in your subject? How do you assess this text? What is the purpose of this writing assignment?" (Flyum & Hertzberg, 2011). These discussions revolved around exploring the meaning of the curriculum statement that basic skills are to be "integrated and adapted to each subject." The authors conclude that the discussions that took place in the project are contrast to the findings from the evaluations of the curriculum referred to earlier; that the implementation was reluctant and partial. Also, it parallels the NORM project in its focus on the significance of making norms on writing explicit and accessible for reflection for the teachers in the subjects. The researchers noted that the initial ambition to develop common norms of writing changed into an investigation of the differences between subjects and why these differences were present. For instance, why was the IMRaD format (Introduction, Method, Results, and Discussion) considered more important by the science teacher than the L1 teacher? Similar initiatives spread to several other schools (Hertzberg & Roe, 2015).

Part of the interest in basic skills in Norway was an interest in the language of science in textbooks (Maagerø & Skjelbred, 2010) and the role of learning materials in students' interactions. Ark & App (Paper & App) (2012–2016) was a research project that focused on the use of learning resources in Norway (Gilje et al., 2016). Twelve case studies comprised the heart of the project, including one case study in social science, English, science, and mathematics in three different grade levels in grades 5–11 (age 10–16). Some of the case studies were conducted in naturalistic settings without researcher intervention; others involved some level of cooperation between researchers and teachers in the design. As this research project focused on the role of learning resources and social interaction in school, it also provides information on how basic skills were part of classroom life in these cases (Rødnes & Gilje, 2016). It is evident that the interactions, including reading, talking, and writing, took place in a mixed ecology of paper-based and digital resources. Furthermore, talking played a significant role in tying different resources together in students' meaning making, whereas writing was significant in tying students' vernacular language to school science patterns of meaning. The findings support the key developments from the Nadderud project, in that the specifics of the school science subject shaped the classroom interactions and made basic skills relevant. However, the project was not designed to focus on whether basic skills were explicitly addressed in teaching.

The WRITE, Nadderud, and NORM projects all focused on the explicit development of basic skills, with an emphasis on writing. These projects were initiated by text researchers from a general literacy perspective; however, the researchers were acutely aware of the tension between perceptions of writing skills across subjects and the particulars of writing practices in different school subjects. The differences between writing practices in science and other subjects were utilized as a vehicle for making different writing norms explicit.

2.2.5 Research Projects Involving Scientific Literacy and Inquiry

We will now present some of the significant undertakings in the science education community in Norway to address basic skills. Many of the initiatives utilized inquiry as a frame for the development of science teaching and sought to develop teacher competence in teacher training courses or in cooperation with schools. We will present three projects where the basic skills were addressed explicitly in the design of the project and in classroom implementation.

The project Students as Researchers in School Science (StudentResearch/ElevForsk) (2007–2012) was funded by the Research Council of Norway and focused on the development of inquiry practices in Norwegian schools. It was one of several projects that focused on inquiry-based science teaching (IBST) in Norway during this time period. It maintained a strong focus on basic skills by incorporating the previously discussed interconnection between the budding researcher and basic skills in the Knowledge Promotion Reform as an explicit criterion for design. In the section on the budding researcher in LK06, the formulation of hypotheses, discussions, critical assessment, argumentation, and presentation are considered key aspects in the processes of science. This is a functional understanding of literacy in science practices that mirrors the following description of basic skills in the Knowledge Promotion Reform: the ability to communicate knowledge and formulate questions, arguments, and explanations, both verbally and in writing. In this very functional literacy sense, basic skills and inquiry practices are mutually constitutive (Knain, 2015). Basic skills are thus tied to inquiry practices and the nature of science, which shapes the relationship between different basic skills and how reading, talking, writing, mathematics, and digital tools (all of which are basic skills) together realize school science practices. One important design challenge of IBST is that teachers need to provide leeway for students' inquiries. More specifically, teachers need to create a space for students to grapple with questions that have some personal and/or social significance and that do not have an immediate answer (at least for the students). This will enable the students to refine their knowledge and learn from their mistakes. Thus, the degree of openness is one of the key design issues to consider in IBST. Activities with a lot of room for students to explore their own ideas and understanding need to be balanced with activities focusing on encounters with authoritative knowledge, synthesis, and critical reflection and assessment (Bjønness & Kolstø, 2015). One support structure for helping students to synthesize and engage in critical reflection was labeled "research meetings" (*Forskermøter*). These meetings were organized as tightly structured events intended to mimic the research seminar one finds in authentic science. In "research meetings," the students were given roles as presenters and as critical and constructive listeners. Each student was provided with prompts and tasks, for instance, asking for clarification or giving advice and suggestions for further inquiry. Furthermore, in a project in which students investigated environmental socio-scientific issues, they were provided templates for writing during the investigation (process document and log) and their report (Byhring & Knain, 2016). The assessment criteria focused on textual aspects linked to the purpose of the text; as a result, the clarity and relevance of the research question, the method, relevance, and proper handling of data, and critical and balanced discussion were tied to disciplinary-valid use of representations, the relevance to arguments, and overall cohesion. Teachers in social science, natural science, and L1 used these criteria to provide formative guidance to their students in a wiki space. In a different sub-project in the StudentResearch project, a focus on basic skills addressed the norms and expectations as the context of students' meaning making. Mestad and Kolstø (2014) demonstrated in a classroom study how strong expectations in the learning context to provide correct explanations—before the students had an opportunity to explain their answers using their own words—implied that students did not engage in a process of reflection. However, the team redesigned the learning context to include the following: (1) students' own ideas and explanations were explicitly required; and (2) the students were informed that their own explanations should be written on the blackboard when requested, but not handed in to the teacher. In the next cycle, they found higher levels of engagement from students in formulating explanatory ideas, possibly because they experienced that their understanding—and not the correct scientific explanation—was what the teacher expected and valued.

The Budding Science and Literacy Project (presented in Chap. 16) focused on exploring how working with an integrated inquiry-based science and literacy approach might challenge and support the teaching and learning of science at the classroom level. The interrelationship between multiple learning modalities (writing, reading, talking, and doing) and phases of inquiry (preparation, data gathering, discussion, communication) was studied by observing six teachers and their students, who were recruited from a professional development course. The teachers tried to implement the Budding Science teaching model taught in the course, the underlying features of which are the synergistic integration of science and literacy and the synergistic effects of teaching them simultaneously. There was a strong focus on reading linked to hands-on inquiry and reading as an inquiry activity. The video analysis demonstrated variations and patterns of inquiry-based science and literacy activities, revealing that the multiple learning modalities were all used in the integrated approach; however, not surprisingly, oral activities dominated. The inquiry phases shifted throughout the students' investigations, but the consolidating phases of discussion and communication were given less space. The data phase of inquiry seemed essential as a driving force for engaging in science learning in consolidating situations. While the multiple learning modalities were integrated in all inquiry phases, this was true to a greater extent in the preparation and data phases (Ødegaard, Haug, Mork, & Sørvik, 2014). The results of the project indicated that literacy activities embedded in science inquiry provided support for teaching and learning science; however, the teachers found it hard to find the time and courage needed to use the discussion and communication phases to consolidate the students' conceptual learning (Haug & Ødegaard, 2014).

Representation and Participation in School Science (REDE) is a recent project started at the University of Oslo funded by the Research Council of Norway. This project takes existing research on representation practices as a point of departure for design-based research with teacher participation. A variety of representations, including verbal language, graphs, diagrams, images, animations, simulations, and equations, are used as both tools for learning and for participation in practices involving science. Representations are integral to disciplines of science and to citizenship. Consequently, all the basic skills are combined through working with representations. This project will also conduct research on students' learning processes, teachers' experiences, and the teaching design that fosters students' competent and critical use of representations. Researchers at ILS are responsible for the design-based research and are cooperating with teachers at two upper secondary schools and one lower secondary school. This project draws on design principles that emphasize that teachers should support students in engaging in inquiries about phenomena and issues using representations as tool; that they explicitly address the benefits and limitations of forms of representations; that the adequacy of specific aspects of representations is critically discussed; and that the role of representations in arenas outside school, including scientists' practices and socio-scientific issues, are addressed. This explicitness makes it possible to connect representation practices to science as a practice that contains social norms.

These projects, that is, Students as Researchers in School Science, the Budding Science and Literacy Project, and Representation and Participation in School Science, all had a predominant focus on school science and approached basic skills through the practices of inquiry. They thus emphasized the intersection between the basic skills and the budding researcher. However, the nature of the basic skills and their theoretical understanding differed among the projects described in this section, as did the degree to which each project addressed the basic skills explicitly. The projects that are described in this section are summarized in Table 2.1.

Table 2.1 Summary of projects addressing scientific literacy after the curriculum reform in Norway, LK06, addressed in the chapter

Project	Grade—age	Goal	Approach
WRITE	K-12	Develop functional writing skills across the curriculum	Focusing purpose and use of texts; functional literacy
NORM	Grades 3–9 (age 8–14)	Develop norms of achievement and intervention across the curriculum	Shared norms on quality of text deduced from teachers reflecting on students' texts; intervention
Nadderud project	Upper secondary (age 15–18)	Making norms of text quality explicit for exploration	Cross-disciplinary group of teachers met regularly for 4 years discussing quality and purpose of writing in subjects
Ark & app	Grades 5–11 (age 10–16)	Study of how educational resources are chosen and used in four school subjects	Various functions educational resources can serve within different ways of organizing teaching
StudentResearch	Grades 8 & 11 (age 13–16)	Develop inquiry-based science teaching integrated by literacy	Basic skills as functional tools for inquiry practices, case— and intervention studies
Budding science and literacy	Grades 1–7 (age 6–12)	Develop inquiry-based science teaching integrated with literacy	Learning modalities (writing, reading, talking, and doing) considered in relation to phases of inquiry
REDE	Grades 9–11 (age 14–16)	Develop visual representations practices in science	Representation practices considered key for learning and participating in science discourses

2.3 Discussion

The Norwegian curriculum reform was an ambitious attempt to focus on basic skills in the Norwegian school system. It was partly spurred by an international ambition to identify key competencies. One of these projects was the DeSeCo project, which had a theoretical underpinning of competencies quite close to functional literacy. However, the basic skills were also influenced by more general notions of *allmenndannelse* (bildung, or liberal education). Thus, the theoretical background of the basic skills could be deemed to meet scientific literacy both in the fundamental and the derived senses.

However, the general picture is that teachers and school leaders have been somewhat hesitant to address basic skills explicitly, even if they embrace the intentions of the reform. Yet, communities of teachers who discuss the purposes and qualities of writing have been successfully established when supported by researchers. It seems that the key to the implementation of the reform is for science teachers to realize the functions of text in science, and consequently, to develop a metalanguage that connects science understanding and knowledge to descriptive aspects of text. Furthermore, the social norms regulating scientific practices must be recognized as text norms.

We have discussed two approaches for the implementation of the basic skills in the LK06 reform. One is based on L1 literacy research and literacy considered as a general competence, with a strong emphasis on writing in the disciplines. The other is from the science education community, which has emphasized scientific literacy for several decades. The introduction of BS in LK06 was an opportunity to (re)direct research on scientific literacy, with a stronger emphasis on classroom implementation. There is a certain amount of overlap in terms of the focus, the teachers involved, and the theoretical basis of the two approaches. Several large and small initiatives have emerged since the reform.

We suggest that the two approaches can be mutually beneficial to science education. The general literacy approach is rooted in textual theoretical traditions that have significantly benefited science education and can also benefit school teachers by offering a theoretical framework for discussing text. These traditions can also help improve science teachers' self-understanding by enabling them to compare their practices with other school subjects. One strong point of science educational research is an embedded understanding and implementation of basic skills from a disciplinary perspective, which makes basic skills functional in the teaching of the products and processes of science. The latter tradition also seems to be more alert to the significance of multimodality.

A new curriculum reform is under way in Norway, and the Ministry of Education intends to include the same basic skills in the future as part of the curriculum. This is fortunate, because there are promising developments currently happening in Norwegian science education in terms of addressing the basic skills.

References

Berge, K. L., & Skar, G. (2015). *Ble elevene bedre skrivere? Intervensjonseffekter på elevers skriveferdigheter og skriveutvikling* [Did the students become better writers? Intervention effects on students' writing skills and development]. Report 2 from the NORM project. Retrieved from http://www.uio.no/studier/emner/uv/ils/PPU3220/h16/ble-elevene-bedre-skrivere.pdf

Bjønness, B., & Kolstø, S. D. (2015). Scaffolding open inquiry: How a teacher provides students with structure and space. *Nordic Studies in Science Education, 11*(3), 223–237.

Byhring, A. K., & Knain, E. (2016). Intertextuality for handling complex environmental issues. *Research in Science Education, 46*(1), 1–19.

Flyum, K. H., & Hertzberg, F. (2011). *Skriv i alle fag! Argumentasjon og kildebruk i videregående skole* [Write in every school subject! Argumentation and use of sources in upper secondary school]. Oslo: Universitetsforlaget.

Gilje, Ø., Ingulfsen, L., Dolonen, J., Furberg, A., Rasmussen, I., Kluge, A., et al (2016). *Med ARK&APP. Bruk av læremidler og ressurser for læring på tvers av arbeidsformer* [Use of learning resources for learning across teaching practices]. Report. University of Oslo.

Haug, B., & Ødegaard, M. (2014). From words to concepts: Focusing on word knowledge when teaching for conceptual understanding within an inquiry-based science setting. *Research in Science Education, 44*(5), 777–800.

Hertzberg, F. (2010). Arbeid med grunnleggende ferdigheter [Working on basic skills]. In E. Ottesen & J. Möller (Eds.), *Underveis, men i svært ulikt tempo. Et blikk inn i ti skoler etter tre år med Kunnskapsløftet*. Oslo: NIFU/STEP Rapport 37/2010.

Hertzberg, F., & Roe, A. (2015). Writing in the content areas: A Norwegian case study. *Reading and Writing, 29*(3), 555–576.
Knain, E. (2015). *Scientific literacy for participation: A systemic functional approach to analysis of school science discourses*. Rotterdam: Sense Publishers.
Maagerø, E., & Skjelbred, D. (2010). *De mangfoldige realfagstekstene: Om lesing og skriving i matematikk og naturfag* [A multitude of science texts. On reading and writing in mathematics and science]. Bergen: Fagbokforlaget.
Matre, S., & Solheim, R. (2015). Writing education and assessment in Norway: Toward shared understanding, shared language and shared responsibility. *L-1 Educational Studies in Language and Literature, 15*, 1–33.
Mestad, I., & Kolstø, S. D. (2014). Using the concept of ZPD to explore the challenges of and opportunities in designing discourse activities based on practical work. *Science Education, 98*(6), 1054–1076.
Ministry of Education and Research. (2004). *Kultur for læring* [Culture for learning]. Stortingsmelding 30 (2003–2004). Oslo: Utdannings- og forskningsdepartementet.
Ministry of Education and Research. (2006). *Kunnskapsløftet - Knowledge promotion*. Information for pupils and parents/guardians. Retrieved from https://www.regjeringen.no/globalassets/upload/kilde/kd/bro/2006/0002/ddd/pdfv/292311-kunnskapsloftet2006_engelsk_ii.pdf.
Mork, S. M. (2013). Revidert læreplan i naturfag. Økt fokus på grunnleggende ferdigheter og forskerspiren 8 [Revised curriculum in science. Icreased attention to basic skills and the Budding researcher]. *Nordic Studies in Science Education, 9*(2), 206–210.
Norris, S. P., & Phillips, L. M. (2003). How literacy in its fundamental sense is central to scientific literacy. *Science Education, 87*(2), 224–240.
Norwegian Directorate for Education and Training. (2012). Framework for basic skills. To use for subject curricula groups appointed by the Norwegian Directorate for Education and Training. *Norwegian Directorate for Education and Training*. Retrieved from http://www.udir.no/contentassets/fd2d6bfbf2364e1c98b73e030119bd38/framework_for_basic_skills.pdf
Ødegaard, M., Haug, B., Mork, S. M., & Sørvik, G. O. (2014). Challenges and support when teaching science through an integrated inquiry and literacy approach. *International Journal of Science Education, 36*(18), 2997–3020.
Official Norwgian Reports - NOU (2003). *I første rekke. Forsterket kvalitet i en grunnopplæring for alle* [First in line. Strenghtened quality in primary and secondary school] (Vol. 16): Statens forvaltningstjeneste.
Rødnes, K., & Gilje, Ø. (2016). *Grunnleggende ferdigheter: På tvers eller i fag?* [Basic skills: Across or within school subjects?] Retrieved from http://www.uv.uio.no/iped/forskning/prosjekter/ark-app/rodnes_gilje_ark_app_grf_2016.pdf
Rychen, D. S., & Salganik, L. H. (Eds.). (2003). *Key competencies for a successful life and a well-functioning society*. Göttingen: Hogrefe & Huber.
Sivesind, K. (2012). Kunnskapsløftet: Implementering av nye læreplaner i reformen. Synteserapport fra evalueringen av Kunnskapsløftet [Knowledge promotion: The implementation of new curricula in the reform]. *Acta Didactica, 2*.
Smidt, J., Solheim, R., & Aasen, A. J. (Eds.). (2011). *På sporet av god skriveopplæring - ei bok for lærere i alle fag* [Developing good teaching of writing—a book for teachers in every school subject]. Bergen: Fagbokforlaget.
Solheim, R., & Matre, S. (2014). Forventninger om skrivekompetanse. Perspektiver på skriving, skriveopplæring og vurdering i «Normprosjektet» [Expectations on writing skills. Perspectives on writing, teaching of writing and assessment in the "Normproject"]. *Viden om læsning, 15*, 76–88.
Sørvik, G. O., Blikstad-Balas, M., & Ødegaard, M. (2015). "Do Books Like These Have Authors?" New Roles for Text and New Demands on Students in Integrated Science-Literacy Instruction. *Science Education, 99*(1), 39–69. doi:10.1002/sce.21143.
The Norwegian Directorate for Education and Training. (2016). Natural science subject curriculum. *Basic skills*. Retrieved from http://www.udir.no/kl06/NAT1-03/Hele/Grunnleggende_ferdigheter?lplang=eng

Chapter 3
But I'm Not an English Teacher!: Disciplinary Literacy in Australian Science Classrooms

Chris Davison and Sue Ollerhead

Abstract In Australia, increasing calls to strengthen the teaching of science in schools, stimulated by a need for higher levels of scientific understanding in the general community and also concerns about falling educational standards, have led to a concerted push to raise scientific literacy. Given the multicultural nature of Australia, government and educators have acknowledged the critical role of identifying and describing the language and literacy demands in all disciplines and supporting learners and teachers to meet them. In science, a range of support material and teacher resources, mostly drawing on systemic functional linguistics, have been developed alongside revisions to national and state-based curricula. In addition, all preservice teacher education programs are required to address literacy as well as the needs of students learning in and through English as an additional language or dialect (EALD) as national priority areas. However, research into the complex language and literacy challenges faced by low literacy and EALD learners suggests that many mainstream secondary school teachers feel inadequately prepared to meet their needs. Thus, some universities are also developing more targeted language and literacy programs for preservice student teachers (PSTs). One example of a language and literacy tutoring program involving science PSTs at a secondary school is described, showing how the student teachers came to understand their key roles not just a disciplinary experts, but as providers of language and literacy development to students both before and after their mentoring placement.

Keywords Science · English language · literacy · curriculum · preservice teacher education

---

C. Davison (✉) · S. Ollerhead
School of Education, University of New South Wales, Sydney, Australia
e-mail: c.davison@unsw.edu.au

S. Ollerhead
e-mail: s.ollerhead@unsw.edu.au

© Springer International Publishing AG 2018
K.-S. Tang, K. Danielsson (eds.), *Global Developments in Literacy Research for Science Education*, https://doi.org/10.1007/978-3-319-69197-8_3

3.1 Introduction

In Australia, as in many other countries, increasing recognition of the need for higher levels of scientific understanding in the general community, combined with concerns about falling educational standards, have led to a new emphasis on raising levels of scientific literacy in schools. According to the Australian Curriculum Assessment and Reporting Authority (ACARA), scientific literacy is defined as a student's ability to apply broad conceptual understandings of science in order to make sense of the world; to understand natural phenomena; and to interpret media reports about scientific issues. It also measures the ability to ask investigable questions, conduct investigations, collect and interpret data, and make informed decisions. This construct evolved from the definition of scientific literacy used by the Organisation for Economic Co-operation and Development (OECD) – Programme for International Student Assessment (PISA), and has been adopted in all Australian curriculum and assessment materials. However, given the multicultural nature of Australia, government and educators have also acknowledged the critical role of identifying and describing the language and literacy demands in all disciplines and supporting learners and teachers to meet them. Hence, in science, a range of support material and teacher resources, most drawing on systemic functional linguistics, have been developed alongside revisions to national and state-based curricula. In addition, all preservice teacher education programs are required to address literacy as well as the needs of students learning in and through English as an additional language or dialect (EALD) as national priority areas. However, research into the complex language and literacy challenges faced by low literacy and EALD learners suggests that many mainstream secondary school teachers feel inadequately prepared to meet their needs. Thus, some universities are also developing more targeted language and literacy programs for preservice student teachers (PSTs).

This chapter will first describe the reasons for a stronger emphasis on language and literacy in Australian curriculum documents and standards and in teacher education, before describing a specific language and literacy tutoring program involving PSTs at a secondary school. This example will show how a particular group of student teachers came to understand their key roles not just as disciplinary experts, but also as providers of language and literacy development to students both before and after a mentoring placement.

3.2 The Call for Greater Scientific Literacy in Australia

Science education is growing in importance in the twenty-first century as more careers require scientific knowledge and skills, and global problems such as climate change demand higher levels of scientific literacy in the broader community. However, international benchmarks of student performance in science such as

PISA show a steady decline over time, exemplified by an increase in the proportion of Australian students in the lowest bands in science literacy and a decrease in higher bands (Thomson, De Bortoli, & Buckley, 2013). According to Australian Chief Scientist (Chubb, 2014):

> We have worrying gaps in the STEM skills pipeline, from primary to tertiary education levels. Australia's relative performance in science has slipped. Of the countries tested in 2006 and 2012, five significantly outperformed Australia in 2012, whereas only three did in 2006. Australian schools also show a decline in the rates of participation in "science" subjects to the lowest level in 20 years. (p. 10)

The lower the socioeconomic background status (SES), the lower the students' performance (Thomson et al., 2013, p. xviii). While only 3% of students in the high SES quartile fell below the 2012 PISA international benchmark in scientific literacy, 22% in the lowest SES quartile failed to reach it (Marginson, Tytler, Freeman, & Roberts, 2013, p. 16). The same pattern can be seen in school leaving results, for example, in the Victorian Certificate of Education chemistry examinations, "the rate of failure soared as the social scale was descended" (Teese, 2013, p. 105). The intersection of language background other than English with SES is a complicating factor. In New South Wales (NSW), the most populous state in Australia, more than 30% (over 240,000) of students are from a language background other than English (LBOTE), and about 20% of all students (over 145,000 students) in NSW government schools are learning English as an additional language or dialect. In NSW schools, a 62% increase in EALD learners over past 3 years has occurred with no extra EALD teacher support; in 2014 alone, 138,487 students needed ESL support, only 91,401 received it.[1]

Internationally, it is now widely recognized that immersing English language learners in English content classrooms by itself is not an adequate solution to language/literacy or cognitive/academic development (Davison & Williams, 2001; Gibbons, 2009, 2014; Hammond, 2014). Learning to use English for academic purposes requires considerably more time than is the case for conversational or social English (Cummins & Early, 2011); simply placing students in English-medium mainstream classes cannot be assumed to provide optimal language learning opportunities.

Similarly, science education researchers (Prain & Tytler, 2012) emphasize that all students need to understand why and how discipline-specific and generic literacies are used to build and validate scientific knowledge. Articulating scientific

---

[1] English an Additional Language/Dialect (EALD) education in NSW is provided in primary and secondary schools and in intensive English language centers to support the English language development of students whose first language is not English. In these schools, EALD programs are delivered in a variety of ways to meet the different needs of EALD students at different stages of learning English, focusing on students learning English in the context of the curriculum they are studying so that they acquire the English language skills relevant to the subject area. Students may receive support from a specialist EALD teacher working with a class teacher or they may be in a separate parallel group for some classes.

knowledge is therefore a critical task, yet in science education, learning about the language of science and how to produce it is a major challenge for students. In Australia, a major factor contributing to declining results in science appears to be the failure of low SES students to develop the disciplinary-specific forms of language and literacy, including graphic and visual literacy, required in progress to senior grades (Macnaught, Maton, Martin, & Matruglio, 2013). For example, Teese (2013) argues that the technical linguistic precision and sophisticated mathematical dexterity required in short answer examination questions in Victorian Certificate of Education Chemistry exceeds the literacy capacities of low SES students.

In the 1980s, Australia was one of the first countries to embrace Halliday's (1978) view that the uses of language are inseparable from its social functions, with language defined in terms of its meaning potential, as a set of linguistic choices to be made, explicitly negating the separation of language and content. The critical role of language in the knowledge building of school disciplinary content was extensively described (Christie & Derewianka, 2008; Freebody, Maton, & Martin, 2008; Halliday & Martin, 1993; Martin, 2013; Unsworth, 1999a, b, c, 2004), stimulating a number of studies which found that enhancing teacher knowledge of meaning-making systems through the use of systemic functional linguistics (SFL) could improve student understandings of academic concepts (Love & Humphrey, 2012; Schleppegrell, 2004). Two key aspects of language and literacy in science were foregrounded through this research. First, the ways in which the verbal, visual, mathematical, and symbolic discourse of science differ significantly from the everyday discourse and literate practice of lowSES and EALD students (Fang, 2005, 2006; O'Halloran, 2003; Unsworth, 2000), and second, how the increasingly technical and abstract scientific language at the higher levels of schooling leaves many of these learners behind. As Fang & Schleppegrell (2008) demonstrate, "the language of science is simultaneously technical, abstract, dense, and tightly knit – features that contrast sharply with the more interactive and interpersonal language of everyday spontaneous speech" (p. 20). Such scientific language builds cumulatively through schooling. Christie and Derewianka (2008) found that although in elementary school Science, the focus is generally on "doing science" through linguistic genres such as procedures and procedural recounts, as students move through school, the focus becomes more abstract and technical, with students learning about "organizing science," describing and classifying information through reports and taxonomies and "explaining science" through articulating sequential, causal, or theoretical relationships. Near the end of compulsory schooling, the focus of science enlarges further to include "challenging science," using exposition to argue and justify a case. As Echevarria et al. (2004) observe, if students do not possess the linguistic skills required to engage in the discussion necessary for scientific enquiry, as is often the case with low SES and EALD learners, they will struggle with academic reasoning. However, research also shows how even a basic knowledge of genres and their key linguistic features in science builds teachers' capacity to make discipline knowledge visible and accessible to their students (Fang & Schleppegrell, 2008;

Martin, 2013; Unsworth, 2000), demonstrate the interplay between language and other meaning-making systems (Georgiou, Maton, & Sharma, 2014; Macnaught et al., 2013), and support the development of increasingly technical and abstract discourse across the school years in ways that enhance student achievement (Gibbons, 2009; Rose & Martin, 2014; Schleppegrell, 2001; Seah, Clarke, & Hart, 2011).

This implies that teachers need a deep understanding of the linguistic features and literacy practices of their subject, especially in countries such as Australia, USA, Canada, UK, and Singapore, which have high numbers of English language learners. Hence, in Australia there is now an even stronger emphasis on language and literacy in state and national curriculum documents and standards, including preservice teacher education, described in the following section.

3.3 A Stronger Emphasis on Language and Literacy in Australian Curriculum Documents and Standards and in Teacher Education

In the USA, calls from industry for more science, technology, engineering, and math (STEM) education and the introduction of new national Next Generation Science Standards have turned the spotlight onto the most effective ways to teach scientific disciplinary content, with specific attention to integrating language and literacy instruction with content knowledge (Echevarria, Vogt, & Short, 2011; Turkan, De Oliveira, Lee, & Phelps, 2014). According to De Jong (2013), this has resulted in a focus on the "nature and quality" of the ways in which preservice teacher education programs prepare their students to teach language and literacy (p. 40). Similarly, in Australia in the last 5 years or so there has been an even stronger focus on identifying and describing the language and literacy demands in all disciplines and supporting learners and teachers to meet them, with new national standards in curriculum and teacher education requiring all teachers, not only English language specialists, address language and literacy, with a particular focus on the needs of learners for whom English is an additional language or dialect (ACARA, 2012).

In science, a range of support material and teacher resources have been developed alongside revisions to national and state-based curricula. The new Australian Curriculum for science (http://www.australiancurriculum.edu.au/science/general-capabilities) emphasizes the subject-specific nature of the language and literacy requirements; the need to "communicate ideas, explanations and processes using scientific representations in a variety of ways, including multi-modal texts" (p. 110). A number of projects have also been initiated to help science text book writers to design comprehensible and engaging texts, and develop strategies to support the increasing numbers of students from EALD and educationally disadvantaged populations in Australian schools.

In addition, all preservice teacher education programs are required to address the Graduate Teacher Standards (AITSL, 2011), with designated National Priority areas in both literacy and teaching students with English as an additional language or dialect (BOSTES, 2014), including literacy across the curriculum and effective teaching and learning strategies for teaching second language learners in the context of the mainstream classroom and the range of key learning areas. It is expected that PSTs will know how to implement pedagogic strategies that facilitate EALD learners' access to scientific knowledge, and to understand which aspects of the science curriculum learners may find challenging and why. For example, where EALD students are expected to provide evidence-based explanations of scientific processes, they may need explicit modelling and teaching of conditional tenses, for example, *if x then y; if x occurred then y would occur; if x had occurred then y would have occurred* (ACARA, 2016). Furthermore, PSTs need to learn how to engage with and respond to EALD learners' cultural diversity, and how to draw upon their valuable "funds of knowledge" (Gonzales, Moll, & Amanti, 2005) as potential learning opportunities for all learners within the classroom. Inviting EALD learners to share their culturally specific examples with their peers will help to recognize and affirm their cultural identities in the classroom (Cummins & Early, 2011).

Not surprisingly, perhaps, Australian teachers are interested in learning about cultural and linguistic diversity, but Hammond (2011) has also found that mainstream content teachers still lack confidence in their ability to incorporate their language and literacy knowledge into their content teaching. Research internationally mirrors this finding; even experienced teachers feel inadequately prepared to meet EALD and low literacy needs, especially at the secondary level (De Jong, 2013; Premier & Miller, 2010; Reeves, 2006). Addressing language and literacy across the curriculum is challenging (Hurst & Davison, 2005; Khong & Saito, 2013; Short & Echevarria, 1999; Turkan et al., 2014), with content specialists immersed in the discourse of their discipline not easily recognizing its language and literacy demands, and/or assuming responsibility for language and literacy development is solely the province of "English" teachers with minimal discussion of language-related problems within subjects (Coady, Harper, & De Jong, 2015; Davison, 2016; Gleeson & Davison, 2016). Davison (2006) and Arkoudis (2007), building on earlier work by Siskin (1994), highlight the sub-communities within each subject discipline which play a critical role in shaping and supporting teachers' identities. Each community has distinct views about the canons of knowledge within the subject discipline, a sense of the importance of their discipline within the institution, and shared assumptions of what needs to be taught and when. This explains one of the main barriers to integrating language development into disciplinary areas at the school level, that is, subject knowledge is viewed as belonging to the teachers in that discipline. Thus, most science specialists see teaching skills such as speaking or grammar as the work of English teachers, not their responsibility. Even experienced science teachers with an understanding of scientific language rarely have well-developed "knowledge about language" (Love, 2010) or metalinguistic awareness – an ability "to extract themselves from the normal use of language and focus their attention on the functions and forms of the language" (Masny, 1997, p. 106). This awareness is needed to teach PSTs how language is used in science (Lee, 2004; Lemke, 1990).

Most attention in the literature on building language and literacy awareness has focused on inservice, especially on methods/techniques to use in the classroom and the analysis of linguistic demands of content areas, with Australian initiatives in this area such as the *ESL in the Mainstream* inservice program being exported to Asia and Europe (Davison, 2006, 2016). Comparatively little attention has been paid to preservice education, with some exceptions (De Jong, 2013; Kibler, Walqui, & Bunch 2014; Love, 2016; Polat & Mahalingappa, 2013), although even student teachers have strong pedagogic beliefs and assumptions about their subject area and what good teaching (and learning) means to them, which evolve with their sense of professional identity (Arkoudis, 2007).

For these reasons in the last decade, many Australian universities have developed language awareness programs in preservice teacher education. One early example was an 18-hour core unit on language in education at the University of Melbourne, utilizing an interactive video-based CD-ROM, consisting of transcripts, various drag-and-drop tasks, and a glossary (Love, Baker, & Quinn, 2008). It was used over 5 years with more than 800 graduate nonliteracy specialists, very effectively introducing them not only to the role of language and literacy in learning, including in EALD, but also to the grammar of technicality, abstraction, density, and coherence. What is also needed, however, are preservice education programs which integrate theory and practice in the field, engaging students in learning-while-doing (Darling-Hammond, 2008), giving prospective teachers the opportunity to integrate the theory and practice of being a teacher. One such initiative in Australia is described below.

3.4 New Directions: Language and Literacy "Mentoring" in Preservice Education

The University of New South Wales has always had a strong focus on the development of language and literacy in its preservice education programs, however, this is now combined with systematic efforts to link theory and practice so PSTs are better prepared to support their students with the language and literacy demands of the subject disciplines, including science (Ollerhead, 2016). In 2015, a mentoring program was established in a low SES secondary school involving 35 PSTs in the second year of their four-year program and 110 secondary students in Year 7–11, identified by their classroom teachers as requiring additional support with language and literacy. Over the course of a 14-week semester, each PST conducted three weekly one-on-one tutoring sessions with three different learners each, totaling 40 contact hours of tutoring. The PSTs were trained to implement pedagogical strategies known to be effective for developing language and literacy in the content areas through five 2-hour-long face-to-face workshops and ongoing online support.

The key idea guiding the training workshops was to provide PSTs with an understanding of the nature of the academic language that EALD and low literacy learners needed to master in order to access and master scientific concepts. Workshops were thus structured around Meltzer's (2002) framework of best practices for supporting academic literacy, including addressing learners' literacy engagement and motivation, implementing research-based literacy strategies and integrating reading and writing across the curriculum (pp. 14–16). Specific strategies addressed in the workshop included making connections to learners' lives through responsive pedagogies, teaching through modelling, providing explicit literacy strategies for engaging with texts, recognizing and analyzing discourse features, understanding text structures, and paying explicit attention to vocabulary development. Thus, the use of genre pedagogy (Christie & Martin, 1997; Derewianka, 1991), where the curriculum cycle is used to plan activities that provide language and literacy scaffolding for learners, underpinned much of the workshop content. PSTs were also taught how to modulate their teacher talk to promote deeper thinking and higher learning, by using techniques such as wait time, recasts, and probing questions to produce extended stretches of speech which could be converted to more expert scientific texts (Gibbons, 2014). Throughout the program, the academic mentor encouraged PSTs to see their role as facilitating learners' capacity to "think like mathematicians, read like historians and write like scientists" by teaching them different ways of reading and writing within each field (Lee, 2004, p. 61).

During each workshop, PSTs were invited to bring a specific example of a science text that their learners were working on, and to discuss the typical generic and linguistic features of the text with the group. For example, PST Dominic presented a task in which one of his learners Jayden had to write a description of the water cycle. PSTs worked collaboratively to analyze the key features needed to produce the text successfully so that it accurately represented the key concepts and processes, working with Gibbons' summary of typical features of school-related genres (2014, pp. 173–178). Together they identified that Jayden needed to produce a "causal explanatory text," which required him to identify the phenomenon, in this case the water cycle, and provide an explanation using a cause-and-effect sequence of events. They brainstormed and agreed upon some typical linguistic features Jayden would need to use, including technical nouns such as "evaporation" and "condensation," classifying adjectives such as "*salt water* intrusion," and time conjunctions such as "once," "after," or "at the second stage." They also discussed that he would need to use the passive voice to foreground the various processes involved in the cycle, as in "salt water intrusion *is shown* to take place" in order to accurately capture the phenomenon. Similarly, he would need to substitute nominalizations such as "evaporation" for descriptions such as "the process whereby water evaporates." Other texts that were collected from science classes discussed in workshops included information reports on bushfires and native plants, and an investigation report into changes in rock formation over time. Through engaging in progressive collaborative enquiries, PSTs gradually began to grow in their confidence to analyze the discourse features required of different texts in order to capture more accurately the scientific concepts and processes.

The nature of the knowledge that PSTs gained through participating in both the training workshops and the fieldwork component, and the ways in which this knowledge contributed toward their identity development, was the object of ongoing research. Interviews with PSTs provided insights into how they positioned themselves with regard to their growing understanding of pedagogical language and literacy knowledge. This afforded an understanding of PSTs' shifting identities, which correspond with Gee's (1996) concept of "ways of being in the world." Data was collected by means of semi-structured interviews conducted with the eight PST participants toward the beginning of the program (Week 3) and at the end of the program (Week 15). Interviews were of approximately 40 minutes duration, and were audio-recorded, transcribed, and analyzed for emergent themes, which were identified and coded. In addition to interview data, PSTs' contributions to an online discussion forum initiated by the academic mentor to provide feedback and advice to learners were also analyzed to yield a clearer picture of their emerging knowledge and shifting identities. The online posts in which PSTs discussed their experiences and growing knowledge during the practicum were collated and subjected to the same thematic analysis as that used with the interview transcripts.

PSTs were asked questions both in the interviews and in the online forums which related to how they conceptualized their role as teachers of language and literacy within both science and their own key learning areas. They were also asked to reflect on how they approached the embedding of language and literacy into curriculum content during their field work placements. They communicated their understandings of the concept of "academic language," and reflected upon on the extent to which they felt they had been prepared to support EAL learners' language and literacy needs within content subjects throughout their ITE program. In addition, students were asked to articulate the key theories underlying specific strategies taught to foster language and literacy skills in the classroom, such as socio-cultural, Vygotskian theory, genre theory, scaffolded support, and so on. They were also asked to reflect upon their knowledge of genre pedagogy, text types, linguistic, structural, and cultural features of the texts within their specific content areas. Finally, they were asked to evaluate the role that they thought academic mentoring played in preparing them to attend to language and literacy needs of EAL learners during the course of the program. Therefore, as PSTs spoke and "authored themselves" (Holland, Lachicotte, Skinner, & Cain, 1998), they created their identities as teachers.

While similar types of questions were asked of PSTs in both round 1 (Week 3) and round 2 (Week 15) of the interviewing phase, to gauge a sense of participants' growing knowledge of language and literacy over time, the first round of interviews was focused on perceptions and initial understandings of language and literacy pedagogy, whereas the second round had a more reflective focus. Students were asked to relate the ways in which their knowledge and confidence had grown over time during their participation in the mentoring program. All of the participants reported that language and literacy was a significant challenge for the learners they had been mentoring. There was an overwhelming sense that both the workshop and practicum components of the program were highly beneficial in helping the PSTs to focus on issues relating to language and literacy in a systematic way and to improve their knowledge and practices. The majority of respondents expressed the view that the program had made salient for them the fact that careful attendance to language and literacy support of all learners, but particularly EALD learners, was a fundamental cornerstone of effective teaching, and stressed their belief that explicit language and literacy pedagogies should be prioritized and foregrounded even more within their teacher education program. They found the Australian Curriculum EALD resource (ACARA, 2016) a rich and helpful resource for making informed decisions about the quality of learners' writing, where they were in their EAL learning progression and where they should be. More specifically, over the course of the semester all the PSTs reported a growing awareness of the linguistic features and patterns of scientific texts, with increasing understanding and knowledge of the nature of disciplinary literacies, including scientific literacy, and how to approach the explicit teaching of text structure, linguistic features, and literacy strategies for planning, researching, and revising, elaborated below.

Initially, the PSTs conceptualized language and literacy support as "understanding the rules of grammar" and how to produce a "good piece of writing," yet they had little understanding of the different genres or text types. Their growing awareness came about gradually as a result of getting to know learners better through recurring weekly meetings, building up rapport and background knowledge which enabled PSTs to identify areas of language and literacy that their learners found challenging. Their increased understanding and knowledge of disciplinary literacies over the course of the program is exemplified in the comment by one of the PSTs, Dominic:

> My first three sessions with Vincent were awful. I was giving him general abstract topics that he wasn't interested in and trying to show him how to go about writing a "good essay" based on the topic. It was only when I actually observed him in a science lesson and saw how he struggled, that I realised I needed to focus on science literacy with him, rather than literacy in general.

Dominic realized that the decontextualized grammar exercises and "rules" about English which he offered to his learners initially did little to help them master the writing tasks needed in their various subject areas. Once he was able to link each literacy task to a specific text type that his students needed to complete in class, he was able to identify certain linguistic features and text patterns that helped his student to master the activity. Dominic also felt that his learners benefited significantly from a more explicit approach to teach planning for literacy tasks, modeling different strategies regarding how to approach and tackle literacy tasks:

> For Fama, ... one of the things that I realised was that nobody ever explicitly told him how to plan approaching a piece of work ... so we went through in quite a lot of detail and we spent a couple of weeks on it without actually producing much work, in terms of going through planning, researching, put this stuff on Post-it notes and putting them in order. (Dominic)

Similarly, PST Jack related how he assisted a reluctant Year 8 writer, Yedica, to approach the writing of an explanatory text about the features, causes, and impacts of bushfires on the environment, in a more thorough, purposeful, and logical way. He commented:

> You need to make everything very explicit for them in their literacy task, whether you like it or not. There needs to be very clear scaffolding and step-by-step guides. Like if we are doing an explanation text about bushfires, we need to give them a pattern, giving them an example of what that kind of text looks like, make salient the vocabulary items, the linguistic structures, that type of thing.

PST Monica reflected upon her understanding of the need to explicitly attend to vocabulary development, and how she worked on encouraging her learners to "talk like scientists" in order to prepare them to "write like scientists." She related how, as a result of a strategy discussed in an earlier workshop, she modelled an "experiment" for her learner, placing a magnetic paper clip on both a metal filing cabinet (where it stuck) and against a wooden door (where it did not), and asked her learner to express what she observed. Through the use of probing questions, recasts, and elicitation of technical and scientific terms, as well as much encouragement, Monica related how she managed to facilitate the student's oral observation from "The magnet sticks to the cabinet but not to the door," to eventually arrive at "magnetic attraction occurs between ferrous metals" (Gibbons, 2009). Monica expressed a profound sense of achievement at having helped her learner to "talk her way to science literacy."

In the same way, PST Cathy discussed the value of dialogic talk in developing learners' oral literacy and building up their technical vocabulary from their everyday commonsense understandings, prior to attempting to complete written tasks in Science. The focus was on helping the students develop a more sound conceptual understanding of the scientific phenomenon which was also more technically correct. She relates helping three of her learners to produce an explanatory text of a variety of plant species growing on the school grounds:

> They had to find a plant that grew on the school grounds and then had to say the scientific name, the indigenous name, the indigenous meaning, the features of it and how that has helped it to survive …. I made sure for them to tell me how they think the feature helped it survive, because a lot of the features of the plants were like … big leaves for rain or shade … and so I tried to make that connection for them for them to see that this feature helped it survive. I made sure they didn't just look up the answers on Google, and like they tried to figure out and reason why they would survive.

Through learning how to address students' language and literacy needs, PSTs also developed their own identities as agents for change and transformation, opening up access for learners and addressing equity in education for EALD learners. As Dominic (Interview, December 2015), commented:

> It's easy to overlook the fact that I have a very privileged access to the education system that some of my learners don't have. So certainly being able to …. I think a critical thing for me, from a social justice perspective, is around how I can be cognizant of that fact and bridge those gaps for people who don't have that kind of access.

The value of exposing or "sensitising" PSTs early in their training to the complexities of language and literacy across the curriculum was also demonstrated succinctly in a comment by Kristina at the end of the program:

> I know more about literacy – what I need to do to teach literacy and that kind of thing and so I feel like this has helped with that a lot and I know now what I need to find out to be able to teach which is also very helpful. I don't like not knowing but at least now I know what I don't know. (Kristina, Interview, December 2015)

All participants reported that they most valued the interlinking of theory and practice, the ability to implement strategies introduced and practised in workshops almost immediately with real leaners. However, they also learnt that attending to literacy and language needs is not just about skills:

> It's about how people learn it's about how people engage, it's about the importance of understanding who you're working with and the nice thing about doing it in a one-on-one environment, is you do have that time to reach out and understand the background and needs and the links and the motivation. It becomes a collaboration, rather than an intervention. (Dominic, December 2015)

This initiative in preservice teacher education demonstrates the value of increased knowledge about language and literacy, but also, perhaps more importantly, the need for explicit, hands-on integration of theory and practice in literacy and language pedagogy and structured opportunities for self-reflection and "re-positioning" or extending of teaching roles.

3.5 Conclusion

Language and literacy development and disciplinary content learning in science can progress "hand-in-hand" (Gibbons, 2002, p. 6), provided there are clear and comprehensive school-level policies, language-infused curriculum and assessment and appropriate support structures and professional learning opportunities. This brief review of developments in Australia helps to illuminate the vital role of such resources in preparing skilled, engaged, and responsive teachers for increasingly multicultural and multilingual classrooms in Australia and internationally.

References

Arkoudis, S. (2007). Collaborating in ESL education in schools. In J. Cummins & C. Davison (Eds.), *International handbook of English language teaching*. Norwell: Springer.

Australian Curriculum and Reporting Authority (ACARA). (2012). *The shape of the Australian curriculum, version 3*. Retrieved from http://www.acara.edu.au/verve/_resources/the_shape_of_the_australian_curriculum_v3.pdf

Australian Curriculum and Reporting Authority. (2016). *Australian curriculum version 8.1 F-10: Science*. Retrieved from http://www.australiancurriculum.edu.au/science/rationale

Australian Institute for Teaching Standards and Leadership (AITSL). (2011). *Australian professional standards for teachers*. Melbourne: Education Services Australia.

Board of Studies, Teaching and Educational Standards (BOSTES). (2014). NSW supplementary documentation: Elaborations in priority areas. Retrieved from http://www.nswteachers.nsw.edu.au/search/?page=2&SearchText=Mandatory%20Areas%20of%20Study

Christie, F., & Derewianka, B. (2008). *School discourse: Learning to write across the years of schooling*. London: Continuum.

Christie, F., & Martin, J. R. (1997). *Genres and institutions: Social processes in the workplace and school*. London: Cassell Academic.

Chubb, I. (2014). Health of Australian science report. Retrieved from http://www.chiefscientist.gov.au/wpcontent/uploads/HASReport_Web-Update_200912.pdf. Accessed 28 November, 2014.

Coady, M. R., Harper, C., & De Jong, E. J. (2015). Aiming for equity: Preparing mainstream teachers for inclusion or inclusive classrooms? *TESOL Quarterly, 50*(2), 340–368.

Cummins, J., & Early, M. (Eds.). (2011). *Identity texts: The collaborative creation of power in multilingual schools*. Stoke-on-Trent: Trentham Books.

Darling-Hammond, L. (2008). Knowledge for teaching: What do we know? In M. Cochran-Smith, S. Feiman-Nemser, D. J. McIntyre & K. E. Demers (Eds.), *Handbook of research on teacher education: Enduring questions in changing contexts*. 3rd ed. (pp. 1316–1323). New York: Routledge, Taylor & Francis Group & the Association of Teacher Educators.

Davison, C. (2006). Collaboration between ESL and content teachers: How do we know we are doing it right? *International Journal of Bilingual Education and Bilingualism, 9*(4), 454–475.

Davison, C. (2016). Collaboration between English language and content teachers: Breaking the boundaries. In A. Tajino, T. Stewart & D. Dalsky (Eds.), *Team teaching and team learning: Collaboration for innovation in language classrooms* (pp. 51–66). New York: Routledge.

Davison, C., & Williams, A. (2001). Integrating language and content: Unresolved issues. In B. Mohan, C. Leung & C. Davison (Eds.), *English as a second language in the mainstream: Teaching, learning and identity* (pp. 51–70). Harlow: Longman Pearson.

De Jong, E. (2013). Preparing mainstream teachers for multilingual classrooms. *Association of Mexican-American Educators (AMAE), Special Invited Issue, 7*(2), 40–48.

Derewianka, B. (1991). *Exploring how texts work*. Sydney: Primary English Teaching Association.

Echevarria, J., Vogt, M., & Short, D. (2004). *Making content comprehensible for English language learners: The SIOP model*. Boston: Pearson Education.

Fang, Z. (2005). Scientific literacy: A systemic functional linguistics perspective. *Science Education, 89*, 335–347.

Fang, Z. (2006). The language demands of science reading in middle school. *International Journal of Science Education, 28*(5), 491–520.

Fang, Z., & Schleppegrell, M. J. (2008). *Reading in secondary content areas: A language-based pedagogy*. Ann Arbor: University of Michigan Press.

Freebody, P., Maton, K., & Martin, J. (2008). Talk, text and knowledge in cumulative, integrated learning: A response to 'intellectual challenge'. *Australian Journal of Language and Literacy, 31*(2), 188–201.

Gee, J. (1996). *Social linguistics and literacies: Ideology in discourses*. 2nd ed. London: Taylor and Francis.

Georgiou, H., Maton, K., & Sharma, M. (2014). Recovering knowledge for science education research: Exploring the "Icarus Effect" in student work. *Canadian Journal of Science, Mathematics and Technology Education, 14*(3), 252–268.

Gibbons, P. (2009). *English learners, academic literacy and thinking: Learning in the challenge zone*. New York: Heinemann.

Gibbons, P. (2014). *Scaffolding language, scaffolding learning. Teaching second language learners in the mainstream classroom*. 2nd ed. Sydney: Primary English Teaching Association (PETA).

Gleeson, M., & Davison, C. (2016). A conflict between experience and professional learning: Subject teachers' beliefs about teaching English language learners. *RELC Journal, 47*(1), 43–57.

Gonzales, N., Moll, L., & Amanti, C. (2005). *Funds of knowledge: Theorizing practices in households, communities and classrooms*. New York: Taylor and Francis.

Halliday, M. A. K. (1978). *Language as social semiotic: The social interpretation of language and meaning*. London: Edward Arnold.

Halliday, M. A. K., & Martin, J. R. (1993). *Writing science: Literacy and discursive power*. London: Falmer.

Hammond, J. (2011). Working with children who speak English as an additional Language: An Australian perspective on what primary teachers need to know. In S. Ellis & E. McCartney (Eds.), *Applied linguistics and primary school teaching* (pp. 32–43). Cambridge: Cambridge University Press.

Hammond, J. (2014). An Australian perspective on standards-based education, teacher knowledge, and students of English as an additional language. *TESOL Quarterly, 48*(3), 507–532.

Holland, D., Lachicotte, W., Skinner, D., & Cain, C. (1998). *Identity and agency in cultural worlds*. Cambridge: Harvard University Press.

Hurst, D., & Davison, C. (2005). Collaborating on the curriculum: Focus on secondary ESOL. In J. Crandall & D. Kaufman (Eds.), *Content-based instruction in primary and secondary school settings* (pp. 41–66). Alexandria: TESOL.

Khong, T. D. H., & Saito, E. (2013). Challenges confronting teachers of English language learners. *Educational Review, 66*(2), 210–225.

Kibler, A. K., Walqui, A., & Bunch, G. C. (2014). Transformational opportunities: Language and literacy instruction for English language learners in the common core era in the United States. *TESOL Journal, 6*(1), 5–23.

Lee, C. (2004). Literacy in the academic disciplines. *Voices in Urban Education, 3*, 14–25. Winter/Spring.

Lemke, J. (1990). *Talking science: Language, learning and values*. Norwood: Ablex.

Love, K. (2010). Literacy pedagogical content knowledge in the secondary curriculum. *Pedagogies: An International Journal, 5*(4), 338–355.

Love, K., Baker, G., & Quinn, M. (2008). *Literacy across the school subjects (LASS) DVD*. University of Melbourne.

Love, K., & Humphrey, S. (2012). A multi-level language toolkit for the Australian Curriculum: English. *Australian Journal of Language and Literacy, 35*(1), 175–193.

Macnaught, L., Maton, K., Martin, J., & Matruglio, E. (2013). Jointly constructing semantic waves: Implications for teacher training. *Linguistics and Education, 24*(1), 50–63.

Marginson, S., Tytler, R., Freeman, B., & Roberts, K. (2013). *STEM: Country comparisons: International comparisons of science, technology, engineering and mathematics (STEM) education*. Final report: Australian Council of Learned Academies.

Martin, J. (2013). Embedded literacy: Knowledge as meaning. *Linguistics and Education, 24*(1), 23–37.

Masny, D. (1997). Linguistic awareness and writing: Exploring the relationship with language awareness. *Language Awareness, 6*(2 & 3), 105–118.

Meltzer, J. (2002). *Adolescent literacy resources: Linking research and practice*. Providence: Northeast and Islands Regional Educational Laboratory at Brown University.

O'Halloran, K. (2003). Intersemiosis in mathematics and science: Grammatical metaphor and semiotic metaphor. In A.-M. Simon-Vandenbergen, M. Taverniers & L. Ravelli (Eds.), *Grammatical metaphor* (pp. 337–366). Philadelphia: John Benjamins.

Ollerhead, S. (2016). Pedagogical language knowledge: Preparing Australian preservice teachers to support English language learners. *Asia-Pacific Journal of Teacher Education*, pp. 1–11, Retrieved from http://dx.doi.org/10.1080/1359866X.2016.1246651

Polat, N., & Mahalingappa, L. (2013). Pre- and in-service teachers' beliefs about ELLs in content area classes: A case for inclusion, responsibility, and instructional support. *Teaching Education, 24*(1), 58–83.

Prain, V., & Tytler, R. (2012). Learning through constructing representations in science: A framework of representational construction affordances. *International Journal of Science Education, 34*(17), 2751–2773.

Premier, J. A., & Miller, J. (2010). Preparing pre-service teachers for multicultural classrooms. *Australian Journal of Teacher Education, 35*(2), 35–48.

Reeves, J. (2006). Secondary teacher attitudes toward including English-language learners in mainstream classrooms. *Journal of Educational Research, 99*, 131–142.

Rose, D., & Martin, J. (2014). Intervening in contexts of schooling. In J. Flowerdew (Eds.), *Discourse in context: Contemporary applied linguistics volume 3* (pp. 273–300). London: Bloomsbury Academic.

Schleppegrell, M. (2001). Linguistic features of the language of schooling. *Linguistics and Education, 12*, 431–459.

Schleppegrell, M. J. (2004). The language of schooling: A functional linguistics approach. *TESOL Journal, 12*(2), 21–27.

Seah, L. H., Clarke, D. J., & Hart, C. (2011). Understanding students' language use about expansion through analyzing their lexicogrammatical resources. *Science Education, 95*(5), 852–876.

Short, D., & Echevarria, J. (1999). *The sheltered instruction observation protocol: A tool for teacher-researcher collaboration and professional development.* Santa Cruz and Washington: Center for Research on Education, Diversity & Excellence.

Siskin, L. (1994). *Realms of knowledge: Academic departments in secondary schools.* London: Falmer.

Teese, R. (2013). *Academic success and social power: Examinations and inequality.* North Melbourne: Australian Scholarly Publishing.

Thomson, S., De Bortoli, L., & Buckley, S. (2013). *PISA 2012: How Australia measures up.* Melbourne: Australian Council for Educational Research.

Turkan, S., de Oliveira, L., Lee, O., & Phelps, G. (2014). Proposing a knowledge base for teaching academic content to English language learners: Disciplinary linguistic knowledge. *Teachers College Record, 116*(1), 1–30.

Unsworth, L. (1999a). Explaining school science in book and CD ROM formats: Using semiotic analyses to compare the textual construction of knowledge. *International Journal of Instructional Media, 26*(2), 159–179.

Unsworth, L. (1999b). Teaching about explanations: Talking out the grammar of written language. In A. Watson & L. Giorcelli (Eds.), *Accepting the literacy challenge* (pp. 189–204). Sydney: Scholastic.

Unsworth, L. (1999c). Developing critical understanding of the specialised language of school science and history texts: A functional grammatical perspective. *Journal of Adolescent and Adult Literacy, 42*, 508–521.

Unsworth, L. (2000). Investigating subject-specific literacies in school learning. In L. Unsworth (Ed.), *Researching language in schools and communities: Functional linguistic perspectives.* London: Cassell.

Unsworth, L. (2004). Comparing school science explanations in books and computer-based formats: The role of images, image/text relations and hyperlinks. *International Journal of Instructional Media, 31*(3), 283–301.

Chapter 4
Meeting Disciplinary Literacy Demands in Content Learning: The Singapore Perspective

Caroline Ho, Natasha Anne Rappa and Kok-Sing Tang

Abstract This chapter examines how systemic language and literacy support for content-area teachers to enhance their students' learning is realised in Singapore with a focus on science at the secondary level. It highlights theoretical underpinnings that inform the perspective of disciplinary literacy guiding this work and describes how disciplinary literacy is contextualised in Singapore against what is broadly understood as effective communication. It unpacks the nature and extent of systemic support for developing literacy in science with specific reference to the professional learning courses and school-based collaborative research. The chapter addresses the challenges encountered and discusses the implications which impact curriculum and pedagogy in the integration of disciplinary literacy practices to meet students' needs in the learning of science.

Keywords Disciplinary literacy · science education · content teaching and learning · science communication

4.1 Introduction

This chapter offers the Singapore perspective to integrating literacy and content language learning in the curriculum, with a specific focus on the science curriculum at the secondary school level. The chapter unpacks the concept of 'disciplinary literacy' and how this has been contextualised to serve the needs of the local teaching

---

C. Ho (✉)
English Language Institute of Singapore, Ministry of Education, Singapore, Singapore
e-mail: caroline_ho@moe.gov.sg

N.A. Rappa
School of Education, Murdoch University, Perth, Western Australia, Australia
e-mail: natashaannetang@gmail.com

K.-S. Tang
School of Education, Curtin University, Perth, Australia
e-mail: kok-sing.tang@curtin.edu.au

© Springer International Publishing AG 2018
K.-S. Tang, K. Danielsson (eds.), *Global Developments in Literacy Research for Science Education*, https://doi.org/10.1007/978-3-319-69197-8_4

fraternity. It outlines the growing attention to the importance of disciplinary literacy in national curricula in the learning of content and preparation for work life in the real world. Specifically, it describes the Ministry of Education (MOE) curricular focus in the Singapore education context which seeks to raise students' literacy levels in content areas on a nationwide level. The rationale and programme specifics of a nationwide initiative driven by the MOE are delineated along with the unpacking of the support model offered to schools. This chapter also highlights an instantiation of collaboration between the National Institute of Education and the MOE to support the development of disciplinary literacy in science. The chapter closes with a consideration of guiding principles that can inform teachers' pedagogic practice with a focus on disciplinary literacy in science.

4.2 Theoretical Foundations

In Singapore, theoretical understandings of disciplinary literacy have drawn largely on the work of scholars such as Fang (2005, 2012), Moje (2007) and Shanahan and Shanahan (2008). Disciplinary literacy refers to the specific ways of talking, reading, writing and thinking valued and used by people in a discipline in order to successfully access and construct knowledge in that discipline (Moje, 2007; Shanahan & Shanahan, 2008). Disciplinary literacy and disciplinary content are mutually constitutive with literate practices being fundamental to engaging in social and cognitive practices that develop and advance disciplinary knowledge (Fang & Coatoam, 2013). Therefore, the aim in developing students' literacy within a discipline is to build students' capacity to engage in literacy skills, strategies and practices, in line with those of content-area experts and as part of the process of socialisation into science discourse (Fang, 2012; Fang & Coatoam, 2013).

Fang and Coatoam (2013, p. 628) observe that there are differences in the way disciplinary content is 'produced, communicated, evaluated, and renovated'. This diversity calls for specificity in literacy practices taught. Moreover, the distinguishing features of scientific language described by Fang (2005) underscore the highly specialised nature of science discourse. It is thus not surprising that advocates of disciplinary literacy argue that literacy instruction should be situated within a given content area so that teachers can use their content-area expertise to give 'explicit attention to discipline-specific cognitive strategies, language skills, literate practices, and habits of mind' (Fang & Coatoam, 2013, p. 628). But how can the existing framework be adapted purposefully and meaningfully to incorporate disciplinary literacy and what kinds of systemic support would teachers need so that they can help students meet the specific challenges of reading, writing, speaking, listening and language in their respective fields? We address these questions in the following section.

4.3 Disciplinary Literacy Through the Lens of Effective Communication in the National Curricula

In Singapore, growing recognition of the importance of disciplinary literacy in the national curricula has led to this literacy being situated within the MOE 21st century competencies framework under the core competency of communication skills: 'Communicating effectively refers to the delivery of information and ideas coherently, in multimodal ways, for specific purposes, audiences, and contexts' (Ministry of Education, 2011, p. 9). Communication is conceptualised as 'the interactive process of sharing concepts, thoughts and feelings between people using the medium of language as a resource' (English Language Institute of Singapore, 2013, p. 1). In addition, this process involves the 'co-construction of meaning' by those involved (English Language Institute of Singapore, 2013, p. 1). Communication, as acknowledged in research literature, can encompass both linguistic skills and non-linguistic skills, such as body language, gestures, facial expressions, as well as cultural and social conventions for interacting with people (Kress, Jewitt, Ogborn, & Tsatsarelis, 2001). To the MOE in Singapore, 'Effective communication occurs when the audience or reader understands a message in the way the communicator intended it to be understood, or when the co-construction of meaning satisfies all parties involved' (English Language Institute of Singapore, 2013, p. 1).

Literacy in a discipline entails the ability to use language appropriately, meaningfully and precisely in a given subject area and this ability requires both proficiency in language and subject knowledge (English Language Institute of Singapore, 2013). Language itself mediates the learning of the concepts, models, theoretical frameworks and skills demanded by each subject (Bailey, Burkett, & Freeman, 2008). Language serves as a window to the content in the subject classroom where it is used to express, create and interpret meanings in the context of the subject. As students progress towards the higher levels in school, they move beyond the basic literacy level of decoding and generic comprehension to acquiring increasingly specialised literacy skills for each subject. Strong early reading skills do not necessarily translate into an ability to deal with the special language requirements required in content-area classrooms. Students have to read to learn, write to learn and talk to learn (Shanahan & Shanahan, 2012) in order to understand and communicate subject-specific content. This is especially so given that curriculum subjects differ in their communicative purposes, their typical text structures and characteristic language features. The linguistic implications are distinctive differences in how texts are organised, how the vocabulary is selected and how grammatical choices are made. Such knowledge and practices constitute the literacy skills and abilities that students need to acquire.

Effective communication by all content-area teachers from this perspective thus implies the skilful use of subject-specific language to help students better understand, process and internalise subject knowledge effectively. This is achieved by explicit instruction of the content as well as explicit attention to the language specifics in teaching the content to help students access the language. As well as conveying subject content through presentational modes of language use, subject teachers can also facilitate thinking and understanding of content through interactional modes of language use in the classroom (Jocuns, 2012). By modelling effective communication, science teachers can raise students' awareness of the norms and conventions of reading, writing, talking and thinking like a scientist (Vacca & Vacca, 2008). As a corollary, students develop the language to understand and effectively explain the concepts of the subjects they are studying which essentially involves disciplinary literacy.

4.3.1 The Implications of Situating Disciplinary Literacy Within the Effective Communication Framework

The perspective on effective communication in Singapore schools is shaped by several contextual factors that have implications for the way teachers conceive disciplinary literacy. First, the desired outcomes for effective communication within the Singapore education context are oriented towards helping students become future-ready—students' communicative skills are intended to help them meet the expectations of employers or Institutes of Higher Learning upon leaving the secondary education system and every student is expected to communicate effectively in social situations with both local and overseas speakers of English (English Language Institute of Singapore, 2012).

The second factor relates to the focus on the specific English language variety used for communicating in the classroom. In multiracial Singapore, English, Mandarin, Malay and Tamil are the official languages. Bilingualism is 'a cornerstone of our education system' (Curriculum Planning and Development Division, 2010, p. 6) with students learning both English and their own Mother Tongue language in school. English is the common language facilitating bonding among the different ethnic and cultural groups. At the global level, English is recognised as 'the lingua franca of the Internet, of science and technology and of world trade' (Curriculum Planning and Development Division, 2010, p. 6). Given that standard English is the medium of instruction for all subjects in Singapore schools except the Mother Tongue languages (Curriculum Planning and Development Division, 2010), effective communication primarily addresses the use of standard English across the curriculum in content-area classrooms.

Finally, specificity in standard English employed in content-area classrooms is delineated by subject-specific notions of communication articulated in the subject syllabuses. We illustrate what they mean by subject-specific understandings of communication with reference to the MOE Secondary Science Syllabus. As the Upper Secondary Science Syllabus is at present undergoing a review, we refer to the Lower Secondary Science Syllabus (Ministry of Education, 2008) which describes 'scientific literacy' largely in terms of cognitive and social practices (Bailey et al., 2008, p. 4). 'Scientific literacy' is outlined as follows:

(i) the capacity to engage in the discipline-specific inquiry process skills of 'identifying questions', 'drawing evidence-based conclusions', 'making decisions' as well as the 'skills and habits of mind' aligned with the aforementioned 21st century competencies such as 'reasoning and analytical skills, decision and problem solving skills, flexibility to respond to different contexts and possessing an open and inquiring mind';
(ii) having an understanding of the key features of scientific inquiry and its impact and;
(iii) having the appropriate ethical and attitudinal disposition.

While there are some overlaps with the notion of disciplinary literacy, Fang & Coatoam's (2013) broader definition of science literacy encompasses not only the linguistic but also the semiotic (Kress & van Leeuwen, 2001) which includes multimodal resources for communication (visual, verbal, gestural). This is reflected in their more encompassing definition of 'habits of mind' as 'ways of reading, writing, viewing, speaking, thinking, reasoning and critiquing' (Fang & Coatoam, 2013, p. 628).

In addition, in the syllabus document, communication is defined as 'the skill of transmitting and receiving information presented in various forms—verbal, tabular, graphical or pictorial' (Bailey et al., 2008, p. 8). While 'communication' in the Lower Secondary Science Syllabus is embedded within science inquiry, it is not viewed as a skill that cuts across and/or underpins the whole inquiry process. Instead, it is conceived as one of the several distinctive features of science inquiry, others being 'question', 'evidence', 'explanation' and 'connections' (Bailey et al., 2008). Moreover, the skill of communicating is confined to contexts where 'students communicate and justify their explanations when they form reasonable and logical argument to communicate explanations' (Bailey et al., 2008, p. 13) and the teacher guidance for communicating is in the form of steps, procedures, guidelines and coaching (Bailey et al., 2008). As have been articulated by others (Adger, Snow, & Christian, 2002; Bailey et al., 2008), working 'side by side with content and grade-level teachers to collaboratively adapt curriculum and classroom instruction to meet the specific needs' (Bailey et al., 2008, p. 19) of students is what those providing support to content teachers can offer. Understanding not only the structure but also 'how language mediates students' access to content, classroom learning processes and assessments' (Adger et al., 2002) is critical. Tang's (2015) deconstructing scientific explanation through the explicit framing of Premise-Reasoning-Outcome (PRO) to help students reason the underlying logic and casual sequencing of an explanation has proved beneficial to students.

By incorporating disciplinary literacy into an existing framework of effective communication in the national curricula, understandings of disciplinary literacy are shaped by, first, the strong emphasis on standard English as the mode of communication and, secondly, the aforementioned entrenched curricular definitions of scientific literacy amongst teachers. The emphasis on standard English is not necessarily at odds with the notion of disciplinary literacy but can potentially detract from the focus on the literacy demands of a discipline. As such, distinctions between the two need to be clearly articulated. The prevailing understanding of scientific literacy, however, presents a more restrictive perspective of literacy. For this reason, it is imperative that the notion of disciplinary literacy is made explicit for science teachers.

4.4 Systemic Support for Developing Literacy in the Content Areas

In Singapore, support for the development of effective communication in all schools is spearheaded by the MOE. In 2012, the Whole School Approach to Effective Communication in English (WSA-EC) was initiated by the MOE English Language Institute of Singapore (ELIS) to enhance the professional standing of teachers as role models of English and to help students become effective communicators in English (English Language Institute of Singapore, 2016a) in line with the emphasis on 21st-century competencies (Ministry of Education, 2016). The emphasis is on content-area teachers modelling good communication skills to communicate subject knowledge more clearly and effectively in every classroom for every subject, providing opportunities for all students to develop these skills, and creating a whole-school environment where effective communication is valued. MOE held the belief that immersion in such an environment would over time help students develop a wide repertoire of communication skills such as questioning, evaluating, explaining, comparing and contrasting, classifying, hypothesising, and distinguish between subject-specific communication skills. The WSA-EC programme has been rolled out to primary, secondary and pre-university institutions in phases.

The support model of the WSA-EC comprises: (i) professional learning courses, (ii) collaborative school-based research, (iii) provision of resources and (iv) interaction with experts (English Language Institute of Singapore, 2016a). For the purposes of this chapter, we focus on the first two components of the support model to illustrate and examine how systemic support impacts the classroom environment.

4.4.1 Professional Learning Courses: Key Features and Challenges

The MOE recognises that the onus of developing a whole-school environment supportive of effective communication and of modelling effective communication skills within the disciplines rests on the teachers. To deepen content-area teachers' understanding of disciplinary literacy, schools on the WSA-EC went through three core curriculum professional learning courses on disciplinary literacy. These courses are targeted at mixed content-area teams from primary and secondary schools comprising teacher leaders (also referred to as 'Champions of Effective communication') in science, mathematics and the humanities. The courses aim to develop a greater awareness and understanding of the strong connection between learning a subject and the language used to convey content and skills in that subject. The courses aim to:

(i) develop teachers' awareness of the importance of language and literacy for teaching disciplinary content and highlight the role language plays when students are learning the concepts, skills and processes of disciplinary content (Language and Literacy in Subject Classrooms);
(ii) examine how talk and interaction can help facilitate deeper learning and engagement for students through a framework for supporting high-quality talk and interaction in content-area classrooms and explore strategies that facilitate productive talk for effective teaching and learning of disciplinary content (Opening Up Talk for Learning in Subject Classrooms);
(iii) examine how talk and writing can be integrated to deepen learning in content-area classrooms through a framework for integrating talk with writing and explore strategies for monitoring and evaluating student learning through talk and writing (Integrating Talk with Writing in Subject Classrooms) (English Language Institute of Singapore, 2016b).

One important feature of the courses is that the content-area teacher leaders representing different disciplines are encouraged to implement the strategies they jointly developed or identified in their content-area group as salient for a given task in their classroom, and reflect on the effectiveness of this implementation before the next session of the course. These teacher leaders, upon completion of the courses, return to school and work towards transferring learning to their colleagues in their specific disciplines. Embedding the implementation within the programme itself gives content-area teachers a platform to share their experiences with one another and develop collective wisdom on literacy instruction that benefited their students the most, identify areas to improve upon and pitfalls to avoid. Moreover, it encourages commitment to developing subject-specific communication skills as an integral part of their subject teaching and proficiency in the use of pedagogy that will enable student to develop these skills. In-course and post-course follow-up by the teacher-leaders allow them to trial and experience the strategies introduced in the course with their own classes. There is further on-site co-facilitation of cascading of learning by content teacher leaders to their subject teams supported by MOE language specialists and subject literacy officers.

Given that these courses are usually facilitated by language specialists, the general approach taken by facilitators has been to highlight key MOE policies and key findings in the research literature, engage content-area teachers in analysing the literacy demands of a given task and elicit from them the strategies they collectively worked out for addressing students' specific literacy needs anticipated for that given task. These courses, therefore, provide a means for teachers to engage in both 'theoretical knowing' and 'experiential knowing' (Nutley, Walter, & Davies, 2007, p. 24) to inform their use of sound pedagogy to facilitate the development of students' literacy in a discipline.

Having participated in the first course, the team of content-area teachers from each school would then develop plans to enhance the development of effective communication skills suited to its particular environment and culture. The strategies outlined in these plans would over time be infused into the school practices and systems. Our focus after teachers return to their schools is on school-based disciplinary literacy instruction in order to meet students' specific literacy needs as this shows how schools transfer the learning to their subject colleagues and the impact the courses might have had.

Having described this systemic support, we are also mindful of challenges content-area teachers face when attending courses on disciplinary literacy. Content-area teachers, as the more proficient and knowledgeable learners and users of the discourse, possess the ability to recognise pertinent texts and how to interact with them. They have a critical role to play in bringing the students along the path of a deeper and broader understanding of curriculum (Draper, Broomhead, Jensen, & Siebert, 2010). To do this, content-area teachers need to conceptualise 'language and literacy practice as an integral aspect of subject area learning, rather than as a set of strategies for engaging with texts' (Moje, 2008, p. 99). This entails literacy being viewed not as generic skills taught in isolation, separately from the content, but contextualised and adapted within their own disciplines to facilitate learning of the content. The challenge, therefore, lies in the fact that a mixed group of content-area teachers representing each school attends these courses together. These teachers need to consider individually and together with their content-area colleagues back in school how they could adopt or adapt literacy strategies they encountered or came up with during the course to help students understand and construct disciplinary content in ways consistent with social norms and ongoing semiotic and cognitive practices. The Champions of Effective Communication work closely with their content-area colleagues through their in-house professional learning and mentoring sessions to consider ways to synergise their efforts during implementation to best meet their students' learning needs.

In science, the specific challenge concerns situating literacy strategies within the inquiry process (Draper & Siebert, 2010) outlined in the MOE Science Syllabus in order to facilitate students interpreting and constructing texts with the distinguishing features of scientific language. The difficulty also arises from identifying and employing literacy strategies that enable students to use and interpret different kinds of representations in the discipline (Draper & Siebert, 2010; Tang, 2011b; Tytler, Prain, Hubber, & Waldrip, 2013). Fang and Coatoam (2013) caution against the problem of generic strategies being re-packaged as discipline-specific ones. As such, content-area teachers need to address the question of which literacy strategies enable students to interpret these representations in ways consistent with norms and recognise nuanced changes in meaning with changes in the mode of representation and the purpose of these different modes of representations.

Further, as content-area teachers develop and refine these literacy strategies for their discipline, they need to heed Draper et al.'s (2010) caution against a general form of literacy applicable only to a school or examination context and is neither useful within the discipline nor in adolescents' lives outside of school. In our view, the first point is problematic only if the inquiry process outlined in the MOE Science Syllabus is not consistent with the real-world practices of scientists and only if assessment methods are not consistent with the inquiry process described. The second point presents the more persistent challenge of developing literacy strategies that draw on adolescents' out-of-school interests and experiences with popular texts and/or hybrid texts on the science topics. This is an important area that warrants further investigation to better inform content-area teachers as they endeavour to develop pedagogical practices that support the development of disciplinary literacy. Some exploratory studies were carried out in this area to investigate the role of out-of-school media representations of science (Parkinson & Adendorff, 2004; Tang, 2013) and the agency of science students across the informal and formal domains (Bell, Lewenstein, Shouse, & Feder, 2009; Rappa & Tang, 2017; Tang, 2011a).

We want to emphasise that what content-area teachers face should not be handled by them alone. According to Fang and Coatoam (2013, p. 629), literacy teachers trained in 'reading instruction, focussed on phonics, vocabulary, fluency and cognitive strategies' lack disciplinary expertise in two areas—they are unfamiliar with the 'content, discourse patterns, literate practices and habits of mind within specific disciplines' and 'they lack knowledge of the big ideas, unifying concepts and key relationships related to the content of the disciplines'. Having said that, Fang and Coatoam (2013, p. 629) also argue that content-area teachers 'lack the necessary language awareness and literacy strategies to help students cope with the specific language and literacy demands of their discipline'.

In the light of the aforementioned view, there is much that ELIS language specialists and the National Institute of Education (NIE) education researchers can do and have done to support content-area teachers. First of all, language specialists have a role to play in bringing to the fore the literacies specific to a discipline (Draper et al., 2010). What this means is that language specialists can begin by helping content-area teachers reflect on the background knowledge and self-questioning practices that support text interpretation, how they go about interpreting texts and the norms for constructing texts (Draper et al., 2010). Second, language specialists can provide support by drawing content-area teachers' attention to instructional frameworks for literacy (Draper et al., 2010). We acknowledge what Draper and Siebert (2010), citing Conley (2008), have highlighted regarding generic strategies, that they 'fit poorly with content-area goals and discipline-specific practices'. One approach lies in adapting instructional frameworks which different disciplines have adhered to by incorporating elements of literacy instruction. Draper and Adair (2010) provide an illustration of how this might be achieved in relation to the 5E Learning Cycle (Bybee et al., 2006), which is popularly known among science teachers. Following this approach, a research collaboration with NIE and two secondary schools developed and tested an integrated literacy-inquiry instructional model that infused literacy elements into the 5E Learning Cycle (Tang & Putra, Chap. 17).

This brings us to the second component of the ELIS support model, which concerns research collaboration with science education researchers at NIE. In the following section, we describe the synergistic relationship between the MOE, NIE and schools as all parties worked towards helping teachers communicate effectively in their subjects.

4.4.2 School-Based Collaborative Research: Impact on Pedagogy

One common form of collaboration between NIE researchers and school teachers is the joint partnership of carrying out design-based research (Collins, Joseph, & Bielaczyc, 2004) with the dual purposes of informing education theory and improving classroom practices situated in the school context. Aligned with the MOE's emphasis on disciplinary literacy, various research studies across a range of school contexts were carried out to integrate some aspects of language and literacy into existing science classroom practices. The range of intervention research includes examining and enhancing primary school teachers' capacity in addressing the language demands of science (Seah, 2016), developing instructional models and strategies for secondary school teachers to explicitly address the language and multimodal demands of science (Tang, 2016a; Tang, Ho, & Putra, 2016), using a genre-based heuristic to support students in constructing scientific explanations at the primary (Seah, 2015) and secondary level (Tang, 2015), harnessing out-of-school media representations of science to foster critical literacy in high school (Rappa & Tang, 2017), and exploring the use of argumentation to foster group discussion in university chemistry (Tan, Lee, & Cheah, In press).

Through design-based research projects situated in science classrooms, the teachers benefited from just-in-time professional development and joint development of resources with researchers with notable changes in the teachers' pedagogical practice. For instance, four teachers in one of the research studies learned a new literacy strategy, called PRO, that was designed to teach students how to construct scientific explanations (Tang, 2015). From classroom observations of their teaching over 2 years, the teachers were able to integrate the PRO strategy into classroom talk in a way that supported logical reasoning and content mastery (Putra & Tang, 2016; Tang, 2015). The teachers were also able to adapt other literacy practices introduced during the professional development session to support classroom talk (Tang et al., 2016). Analysis of the students' writing suggests a positive impact in the use of the PRO strategy to improve the quality of the students' written explanations (Tang, 2016a).

4.5 Implications

Subject-based mixed teams of content teacher leaders applying their learning acquired from courses to their school-based subject teams and co-facilitation professional learning sessions with MOE officers present opportunities for empowering content teachers at various levels. This can also pave the way for more ground-up collaborative school-based research partnerships with the MOE officers and NIE faculty inquiring into identified areas of concern or challenge in the process of seeking solutions collaboratively to enhance students' content learning. At the same time, there is a need to ensure initiatives made to support subject teachers' professional learning and facilitate collaborative research are meeting targeted needs, particularly where students' learning is concerned. Indeed, it is acknowledged that the 'integration of language, subject content, and thinking skills requires systematic planning and monitoring' (Gibbons, 2002, p. 6). These can be framed around the following aspects which have surfaced as not only necessary but also critical in subject-specific learning contexts: Coherence, Contextualisation and Cascading.

4.6 Coherence

Implementation and monitoring of effective communication skills across the whole school must be easily integrated into existing school practices to reinforce current initiatives. There is a need for coherence in literacy programmes/initiatives in order to examine the impact on different stakeholders at different levels and to differing degrees. In education, features of programme design and research initiatives resembling coherence have been advocated under other names—such as integration, articulation. 'Coherence' denotes 'connectedness which, in turn, suggests consistency and accord among elements' (Buchmann & Floden, 1991). The move towards connection among various components is epitomised by Tyler's (1949) seminal work with the consequent continuity, sequence and integration that would ensue. Each of these qualities is a form of connectedness. Continuity means having links between one component and another in the system. Sequence extends the idea of continuity, requiring that links over time—'vertical' relations (Tyler, 1949)—involve a broadening and deepening of what is examined or focused, rather than mere repetition. Integration refers to connections across different aspects in different subjects—'horizontal' relations. Connectedness is required given that haphazard, isolated experiences are unlikely to ensure intended learning.

Coherence extends to the links across teacher facilitation, school leadership, subject teaching and student learning, and how these support and reinforce each other. Decisions about professional learning and development must be based on a good understanding of the relationship between the different layers. For example, if the students' needs-analysis identifies students' content vocabulary as a common 'gap', it would be important to understand how current teaching impacts student content vocabulary learning, and how current leadership and organisational practices contribute to that pattern of teaching through channels such as professional learning communities focused on evidence of teaching and learning (Ministry of Education NZ, 2013, p. 14). The strong school leadership support from the key personnel provided a foundation for the alignment of disciplinary literacy initiatives adopted by the content teachers in the school context with the overall thrust of schools' strategic goals for effective classroom pedagogical practice. In particular, quality academic experience, staff engagement and development, and partnerships were identified among the strategic thrusts in one school as critical considerations in driving initiative and programmes implemented in the school. In another, disciplinary literacy initiatives reinforced the schools' strategic thrusts that included academic excellence in terms of customised instructional approaches catering to diverse students' needs and developing skilful teachers to be curriculum leaders and reflective practitioners in their disciplines.

4.7 Contextualisation

There is a need for the contextualisation of literacy skills to meet the specific demands and requirements of learning environments and particular curricular contexts. The extent to which initiatives mediated by language and literacy facilitate content learning can be adapted or modified to aid transferability to similar or related contexts or settings must be considered in any whole-school implementation of a disciplinary literacy-based programme. At the same time, there is a need to address on-the-ground realities, contextual constraints in order to support science teachers and students in their learning endeavour. This will ensure a more targeted approach in supporting students to acquire the relevant disciplinary literacy skills required. Systematic scaffolding as realised in specific disciplinary literacy practices outlined in Tang (2015) is attentive to students' needs and aligned with their ability level, and seeks to address specific challenges in constructing scientific explanation.

Uncovering the critical aspects in authentic contexts or actual settings can inform the science learning experience which the targeted research is addressing. Understanding the learning context can reveal much more with an enhanced understanding of general and specific participant behaviours and decisions taken to provide the most relevant, engaging experience possible for students and teachers. Important insights as to what works and should be sustained and what may need further refinement can be gleaned from the research process, and the learning experiences and expectations of those involved. Such information is essential if meaningful analyses are to be provided. There is also a need to extend beyond 'surface manifestations (discrete activities, materials, or classroom organisation)' to inculcating in teachers an enhanced awareness of deeper pedagogical principles' (Century & Levy, 2002, p. 4). This could mean that the underlying principles of the literacy initiatives or programmes and the associated teacher beliefs and expectations of students are maintained over time.

The need for contextualising what is investigated within appropriate disciplinary discourses and paradigms cannot be overemphasised. This will develop teachers' capacity to recognise and contextualise research questions or hypotheses within specific disciplinary frameworks, and provide them with the opportunity to explore theoretical frameworks and methodologies in relation to their particular contexts. Initiatives adopted must be interpreted in the context of the specific classroom setting and examined with attention given to the on-the-ground realities, contextual constraints and implementation challenges.

4.8 Cascading

For any literacy initiative or programme implemented to support science teachers, the need for sustainability over a period of time is not to be overlooked. Initial efforts taken to implement literacy strategies to support content learning deserves attention to maintaining scaling up through the transfer of learning and cascading disciplinary literacy practices that have proven to be worthwhile. The central question to be addressed is: how does one ensure that literacy initiatives or programmes implemented will last? This question begs another: which specific aspects of literacy initiatives or programmes would be lasting in 1, 5 or 10 years' time? Research has shown that 'the programme or pedagogical approaches that were promoted through the professional learning/development experience' (Timperley, Wilson, Barrar, & Fung, 2007, p. 218) are the ones that may have a great lasting impact. A key criterion identified for judging sustainability appears to focus on 'continued, improved, worthwhile student outcomes' (Timperley et al., 2007, p. 218). The belief is that 'the conditions for sustainability are set in place during the professional learning experience as much as after it' (Timperley et al., 2007, p. 218).

Sustainability under these circumstances requires 'sufficient depth of principled knowledge for teachers to be able to recognise what is consistent and inconsistent with the changed practice being promoted' (Timperley et al., 2007, p. 219). Indeed, earlier work in the local context has surfaced the need for sustained, focused professional learning over a period of time rather than 'just-in-time' feedback for instructional planning and resources (Tang, 2016b) in order for scaling up pedagogic efforts and initiatives that support disciplinary literacy practices.

The following guiding principles for infusing disciplinary literacy practices into subject-specific pedagogy could inform schools that are focused on strategising disciplinary literacy practices to support content learning:

(i) Which aspects of the professional learning (e.g., specific expectations, principles, theories) are expected to be sustained (if stated or implied)?
(ii) At what level is the implementation (e.g., classroom/level/school-wide) expected to be sustained?
(iii) What kind of conditions created for sustainability was evident during the professional development? (tools for evidence-informed study, focus on theory/principle, other conditions)
(iv) What kind of conditions created for sustainability was evident after the professional development? (integration of implementation efforts that are coherent with school curriculum policy/framework, institutionalisation of implementation through school restructuring/re-culturing?) (Adapted from Timperley et al., 2007, pp. 219–220).

4.9 Conclusion

This chapter has examined how focused planning and strategic design of professional learning programmes and collaborative school-based research have reinforced the concretising of disciplinary pedagogic practices that draw on literacy support to meet students' learning needs in the science curriculum. Purposeful strategising at the systemic level informed by curricula focus that clearly delineates the parameters for the integration of content and language-specific tasks and processes. With the support of key partners (MOE, NIE) in collaboration with school partners, this has facilitated the school-level implementation of disciplinary literacy initiatives that is extending to more of the unreached among schools in the local context. At the same time, the cascading of learning to school-based subject teams at various levels has been set in place with structures supporting the co-facilitation of subject teacher leaders working closely with language specialists and subject literacy officers to infuse disciplinary literacy practices into content teaching at the classroom level.

Further work necessitates ongoing monitoring of the impact of disciplinary literacy initiatives adopted and adapted by schools to enable the necessary adjustments and modifications based on what is or is not enhancing student improvements in learning (Kaufman, Grimm, & Miller, 2012). Attempts to assess disciplinary literacy, as Fang and Coatoam (2013, p. 630) remind us, necessitate collaboration between language specialists and content subject teachers on identifying core skills for developing content and habits of mind, selecting relevant and significant texts, and designing authentic tasks and experiences. More studies along this line will contribute to a comprehensive picture of how perspectives on disciplinary literacy practice in the science curriculum are enacted to support students in reading, thinking, writing and speaking science the way scientists do.

References

Adger, C. T., Snow, C. E., & Christian, D. (Eds.). (2002). *Teachers need to know about language*. Washington: Center for Applied Linguistics.
Bailey, F., Burkett, B., & Freeman, D. (2008). The mediating role of language in teaching and learning: A classroom perspective. In B. Spolsky & F. M. Hult (Eds.), *The handbook of educational linguistics* (pp. 606–625). Hoboken: Wiley-Blackwell.
Bell, P., Lewenstein, B., Shouse, A., & Feder, M. (Eds.). (2009). *Learning science in informal environments: People, places and pursuits*. Washington: National Academy Press.
Buchmann, M., & Floden, R. E. (1991). Program coherence in teacher education: A view from the United States. *Oxford Review of Education, 17*, 65–72.
Bybee, R. W., Taylor, J. A., Gardner, A., Van Scotter, P., Powell, J. C., Westbrook, A., et al. (2006). *The BSCS 5E instructional model: Origins and effectiveness*. Colorado Springs: BSCS.
Century, J. R., & Levy, A. J. (2002). Sustaining your reform: Five lessons from research. *Benchmarks: The Quarterly Newsletter of the National Clearinghouse for Comprehensive School Reform, 3*(3), 1–7.

Collins, A., Joseph, D., & Bielaczyc, K. (2004). Design research: Theoretical & methodological issues. *Journal of Learning Sciences, 13*(1), 15–42.

Conley, M. (2008). Cognitive strategy instruction for adolescents: What we know about the promise, what we don't know about the potential. *Harvard Educational Review, 78*(1), 84–106.

Curriculum Planning and Development Division. (2010). *English language syllabus 2010*. Singapore: Curriculum Planning and Development Division, Ministry of Education.

Draper, R. J., & Adair, M. (2010). (Re)Imagining literacies for science classrooms. In R. J. Draper, P. Broomhead, A. P. Jensen, J. D. Nokes, & D. Siebert (Eds.), *(Re)Imagining content-area literacy instruction* (pp. 127–143). New York: Teachers College Press.

Draper, R. J., Broomhead, P., Jensen, A. P., & Siebert, D. (2010). Aims and criteria for collaboration in content-area classrooms. In R. J. Draper, P. Broomhead, A. P. Jensen, J. D. Nokes, & D. Siebert (Eds.), *(Re)Imagining content-area literacy instruction* (pp. 1–19). New York: Teachers College Press.

Draper, R. J., & Siebert, D. (2010). Rethinking texts, literacies, and literacy across the curriculum. In R. J. Draper, P. Broomhead, A. P. Jensen, J. D. Nokes, & D. Siebert (Eds.), *(Re)Imagining content-area literacy instruction* (pp. 20–39). New York: Teachers College Press.

English Language Institute of Singapore. (2012). *Position statement, attainment levels framework and key implementation strategies for effective communication*. (EDUN: N07-08-069).

English Language Institute of Singapore. (2013). *Effective communication across the curriculum: The importance of paying attention to subject literacy*. (EDUNN07-08-069 V6). Singapore: Ministry of Education.

English Language Institute of Singapore. (2016a). Retrieved from http://www.elis.moe.edu.sg/

English Language Institute of Singapore. (2016b). Professional learning opportunities @ ELIS. Retrieved from http://www.elis.moe.edu.sg/elis/slot/u54/news-n-events/publications/prospectus/2016-ELIS-prospectus.pdf

Fang, Z. (2005). Scientific literacy: A systemic functional linguistics perspective. *Science Education, 89*(2), 335–347.

Fang, Z. (2012). Language correlates of disciplinary literacy. *Topics in Language Disorders, 32*(1), 19–34 https://doi.org/10.1097/TLD.1090b1013e31824501de

Fang, Z., & Coatoam, S. (2013). Disciplinary literacy: What you want to know about it. *Journal of Adolescent & Adult Literacy, 56*(8), 627–632. https://doi.org/10.1002/jaal.190

Gibbons, P. (2002). *Scaffolding language, scaffolding learning: Teaching second language learners in the mainstream classroom*. Portsmouth: Heinemann.

Jocuns, A. (2012). Classroom discourse. In C. A. Chapelle (Ed.), *The encyclopedia of applied linguistics* (pp. 620–625). Oxford: Blackwell Publishing.

Kaufman, T. E., Grimm, E. D., & Miller, A. E. (2012). *Collaborative school improvement*. Cambridge: Harvard Education Press.

Kress, G., Jewitt, C., Ogborn, J., & Tsatsarelis, C. (2001). *Multimodal teaching and learning: The rhetorics of the science classroom*. London: Continuum.

Kress, G., & van Leeuwen, T. (2001). *Multimodal discourse: The modes and media of contemporary Communication*. Oxford: Oxford University Press.

Ministry of Education. (2008). Science syllabus (lower secondary). Retrieved from http://www.moe.gov.sg/education/syllabuses/sciences/files/science-primary-2008.pdf

Ministry of Education. (2011). *Standards and benchmarks for 21st century competencies*. Singapore: Curriculum Policy Office, Ministry of Education.

Ministry of Education. (2016). 21st century competencies. Retrieved from https://www.moe.gov.sg/education/education-system/21st-century-competencies

Ministry of Education NZ. (2013). *Research into the implementation of the Secondary Literacy Project (SLP) in schools*. New Zealand.

Moje, E. B. (2007). Developing socially just subject-matter instruction: A review of the literature on disciplinary literacy teaching. *Review of Research in Education, 31*, 1–44.

Moje, E. B. (2008). Foregrounding the disciplines in secondary literacy teaching and learning: A call for change. *Journal of Adolescent & Adult Literacy, 52*(2), 96–107.

Nutley, S., Walter, I., & Davies, H. (2007). *Using evidence: How research can inform public services*. Bristol: The Policy Press.

Parkinson, J., & Adendorff, R. (2004). The use of popular science articles in teaching scientific literacy. *English for Specific Purposes, 23*(4), 379–396.

Putra, G.B.S., & Tang, K.S. (2016). Disciplinary literacy instructions on writing scientific explanations: A case study from a chemistry classroom in an all-girls school. *Chemistry Education Research and Practice, 17*(3), 569–579. https://doi.org/10.1039/c6rp00022c

Rappa, N. A., & Tang, K. S. (2017). Student agency: An analysis of students' networked relations across the informal and formal learning domains. *Research in Science Education, 47*(3), 673–684. https://doi.org/10.1007/s11165-016-9523-0

Seah, L. H. (2015). Understanding the conceptual and language challenges encountered by grade 4 students when writing scientific explanations. *Research in Science Education*, 1–25. https://doi.org/10.1007/s11165-015-9464-z

Seah, L. H. (2016). Elementary teachers' perception of language issues in science classrooms. *International Journal of Science and Mathematics Education, 14*(6), 1059–1078. https://doi.org/10.1007/s10763-015-9648-z

Shanahan, T., & Shanahan, C. (2008). Teaching disciplinary literacy to adolescents: Rethinking content-area literacy. *Harvard Educational Review, 78*(1), 40–59.

Shanahan, T., & Shanahan, C. (2012). What is disciplinary literacy and why does it matter? *Topics in Language Disorders, 32*, 1–12.

Tan, A.-L., & Lee, P. P. F., & Cheah, Y. H. (In press). Educating science teachers in the twenty-first century: Implications for pre-service teacher education, *Asia Pacific Journal of Education*. https://doi.org/10.1080/02188791.2017.1386092

Tang, K. S. (2011a). *Hybridizing cultural understandings of the natural world to foster critical science literacy*. Doctoral dissertation, University of Michigan, Ann Arbor. Retrieved from ProQuest Dissertations and Theses database. UMI No. 3476796.

Tang, K. S. (2011b). Reassembling curricular concepts: A multimodal approach to the study of curriculum and instruction. *International Journal of Science and Mathematics Education, 9*, 109–135.

Tang, K. S. (2013). Out-of-school media representations of science and technology and their relevance for engineering learning. *Journal of Engineering Education, 102*(1), 51–76. https://doi.org/10.1002/jee.20007

Tang, K. S. (2015). The PRO instructional strategy in the construction of scientific explanations. *Teaching Science, 61*(4), 14–21.

Tang, K. S. (2016a). Constructing scientific explanations through premise—reasoning—outcome (PRO): An exploratory study to scaffold students in structuring written explanations. *International Journal of Science Education, 38*(9), 1415–1440. https://doi.org/10.1080/09500693.2016.1192309

Tang, K. S. (2016b). How is disciplinary literacy addressed in the science classrooms? A Singaporean case study. *Australian Journal of Language and Literacy, 39*(3), 220–232.

Tang, K. S., Ho, C., & Putra, G. B. S. (2016). Developing multimodal communication competencies: A case of disciplinary literacy focus in Singapore. In M. Mcdermott & B. Hand (Eds.), *Using multimodal representations to support learning in the science classroom* (pp. 135–158). New York: Springer.

Timperley, H., Wilson, A., Barrar, H., & Fung, I. (2007). *Teacher professional learning and development: Best evidence synthesis iteration*. Wellington: Ministry of Education.

Tyler, R. (1949). *Basic principles of curriculum and instruction*. Chicago: University of Chicago Press.

Tytler, R., Prain, V., Hubber, P., & Waldrip, B. (2013). *Constructing representations to learn in science*. Rotterdam: Sense Publishers.

Vacca, R. T., & Vacca, J. A. (2008). *Content area reading: Literacy and learning across the curriculum*. 9th ed. Boston: Allyn & Bacon.

Part 2
Content and Language Integrated Learning (CLIL) in Science

Chapter 5
Learning Language and Intercultural Understanding in Science Classes in Germany

Silvija Markic

Abstract This study discusses a collaborative research and development project consisting of science teachers, German as a Second Language (GSL) teachers, and science educators. The project follows the model of Participatory Action Research in science education. It focuses on the development of teaching modules for early lower secondary school science lessons in grades 5–8 (age ranges roughly from 10 to 11 and 13 to 14, respectively) on different topics. The lesson modules implement the integration of content and language with the help of the Content and Language Integrated Learning (CLIL) approach. All lessons are structured using cooperative and autonomous learning methods. Over the last 2 years, the group has included intercultural understanding in its teaching materials. The accompanying research attempts to answer the following question: to what extent is it possible for students to learn science content, scientific terminology, and the German language simultaneously in an intercultural context, while working in cooperative learning settings and developing their intercultural understanding? Data were collected from classroom observations, student feedback questionnaires, cognitive tests, and teacher feedback. The initial results show that it is possible to successfully combine science content, language, and intercultural factors in the same lesson module. Students were highly motivated and the lesson modules showed great potential for improving students' learning about the science subject matter. The lessons simultaneously contributed to improvements in the students' German language skills and intercultural understanding. The findings reflect both the potential benefits and consequences of the language and intercultural understanding aspects selected for this lesson module. Conclusions from the results and further ideas are also addressed.

Keywords German language · CLIL · intercultural understanding · participatory action research · lower secondary school

---

S. Markic (✉)
Institute for Science and Technology – Chemistry Department, University of Education Ludwigsburg, Ludwigsburg, Germany
e-mail: markic@ph-ludwigsburg.de

© Springer International Publishing AG 2018
K.-S. Tang, K. Danielsson (eds.), *Global Developments in Literacy Research for Science Education*, https://doi.org/10.1007/978-3-319-69197-8_5

5.1 Introduction

Teaching for diversity and heterogeneity is a new challenge for most of the teachers in general and science subject areas in particular. However, there is no *general* heterogeneity. Students in our schools differ in all of the dimensions of the diversity wheel: in their knowledge, immigration background, social background, culture (religion, tradition, national origin, etc.), cognitive skills, personal (special) needs, in their mother language, and much more (Markic & Abels, 2016).

Conventionally, language has been understood as a simple vehicle for the transfer of information (Fang, 2006; Ford & Peat, 1988). However, in science education research, language and its role in the teaching and learning process did not play a prominent role for a long time. An examination of science education literature over the last few decades suggests that the topic of language and its importance within science education has been poorly represented. However, there has been a dramatic change when it comes to research on language in science education and science teaching. At present, language is considered to be one of the central issues that fosters or hinders learning in general and in the science classroom in particular (Osborne, 2002). Science education researchers have become increasingly aware of the fact that students' linguistic abilities do interact with learning in general and science learning in particular. For example, Lee (2005) and Lee and Fradd (1998) showed that students' lack of linguistic skills and unfamiliarity with asking questions, investigating, and reporting results using scientific language can cause students to lose interest in science lessons, which, in turn, causes lack of understanding of science as a subject.

Furthermore, learning and the proper use of scientific language is necessary for both communication among the students and communication between the students and the teacher within the science classroom. Yet, it is no secret that students have problems employing scientific language. This difficulty does not depend on familial and/or social background. Scientific language can be regarded as a new language for all the students. In addition, students with migration backgrounds have further difficulties when it comes to correctly using the official language of their country of residence. For most of these students, a science lesson is, in fact, a bilingual lesson with specialized scientific language. In contemporary times, there has been a large influx of refugees and other migrants in a lot of countries all over the world. Due to these factors, the issue of communication is taking on increasing importance for the overall success of national, regional, and local education programs (Childs, Markic, & Ryan, 2015).

To support the argumentation and starting from the aims of *Scientific Literacy* more than 15 years ago, Phillips and Norris (1999) mentioned that one of the main features of *Scientific Literacy* was the skill of text understanding, including the capability of employing information rationally during discourse or decision-making in science-related, personal, and social issues. Furthermore, Scherz, Spector-Levy, and Eylon (2005), in their study on the impact of explicit instruction of literacy and communication abilities in middle school science lessons, show that significant improvements could be observed in the intervention group for communication skills, when this group was compared with the control group, in which no explicit instruction of communication and literacy skills had taken place.

An additional issue is that problems surrounding language in science education are not solved through one-sided action, for example, through changes on the part of the students. Talking and understanding a foreign language properly takes years, depending on the age of the student (Collier, 1987). Furthermore, it is widely known that the teacher is a key factor when it comes to the implementation of the new ideas, changes in the activities in the classroom, and reforms in science lessons (Markic & Eilks, 2008; Nespor, 1987). However, there is still a perceived dichotomy between language and science amongst both teachers and students. There is a crucial need for well-prepared teaching materials, which incorporate language learning, and for teachers' knowledge on how to create an environment, which might include specific teaching materials, that promotes integration of literacy and science learning. Not only teachers, but also science researchers alike, consider culture, in addition to language, as an obstacle to learning science (Carter, 2007; Grosser & Glombard, 2008; Nieto, 2000; Roth & Tobin, 2009). Thus, language and cultural prerequisites have been seen as a challenge in modern science education and have influence on science teaching and learning.

The question remains, are language and students' linguistic skills and their different cultural backgrounds only to be seen as challenges in science classrooms or can we see language and culture as an opportunity for science education? Answering the question positively would mean seeing the given situation in science classes from a different angle. To do so, first it is necessary to clarify vocabulary. The words "heterogeneity," "diversity," "integration," and "inclusion" are often used synonymously but different researchers have different definitions of these terms. This chapter focuses on the definition of Sliwka (2010) where heterogeneity is understood as adjustments made to come to the terms with students' different needs. With this in mind, integration may be interpreted as viewing students' differences as challenges that need to be dealt with in the classroom. On the other hand, diversity means that differences serve as a resource for both student's and peers learning, where development and inclusion involves seeing differences as an asset and opportunity in classes (see also Fig. 5.1).

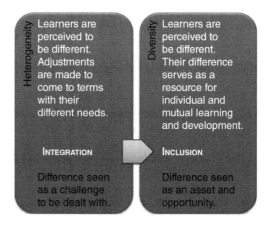

Fig. 5.1 From heterogeneity to diversity (Sliwka, 2010, p. 214)

Starting from these definitions in this chapter, the expressions "*linguistic heterogeneity*" and "*cultural diversity*" will be used to describe the current research and development project.

5.2 Theoretical Background

A good overview of German science education research is given in a special issue "*Traditions and trends in German mathematics, science and technology education*" by Eilks and Markic (2014). In alignment with international studies such as PISA, IGLU, and international science education researchers (Bryan & Atwater, 2002; Cassels & Johnstone, 1983; Johnstone & Selepeng, 2001; Rodrguez & Kitchen, 2005), Markic and Abels (2014) point out in this issue that multilingualism and multiculturalism are primarily viewed as a disadvantage in German education. This is considered to be one of the major problems with the German educational system in general and in German science teaching in particular. The majority of German students with migration backgrounds only begin to learn German at the age of 6–7 after entering primary school (Brandenburger, 2007). Most of those students never attended German kindergarten and mainly have contact with children from the same migrant background until they start school in the German school system. Even then, outside of school, they almost exclusively speak their mother tongue with their families and friends. This translates, quite often, into students with migrant backgrounds achieving overall lower educational levels than native German speakers due to a lack of German language skills. Reich and Roth (2002) discovered that it is only in a few cases that bi- or multilingual students ever reached the language standard of native speakers. Furthermore, the official school language during lessons, German, is a huge challenge for such students for two reasons: (i) they often do not know the grammatical rules of either their own spoken language or German (Maas, 2005) and (ii) explicit instruction in their mother tongue is not offered in school. However, Riebling and Bolte (2008) found that multilingual students in the German context have high metalinguistic competencies in comparison with monolingual, native speakers. This is because they have already been actively exposed to learning more than one language system. Unfortunately, this skill has not been used. Students with migrant backgrounds proved to be more attentive with respect to the language used in chemistry lessons. Riebling and Bolte (2008), in addressing the "hidden linguistic issues," said that answers, statements, and questions given by these students also tended to be much shorter and less complex, with less usage of specific, scientific terminology. Their answers do not reflect a high level of complexity of using elements of scientific language in German. Furthermore, those students have relatively few possibilities to participate actively and productively in regular classroom settings. They experience the new language receptively and have less opportunity to actively use it in order to develop their linguistic competency. Furthermore, the lack of mastery of the German language makes content learning in science lessons difficult for nonnative speakers. Thus, regular science lessons often turn into a bilingual minefield for such learners. They not only have to assimilate the basic content presented in the lesson, but also have to understand and learn the specific scientific language (Leisen, 2004). Due to this combination, these students lack the language competencies necessary to communicate and to actively participate in the lesson (Phillips & Norris, 1999).

Additionally, German science teachers often do not accept the teaching of language as a necessary goal within their own science lessons (Markic, 2010). In many cases, they attempt to relegate it to a secondary position, as an issue which should be addressed by other subjects and by other teachers, for example, in German lessons and German teachers (Tajmel, 2010). Although the importance of the work on linguistic heterogeneity and dealing with it in science classes is already known in the German context, studies concerning science teachers' beliefs about dealing with the linguistic heterogeneity in science classes are rare in our country. As such, Riebling and Bolte (2008) propose that science teachers pay greater attention to students' language and display sensitivity when they are teaching in linguistically heterogeneous classes. This is important, so that teachers can realize the problem, deal with it in their lesson planning, and try to address this issue in their teaching. However, Benholz and Iordanidou (2004) noticed that this is especially difficult for science teachers that are monolingual. The authors showed in their study that, in particular, monolingual teachers have problems noticing linguistic heterogeneity of their classes. Thus, science teachers plan their lesson and teaching for monolingual classes.

The issue of linguistic heterogeneity, however, is not only an issue connected to migration. Increasingly, native-speaking students have less developed language abilities in many countries (Tajmel, 2010). The reasons lie in the special needs of some of the students. Problematic familial and social backgrounds can also lead to lower levels of linguistic abilities, which directly influence the student's potential learning success in any domain of school education. This is why this issue should not be seen as a problem just for students with migrant backgrounds, but more as arising from the linguistic heterogeneity of the students.

Additionally, in their classrooms students are confronted with different cultures, which often have different belief systems and attitudes (Mamlok-Naaman, Abels, & Markic, 2015). To work with other students, it is necessary to understand their behavior, as well as to speak the same language. Thus, intercultural understanding needs to become an important part of science lessons. Intercultural understanding is seen as a fundamental, or even essential, part of international education (Walker, 2004). It helps students appreciate the richness and diversity of other cultures and recognize that there are different ways of seeing the world (Bredella, 2003). It requires the development of (i) specific knowledge – awareness of cultural differences, (ii) attitude – raising awareness of the attitudes which inform how we react and the development of the ability to adjust our own behavior, when required, and (iii) building rapport – understanding the way we need to act and react to respond in an appropriate and respectful way (Van Oord & Corn, 2013). This research field is, unfortunately, underrepresented in German science education. From only a few scholars do we see the realization of a strong connection between cultural diversity and linguistic heterogeneity. Some of this work is presented by Tajmel (2010), although the focus lies on research of students' linguistic skills. The present project is aiming to develop language- and cultural-sensitive teaching materials for science classes.

5.3 Rationale of the Project

Starting from the present situation in Germany, this research and development project aims to develop teaching methods and learning materials for linguistically heterogeneous and culturally diverse classes, including research on their effect on teaching and learning. Thus, from one perspective, the lesson modules should help students to develop a linguistic basis for scientific language and to avoid learning incorrect scientific language. Using the lesson modules should help teachers to support communication between students, not only help them express themselves in good German, but also use proper scientific language. Consequently, the lesson approach selected and the learning materials developed combine both content and language using Content and Language Integrated Learning (CLIL), along with cooperative and autonomous learning. From another perspective, the lesson module seeks to develop students' intercultural understanding.

From this initial point the main research question is:

> To what extent is it possible for students to learn science content, scientific terminology and the German language simultaneously in an intercultural context, while working in cooperative learning settings and developing their intercultural understanding?

5.4 Research Methods

5.4.1 Participatory Action Research

This project is based on the Participatory Action Research (PAR) model of science education (Fig. 5.2; Eilks & Ralle, 2002). PAR is a joint effort between teachers and science educators for curriculum development, educational research, and classroom innovation. By using this approach, different competencies melt together into development of teaching practice.

This chapter describes the work of a group of nine chemistry/science teachers and three teachers of German as a Second Language (GSL) from different secondary schools, who are collaborating with a university researcher (Fig. 5.3). The group meets regularly every 3–4 weeks and has been developing lesson modules concerning CLIL and intercultural understanding for about 2 years. Also before, the same group was working on a development of language sensitive teaching materials for science classes. At the group meetings, changes in teaching practices

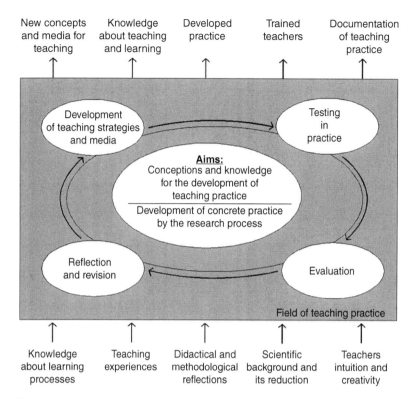

Fig. 5.2 PAR within science education (Eilks & Ralle, 2002)

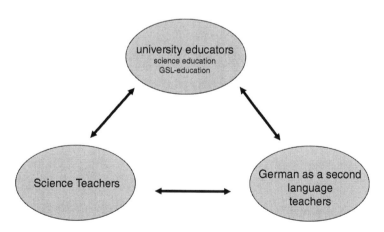

Fig. 5.3 PAR group in the present project

Table 5.1 Development and evaluation of a lesson module "Staying Healthy"

August 2014	Analysis of relevant literature; collecting ideas for methods and experiments; first provisional structuring of the lesson module
End of September 2014 (Meeting of the group of teachers)	Presentation of the provisional lesson module; negotiating and restructuring the first part of the lesson module; collecting ideas for structuring the second half
October to November 2014	Revising the lesson module; testing of the lesson module in two student learning groups; observation of the lessons by one university researcher and teacher self-reflection after each lesson
End of November 2014 (Meeting of the group of teachers)	Reflection on first experiences with the whole group of teachers; negotiating the test and student questionnaires
November to December 2014	Testing occurs in another learning group; test and student questionnaires
Mid of December 2014 (Meeting of the group of teachers)	Reflection in the whole group
January to June 2015	Testing in another three learning groups occurs; test and student questionnaires

are proposed, negotiated, and refined, so that the resulting structures can be tested and applied in classroom situations before being reflected upon and improved.

Two different lesson modules have been developed using this model and focusing on both linguistic heterogeneity and cultural diversity. Table 5.1 offers an overview of the development and evaluation process. This example of the development of the lesson module "Staying Healthy" is representative for the work of the group.

To answer our research question, multidimensional triangulation was done. All of the student groups that implemented the lesson module were continuously accompanied and observed by university researchers. Furthermore, after each lesson a self-reflection by the teachers was completed and written down. These experiences were regularly discussed by the entire PAR group. Finally, students were asked to write a short cognitive test which was developed by the teacher group, based on their experiences and knowledge. Additionally, a student feedback tool was collected, which is a combination of an open and a Likert-type questionnaire.

5.4.2 Lesson Module "Staying Healthy"

In the first phase of the lesson module "Staying Healthy," students begin by working on a worksheet which presents different food products that are not typical in German diet (e.g., cooked bananas, mango, okra, soy). The goal in starting this discussion is to help students analyze what it means to eat healthy and what criteria can be used to label food as healthy. Following this idea, the students worked in groups by using different worksheets to clarify these questions. Using this knowledge, in the following experimental phase students researched different food products containing sugar, fat, etc. Then the students worked on a research folder. The first page listed all the materials needed to carry out the experiments. The experiments are about testing the food on sugar, fat, protein, and starch. In addition, German vocabulary and definitions were provided, as well as the definite and indefinite articles for German masculine, feminine, and neuter nouns in both singular and plural forms. Students worked in pairs, using the learning station method. Every station was based on an experiment and contained exercises about the topic assigned to the station. However, every exercise not only aimed to repeat and build knowledge, but also aimed to improve the students' knowledge of the German language. Therefore, in addition to every exercise, students had to work on an exercise in German language. The experiments were mainly presented as drawings or a sequence of pictures. To acquire the knowledge of writing a laboratory report, the students had aid in the form of "Help Cards" (that offer help on different students' language levels) at nearly all stations. By doing so, students were able to decide what they needed help with and on what kind of level they needed it. The next step involved students reflecting on their own eating habits. As a homework, they monitored their eating habits by recording them in writing. They analyzed the results of the observations of their eating habits in the duration of 1 week and interpreted the result based on the information they have gathered. To make students sensitive toward different cultures and give them knowledge about different cultures, students dealt with the food from the beginning of the lesson module and with fictional letters from students from different countries (China, India, South Africa, Peru). They analyzed the food from these countries and learnt more about the eating habits in those countries. Following this, students wrote a letter back to one of the fictional students. At this point, the exercises were merged with exercises for the German language. In this phase students could rely on the "Help Cards" that were offered. It was important that students were aware of the help, but they were not forced to use it.

Different tools and methods for the acquisition of GSL were used in the lesson module. From this vast repertoire, some are named here:

- *Simple phrasing* (one-sentence-constructions) – the sentences were written as easy as possible. The focus was on only full information block without subordinate clauses.
- *List of vocabulary* (with articles, and plural forms) – glossary of the new words was written, containing their explanations and definitions.
- *Words for helping to write the observation and discussion* – a list of word was given that students can use to build a sentence describing their observation or discussion.
- *Beginning of the sentence* – the first two to three words of a sentence were named.

- *Connecting the parts of the sentences* – the sentence was cut into different parts and students had to connect the parts to build a correct sentence.
- *Sample sentences* – one sentence was given as an example. Students should use this as a pattern for writing their own sentences.
- *Drawings as explanation* – explicitly while describing the experiments, some parts of the experiment were presented as a drawing, instead of an explanation. For some students, it is easier to copy the drawing than to understand the explanation properly.
- *Cloze* – the parts of the text are left out. Students need to fill in the gaps and build the correct text. Usually, the scientific words are left out.

More methods are presented in Markic, Broggy, and Childs (2012).

5.4.3 Sample

The testing and evaluation phases were carried out using six learning groups (grade 7; age range 11–13) with a total of 144 students for the lesson module on "Staying Healthy." The lesson module was tested in different schools in the city-state of Bremen, Germany. All of the schools which took part in the study are located in the suburbs of Bremen. The population of these suburbs is, typically, comprised of individuals who have a lower than average social and educational background, including a number of residents with a migrant background. Table 5.2 presents some of the characteristics taken from the sample.

When looking at Table 5.2, it is clear that the students predominantly come from a migrant background and that a very high percentage of students do not speak the German language at home. Information about their competencies in German language has been provided by the science teacher in cooperation with the German language teacher. The students who took part in our studies are generally poor in their German language proficiency, particularly when it comes to expressing their own knowledge in writing and creating proper sentences. The students in this study mainly speak Turkish or Arabic as their first language. Also, different Slavic languages are spoken by students in this study as mother languages. Only a few of the students speak Pakistani, Tamil, or English as their mother tongue.

Table 5.2 Characteristics of sample population

Characteristic		Staying Healthy ($N = 144$)
Sex	Female	83 (57.6%)
	Male	61 (42.4%)
Students with a migrant background		130 (90.3 %)
German not spoken as the home language		90 (62.5 %)

All the participants (teachers and students) voluntarily participated in the study. Everything was performed in compliance with the relevant laws and institutional guidelines. Prior to the study, the school principals were informed about the purposes and the duration of the study. For students' data, codes chosen by students were used.

5.5 Results

The cognitive test was developed by the teachers according to their personal teaching experiences. The scoring of the test was based on the prestructured pattern for evaluating the test. The focus was on students' decision-making, argumentation, and content knowledge. The majority of students passed the test successfully, achieving scores higher than 50% of the total available points. A high percentage of all student groups had scores of "good" or "very good." A total of 75% of the participants achieved more than 80% of the total points possible. Such achievement was considered to be high and quite a remarkable factor by the teachers.

From the teachers' reflection, it was noted that they were happy with the outcome, with the openness of the lessons, and with the overall motivation of their students. Furthermore, they reported that students needed less support in understanding and writing while working on their materials compared to the other lessons. They also said that they were surprised how easy it was for some students to deal with the topic. It was interesting for the teachers to see that some students were more motivated during the work and also started to show their results proudly to the teacher, emphasizing their language and "… writing a … good sentence …" (student). This reaction was consistent both with the feedback given by the students and the classroom observation. The results show that the students were able to learn autonomously and liked to work cooperatively in smaller groups of 2–3 persons (see Fig. 5.4). About 80% of the students do not agree with the statement that it was difficult for them to understand the materials. The learners judged the lessons to be remarkably good, especially concerning aspects such as help in the verbalization of their own ideas and knowledge, the autonomy of learning, and structured cooperation and communication. In particular, they mentioned that the working materials had helped them better understand the topic both by themselves and within their peer group (see next to last and last statement in Fig. 5.4). During the lesson module it was easy to observe that students were proud of themselves and of their own work. They also agreed that their ability to express their own ideas and results in proper German had grown commensurately (compare also Markic, 2011, 2012).

Though it was not explicitly measured and evaluated but, by combining the results from the students' cognitive test and their questionnaire, teachers' reflection and classroom observation, it can be said that the language support given in this lesson does support the development of students' scientific literacy as well. We can see that, with such language support, students were able to understand and

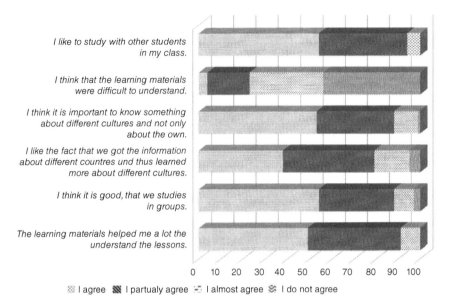

Fig. 5.4 Presentation of the results from the student Likert-questionnaire

experiment scientific facts and their meaning. Furthermore, it seemed easy for them to describe, explain, and interpret scientific facts, which was the focus of the cognitive test.

Finally, the result was that students were very open to and interested in different cultures. They were interested in gaining more knowledge about other cultures and wanted to exchange their ideas in the classroom, as well (see third statement in Fig. 5.4). One noticeable fact was that participants in the tested classes became more open and aware of different cultures in their own classes. Students started to talk about their own eating habits at home and both how and what they ate during the day. Some of the students even talked about their eating habits when they visited the countries where their families had come from. In addition, one class organized a day for international healthy cooking in cooperation with the students' parents. It seems that the lesson module was a step in the right direction for developing and supporting students' intercultural understanding.

5.6 Conclusions and Implications

The teaching and learning methods described above started a wave of strong developments and changes in many specific pedagogies in Germany, in their effort to reduce the difficulties students had with scientific language in German schools (Busch & Ralle, 2012; Leisen, 2004; Markic, 2012). Special teaching methods and different tools for dealing with students' linguistic heterogeneity in science classes are under development (Leisen, 2004; Markic, 2011). Methods are being devised for supporting linguistically heterogeneous classes and encouraging linguistically sensitive teaching and learning of science. The evidence supports the claim that instruction and evaluation of practical work in linguistically heterogeneous classes needs to be assisted by language-activating and supporting tools. This allows for the active integration of more students in practical and experimental tasks and contributes to better levels of achievement. However, the use of language tools as a supporting measure for promoting lab work in classes that are linguistically heterogeneous is a relatively new field in German science education. Research regarding good practices and their effects in this area, therefore, is still quite lacking. The present project tries to bridge this gap.

Although the cognitive test in the present study is limited in its scope in terms of judging long-term learning effects, the short-term results provided a good baseline for measuring whether students can understand topics on their own. Students' understanding of topics includes their ability to express themselves more easily and correctly through the German language. The initial data seems very promising for implementation of further lesson modules which combine the learning of scientific knowledge, German language skills, intercultural understanding, and cooperative learning methods.

Despite the process of collaborative development being new for teachers and students alike, each group dealt with it in an autonomous fashion, aided by the newly created teaching materials for the lesson modules. This also held true for the aspects focusing on the teaching of the German language and teaching methods employed. The students were able to cooperatively manage the lesson module, despite initial doubts expressed by some of the teachers. The expectations of the teachers, which had been set down in the form of a prestructured test, were exceeded by the students, most of whom achieved unexpectedly positive cognitive results.

The cooperative efforts between teachers of science and teachers of GSL appear to have offered attractive possibilities for developing new teaching materials which further linguistic heterogeneity and support intercultural understanding in science lessons. The researchers also had a chance to exchange their personal experiences of linguistic difficulties, knowledge of the students, and any pertinent interdisciplinary information, including methodologies with the teachers. Furthermore, cooperation between experts stemming from multiple disciplines offers a promising path for creating, motivating, and highly attractive learning environments. Finally, this cooperation provides the teachers with opportunities for Continuous Professional Development and extending their professional capabilities.

The results of the present study indicate that there is a definite way of answering the question of seeing students' differences as an opportunity with an "yes." It is a longer but, nevertheless, an optimistic way for the German school system to establish and implement the developed suggestions and changes in science classes. It is also a way to change teaching for heterogeneity into *an opportunity* for most of the teachers in general and science teachers in particular.

References

Benholz, C., & Iordanidou, C. (2004). Fachtexte im Deutschunterricht der Sekundarstufe I. 5.- 8. Jahrgangsstufe. *Deutschunterricht. Sonderheft: Das mehrsprachige Klassenzimmer, 4*, 19–27.

Brandenburger, A. (2007). Fachunterricht ist Sprachförderung ... Selbstständiger und kompetenter Umgang mit Lesetexten und Fachsprache. *Pädagogik, 6*, 29–32.

Bredella, L. (2003). For a flexible model of intercultural understanding. In G. Alred, M. Byram & M. Fleming (Eds.), *Intercultural experience and education* (pp. 31–49). Clevedon: Multilingual Matters.

Bryan, L. A., & Atwater, M. M. (2002). Teacher beliefs and cultural models: A challenge for science teacher preparation programs. *Science Education, 86*, 821–839. https://doi.org/10.1002/sce.10043

Busch, H., & Ralle, B. (2012). Special language competences – diagnosis and individual support. In S. Markic, D. di Fuccia, I. Eilks & B. Ralle (Eds.), *Heterogeneity and cultural diversity in science education and science education research* (pp. 11–22). Aachen: Shaker.

Carter, L. (2007). Sociocultural influences on science education: Innovation for contemporary times. *Science & Education, 92*, 165–181. https://doi.org/10.1002/sce.20228

Cassels, J. R. T., & Johnstone, A. H. (1983). Meaning of the words and the teaching in chemistry. *Education in Chemistry, 20*, 10–11.

Childs, P. E., Markic, S., & Ryan, M. C. (2015). The role of language in teaching and learning of chemistry. In J. Garcia-Martinez & E. Serrano-Torregrosa (Eds.), *Chemistry education: Best practice, innovative strategies and new technologies* (pp. 421–446). Weinheim: Wiley.

Collier, V. (1987). Age and rate of acquisition of second language for academic purposes. *TESOL Quarterly, 21*, 617–641. https://doi.org/10.2307/3586986

Eilks, I., & Markic, S. (Eds.). (2014). Traditions and trends in German mathematics, science and technology education. *Eurasia Journal of Mathematics, Science and Technology Education, 10*(4), 229–396.

Eilks, I., & Ralle, B. (2002). Participatory action research in chemical education. In B. Ralle & I. Eilks (Eds.), *Research in chemical education – what does this mean?* (pp. 87–98). Aachen: Shaker.

Fang, Z. (2006). The language demands of science reading in middle school. *International Journal of Science Education, 28*, 491–520. https://doi.org/10.1080/09500690500339092

Ford, A., & Peat, F. (1988). The role of the language in science. *Foundation of Physics, 18*, 1233–1241.

Grosser, M., & Glombard, B. (2008). The relationship between culture and the development of critical thinking abilities of prospective teachers. *Teaching and Teacher Education, 24*, 1364–1375. https://doi.org/10.1016/j.tate.2007.10.001

Johnstone, A. H., & Selepeng, D. (2001). A language problem revisited. *Chemistry Education Research and Practice, 2*, 19–29. https://doi.org/10.1039/B0RP90028A

Lee, O. (2005). Science education with English language learners: Synthesis and research agenda. *Review of Educational Research, 75*, 491–530. https://doi.org/10.3102/00346543075004491

Lee, O., & Fradd, S. H. (1998). Science for all, including students from non-English language backgrounds. *Educational Researcher, 27*, 12–21. https://doi.org/10.3102/0013189X027004012

Leisen, J. (2004). Der bilinguale Sachfachunterricht aus verschiedenen Perspektiven – Deutsch als Arbeitssprache, als Lernsprache, als Unterrichtssprache und als Sachfachsprache im Deutschsprachigen Fachunterricht (DFU). *Fremdsprache Deutsch, 30*, 7–14.

Maas, U. (2005). Sprache und Sprechen in der Migration im Einwanderungsland Deutschland. In Instituts für Migrationsforschung und Interkulturelle Studien (IMIS) der Universität Osnabrück (Ed.), *Sprache und Migration* (pp. 89–134). Bad Iburg: Grote Druck.

Mamlok-Naaman, R., Abels, S., & Markic, S. (2015). Learning about relevance concerning cultural and gender differences in chemistry education. In I. Eilks & A. Hofstein (Eds.), *Relevant chemistry education – from theory to practice* (pp. 219–240). Rotterdam: Sense.

Markic, S. (2010). Umgang mit sprachlichen Defiziten von Schülerinnen und Schülern im Chemieunterricht. In D. Höttecke (Ed.), *Entwicklung naturwissenschaftlichen Denkens zwischen Phänomen und Systematik* (pp. 496–498). Münster: LIT-Verlag.

Markic, S. (2011). Lesson plans for student language heterogeneity while learning about "matter and its properties". In C. Bruguiere, A. Tiberghien & P. Clement (Eds.), *Science learning and citizenship. Proceeding of the ESERA 2011 Conference. Part 3* (pp. 115–121). Lyon: European Science Education Research Association.

Markic, S. (2012). Lesson plans for students language heterogeneity while learning science. In S. Markic, D. di Fuccia, I. Eilks & B. Ralle (Eds.), *Heterogeneity and cultural diversity in science education and science education research* (pp. 41–52). Aachen: Shaker.

Markic, S., & Abels, S. (2014). Heterogeneity and diversity – a growing challenge or enrichment for science education in German schools? *Eurasia Journal of Mathematics, Science and Technological Education, 10*(4), 271–283. https://doi.org/10.12973/eurasia.2014.1082a

Markic, S., & Abels, S. (2016). Science education meets inclusion. In S. Markic & S. Abels (Eds.), *Science education towards inclusion* (pp. 1–6). New York: Nova Publishing.

Markic, S., Broggy, J., & Childs, P. (2012). How to deal with linguistic issues in chemistry classes. In I. Eilks & A. Hofstein (Eds.), *Teaching chemistry – a studybook* (pp. 141–166). Rotterdam: Sense.

Markic, S., & Eilks, I. (2008). A case study on German first year chemistry student teachers' beliefs about chemistry teaching and their comparison with student teachers from other science teaching domains. *Chemistry Education Research and Practice, 8*, 25–34. https://doi.org/10.1039/B801288C

Nespor, J. (1987). The role of beliefs in the practice of teaching. *Journal of Curriculum Studies, 19*, 317–328. https://doi.org/10.1080/0022027870190403

Nieto, S. (Ed.). (2000). *Affirming diversity: The sociopolitical context of multicultural education*. 3rd ed. New York: Longman.

Osborne, J. (2002). Science without literacy: A ship without a sail? *Cambridge Journal of Education, 32*(2), 203–218. https://doi.org/10.1080/03057640220147559

Phillips, L. M., & Norris, S. P. (1999). Interpreting popular reports of science: What happens when the readers' world meets the world on paper? *International Journal of Science Education, 21*, 317–327. https://doi.org/10.1080/095006999290723

Reich, H. H., & Roth, H.-J. (Eds.). (2002). *Spracherwerb zweisprachig aufwachsender Kinder und Jugendlicher*. Hamburg: Behörde für Bildung und Sport.

Riebling, L., & Bolte, C. (2008). Sprachliche Heterogenität im Chemieunterricht. In D. Höttecke (Ed.), *Kompetenzen, Kompetenzmodelle, Kompetenzentwicklung. Gesellschaft für Didaktik der Chemie und Physik – Jahrestagung in Essen 2007* (pp. 176–178). Münster: LIT.

Rodrguez, A., & Kitchen, R. S. (2005). *Preparing prospective mathematics and science teachers to teach for diversity: Promising strategies for transformative action*. Mahwah: Lawrence Erlbaum & Associates.

Roth, W., & Tobin, K. (2009). Solidarity and conflict: Aligned and misaligned prosody as a transactional resource in intra – and intercultural communication involving power differences. *Cultural Studies of Science Education, 5*, 807–847. https://doi.org/10.1007/s11422-010-9272-8

Scherz, Z., Spector-Levy, O., & Eylon, B. (2005). "Scientific communication": An instructional program for high-order learning skills and its impact on students' performance. In B. K. M. Goedhart, O. de-Jong & H. Eijkelhof (Eds.), *Research and quality of science education* (pp. 231–243). Dordrecht: Springer.

Sliwka, A. (2010). From homogeneity to diversity in German education. In OECD (Eds.), *Effective teacher education for diversity: Strategies and challenges* (pp. 205–217). Paris: OECD.

Tajmel, T. (2010). DaZ-Förderung im naturwissenschaftlichen Fachunterricht. In B. Ahrenholz (Ed.), *Fachunterricht und Deutsch als Zweitsprache* (pp. 167–184). Tübingen: Narr Verlag.

Van Oord, L., & Corn, K. (2013). Learning how to "swallow the world": Engaging with human differences in culturally diverse classrooms. *Journal of Research in International Education, 12*(1), 22–32. https://doi.org/10.1177/1475240913478085

Walker, G. (2004). *To educate the nations: Reflections on an international education*. Woodbridge: John Catt.

Chapter 6
Supporting English-as-a-Foreign-Language (EFL) Learners' Science Literacy Development in CLIL: A Genre-Based Approach

Yuen Yi Lo, Angel M. Y. Lin and Tracy C. L. Cheung

Abstract In recent years, the practice of using a second/foreign language to teach non-language content subjects (e.g. science) has become increasingly popular, especially in English-as-foreign-language (EFL) contexts. Such a practice can be categorised as 'Content and Language Integrated Learning' (CLIL). However, EFL learners may encounter difficulties in accessing content knowledge through English, especially with regard to the subject-specific academic literacy. Hence, content subject teachers may need to provide more scaffolding to help EFL learners in CLIL bridge the gap between everyday language and academic literacy. A genre-based approach, which emphasises contextualised language learning and use, serves as a useful framework to integrate content and language teaching in CLIL. This chapter shares an example of how university language specialists collaborated with science teachers in one secondary school in Hong Kong, where CLIL is practised. Drawing insights from the genre-based approach, the university-school collaborative team designed and tried out a set of materials to help grade 8 students write a piece of sequential explanation text. Lesson observations revealed how teachers integrated language scaffolding into their science lessons, and students' sample work and teachers' reflection showed that the materials were useful for helping the students develop science literacy. These findings yield significant implications for literacy development in science education and language across the curriculum in CLIL.

Keywords Content and language integrated learning (CLIL) · genre-based pedagogy · English medium education (EMI) · language across the curriculum · academic literacies in science

---

Y.Y. Lo · A.M.Y. Lin (✉) · T.C.L. Cheung
Faculty of Education, The University of Hong Kong, Pokfulam, Hong Kong
e-mail: yeonmia@gmail.com

Y.Y. Lo
e-mail: yuenyilo@hku.hk

T.C.L. Cheung
e-mail: tracyccl@hku.hk

© Springer International Publishing AG 2018
K.-S. Tang, K. Danielsson (eds.), *Global Developments in Literacy Research for Science Education*, https://doi.org/10.1007/978-3-319-69197-8_6

6.1 Introduction

In the era of globalisation, learning and using (an) additional language(s) apart from one's mother tongue is becoming the norm internationally. Hence, identifying ways to learn a second language (L2) more effectively is a topic of considerable research interest. English, often referred to as the lingua franca, is clearly the most popular L2 in English as a Foreign Language (EFL) contexts (Dalton-Puffer, Nikula, & Smit, 2010). Content and Language Integrated Learning (CLIL), with its principle of learning content subjects such as science and history through an L2, holds great potential to facilitate L2 learning and is gaining popularity around the world, including such EFL contexts as Europe and Asia. However, researchers have observed notable differences between academic registers and conversational registers, which poses tremendous difficulties for L2 learners who have to master academic literacy and content knowledge at the same time (Gibbons, 2009). How to integrate content and language teaching in content subject lessons remains a huge challenge for CLIL teachers (Davison & Williams, 2001), especially those subject specialists who may lack the knowledge and pedagogical skills of language teaching.

Meanwhile, genre theory and genre-based pedagogy have become popular in the field of English language teaching (Derewianka, 2003). Given its principles of contextualising language learning and use for particular social purposes, genre-based pedagogy has also recently attracted the attention of CLIL researchers (Llinares, Morton, & Whittaker, 2012; Lorenzo, 2013), who suggest that it can constitute an effective pedagogical framework for CLIL. This chapter describes how genre-based pedagogy can be applied in CLIL science lessons to integrate content and language teaching and how it can help EFL learners develop science literacy. Here, it is necessary to define 'science literacy', as it is used in this study. Since our project focuses on helping students acquire the academic language that is needed to 'express concisely and precisely the complex ideas and concepts that are embedded in the content of a subject and that are essential for learning in that subject' (Gibbons, 2009, p. 5), we are primarily concerned about the 'fundamental sense' of science literacy, which refers to the specific ways of using language when doing science, understanding science and communicating science (Hand et al., 2003; Norris & Phillips, 2003).

6.2 Science Literacy and Challenges Imposed on EFL Learners

Knowledge is construed largely through language, and how different academic disciplines construct or construe their disciplinary knowledge is different (Llinares et al., 2012). For instance, in school science texts, it is common to find the prevalence of abstract subject-specific terms, high lexical density, complex noun groups (with nominalisation) and the use of passive voice and implicit logical reasoning (de Oliveira, 2010; Fang, Lamme, & Pringle, 2010). It is widely agreed that mastery of the content of a discipline is in large part mastery of the discipline's specific ways of using language, or discipline-specific literacy (Luke, Freebody, & Land, 2000). At the same time, discipline-specific literacy can be understood with the notion of 'genre', which can be defined as socially recognised ways of using language to achieve certain communicative goals (Hyland, 2007). For example, in school science, the basic genres include procedures, procedural recounts, reports and explanation texts (Rose & Martin, 2012), and each of these genres has distinctive discourse structure and lexicogrammatical features to achieve its communicative goal. Take the genre of 'sequential explanation' as an example. Its purpose is to explain a sequence of events and it usually starts with a general description of the 'phenomenon', followed by the 'explanation' in a series of 'steps' (Rose & Martin, 2012). Such a way of using language in science (i.e. science literacy) is apparently different from how language is used in everyday life, and this presents difficulties to students, particularly those L2 learners in CLIL who have to learn content subjects through an L2. As students are very often expected to express their understanding of subject knowledge through academic literacy in high-stakes assessments, there have been calls for more explicit instruction of subject-specific genres so as to help students master academic literacy (Gibbons, 2009).

6.3 Dilemmas Facing Science Teachers in CLIL

Although there is call for more explicit instruction of academic literacy in content subject lessons, content subject teachers may not share similar views. Content subject teachers tend to construct their identity as 'content subject teachers' *only*, and they may not believe that it is their role to teach 'language' (Lo, 2014; Tan, 2011). Even though some content subject teachers are willing to take on the responsibilities for language teaching, owing to their professional training as subject specialists, they may lack the knowledge and pedagogical strategies to do so (Koopman, Skeet, & de Graaff, 2014). This may be particularly true for science teachers, who have been trained to adopt inquiry-based instruction, in which students are actively engaged in the process of exploring and constructing knowledge (Weinburgh, Silva, Smith, Groulx, & Nettles, 2014). Such inquiry-based pedagogy may not be compatible with language teaching pedagogy. For instance, the well-known Sheltered Instruction Observation Protocol (SIOP) aims at making content accessible to English language learners in mainstream schools in the USA (Echevarria, Vogt, & Short, 2010). It emphasises the provision of language scaffolding in content subject lessons, but its approach of 'frontloading' or 'foregrounding' language teaching has been challenged by science education researchers (Weinburgh et al., 2014). Thus, one prominent challenge faced by content subject teachers in CLIL is how they can systematically integrate content and language teaching in their lessons (Davison & Williams, 2001). In some educational contexts, CLIL teachers are also responsible for preparing students to sit for high-stakes public examinations in content subjects, and may thus be unwilling to spare time in their lessons to teach language (Lo, 2014; Tan, 2011). All these place CLIL content subject teachers in a difficult position when planning and delivering their lessons.

6.4 Genre-Based Pedagogy – A Possible Solution to the Problem of How to Integrate Content and Language Teaching

Researchers have been looking for a systematic pedagogical framework for CLIL and some have realised the potential of genre-based approaches, as such approaches can provide a contextualised language learning experience by pulling together language, content and context (Hyland, 2007). Through identifying subject-specific genres and their linguistic features, teachers can develop a better idea of how to incorporate language teaching during the process of knowledge construction in content subject lessons (Lorenzo, 2013).

There is no single pedagogy associated with genre theory (Derewianka, 2003), and the genre-based pedagogy that our project adopted is the Sydney School's 'teaching/learning cycle' (Rose & Martin, 2012), which consists of three stages, namely deconstruction, joint construction and independent construction. During the deconstruction stage, the teacher builds up field knowledge (such as science concepts in science lessons) and prepares students to read. The teacher then reads the academic text together with the students. During the reading process, the teacher deconstructs the text and draws students' attention to the linguistic features of the genre in question. With such language awareness, the teacher then co-constructs a piece of writing (be it a few sentences, short paragraphs or texts) with students. After guided practice, students will then be able to write another short text on their own during the independent construction stage. Hence, genre-based pedagogy integrates top-down and bottom-up strategies in helping students to read and produce subject-specific genres, thereby helping them to develop academic literacy. Such a pedagogical framework is grounded on the principle that 'successful learning depends on guidance through interaction in the context of shared experience' (Rose & Martin, 2012, p. 58) and also Vygotsky's socio-cultural theory, which suggests that learning is facilitated when learners obtain scaffolding through interacting with the teacher and peers (Vygotsky, 1978).

In Anglophone countries, including Australia and the USA, genre-based pedagogy has been adopted to assist English language learners to master school genres (de Oliveira & Iddings, 2014; Rose & Martin, 2012). In CLIL, there have also been attempts to adopt genre-based pedagogy. For instance, Fan and Lo (2016) evaluated the effectiveness of genre-based pedagogy in facilitating students' science literacy development in Hong Kong. The grade 7 students in their study showed significant improvement in their writing of classifying reports and consequential explanation texts. However, in their study, the genre-based pedagogy was implemented by an English language teacher in after-school classes. Hence, their findings may not be able to illuminate how language scaffolding can actually be implemented by science teachers in science lessons. Further application and evaluation of the impact of genre-based pedagogy on students' academic literacy in CLIL is thus necessary.

It is against this background that this project was conducted to help EFL learners develop their science literacy. This chapter aims to illustrate how genre-based pedagogy can be applied in CLIL science lessons to integrate content and language learning, and to examine the potential impact of adopting such pedagogy on students' science literacy development. The specific research questions are:

1. How can genre-based pedagogy be adopted in science lessons to integrate content and language teaching?
2. How do science teachers perceive the effectiveness of genre-based pedagogy?

6.5 The Project

This project was commissioned by the Education Bureau of Hong Kong, with the aim of promoting talking and writing in junior secondary science classes[1] (grades 7–9, students aged between 13 and 15). In Hong Kong, which is a former British colony and now a special administrative region of China, Chinese and English are the co-official languages. Owing to the colonial history and economic development of Hong Kong, English is highly valued in the society (Poon, 2010). With a view to increasing students' exposure to English and enhancing their English proficiency, English is adopted as the medium of instruction for some or all content subjects in most secondary schools in Hong Kong.[2] In other words, CLIL is implemented in most secondary schools, though to different extents. Realising the difficulties that students may encounter when learning content subjects in English, the Education Bureau of Hong Kong has been providing different kinds of support for schools and teachers, including professional training and school support programmes like the one this chapter reports. In this 2-year project, the university

---

[1]This 2-year project, *'Promoting Talking and Writing in Science at Junior Secondary Level Using English as the Medium of Instruction'* was commissioned by the Science Education Section, the Education Bureau of Hong Kong.

[2]Before 1998, secondary schools in Hong Kong could choose their own medium of instruction, whereas the compulsory mother-tongue policy was implemented in 1998. Since 2010/2011, the government 'fine-tuned' the mother-tongue policy by allowing secondary schools to choose their medium of instruction according to some criteria concerning teachers, students and resources available (see Poon, 2010, for a review of the development of medium of instruction policy in Hong Kong).

team organised professional development workshops for teachers from several secondary schools and provided on-site support through collaborative planning meetings, materials development, lesson observations and reflections. This chapter focuses on the implementation of the project in one of the project schools.

The project school adopted English as the medium of instruction for most content subjects (e.g. science, geography) at all grade levels. Most students of the school belonged to Band 1, the top tier of the three-tier categorisation system of primary school students in Hong Kong. Hence, it is reasonable to assume that students in the school possessed an above average level of academic ability and English proficiency. The students had learned English as a foreign language for at least 6 years in primary school and they were also learning English in their English language lessons in the secondary school. The school participated in the project mainly because they desired to receive more support from university educators to help their students further develop their academic literacy.

The school decided to focus on grade 8 students (aged around 13 years old), and hence the three science teachers teaching grade 8 classes were involved in the project, with the vice principal being the coordinator. All three science teachers were experienced in teaching science through English. Two English teachers were also involved in the project, as the school regarded this as part of the language across the curriculum initiative. They were invited to attend co-planning meetings and comment on the materials designed by the university team, but they did not participate in the implementation and evaluation stage.

The university team, comprised of four English language educators and one science educator, first met with the teachers to discuss the topic to focus on and their objectives. They decided to focus on the topic 'breathing mechanism', with the aim that the grade 8 students would be able to write a short sequential explanation text to explain the processes of 'breathing in' and 'breathing out'. The university team and science teachers worked together to design a set of materials for the topic (more details in the next section) and the materials were then tried out by the three teachers in four grade 8 classes. During the try-out, the university team observed and video-recorded at least one lesson of each class. After that, the university team and teachers reflected on the effectiveness of the materials.

6.6 Data Collection and Analysis

Multiple sources of data were collected to address the research questions. These included the materials developed in collaboration with the teachers, four video-recorded lessons, teachers' reflection and some students' sample work collected by the teachers. The recorded lessons were transcribed and analysed to identify the episodes where the teachers attempted to integrate content and language teaching. The strategies that the teachers used were coded and categorised according to the language features being focused on (e.g. 'teaching subject-specific words using syllabication' and 'teaching common word roots or affixes' are 'vocabulary' level strategies; 'providing some useful connectives for students to link up their ideas in sentences' belong to 'sentence' level strategies). Some illustrative episodes will be presented in this chapter to demonstrate the application of genre-based pedagogy and some useful pedagogical strategies in CLIL science lessons. Teachers' reflections were also transcribed and coded to identify themes related to the usefulness and limitations of the materials. In particular, the researchers were interested in exploring whether and how the teachers found the materials useful in helping students write sequential explanations, how they incorporated the materials into their science lessons, possible difficulties they may have encountered and any limitations of the materials. These, together with some students' sample work provided by the teachers, can reveal the teachers' perceptions of the effectiveness of the materials.

6.7 Design of the Materials Based on Genre-Based Pedagogy

Based on the principles of genre-based pedagogy, the materials designed for the project consisted of four main parts. Activity 1 drew students' attention to subject-specific words, which referred to the different parts of body involved in the breathing mechanism and their collocation with certain verbs (e.g. intercostal muscles *contract* or *relax*; diaphragm *becomes flattened* or *returns to dome shape*). After grasping those key words and phrases, students' understanding of the processes of breathing in and breathing out was further consolidated in Activity 2, which required students to complete two flowcharts showing the breathing mechanism. The use of graphic organisers (flowcharts), together with symbols like arrows or '<', serves as another semiotic resource to complement language. The use of multimodality here was particularly useful, as students were required to thoroughly understand the processes and their sequence involved in the breathing mechanism. When the students completed the flowcharts and shared answers with the teachers, they were already engaged in the 'joint-construction' process, though they did so mainly orally. This then led to Activity 3, which asked students to first rearrange the order of words/phrases in sentences and then further rearrange the order of those sentences, which would then form the target sequential explanation text. After this joint-construction stage, students should be well-prepared for the final task (Activity 4), in which they were asked to write two short texts (or paragraphs) describing the processes of breathing in and breathing out respectively.

In short, the different tasks of the set of materials follow closely the various stages in the teaching/learning cycle, providing scaffolding at the word, sentence and text levels, and progressing from receptive to productive skills. Multimodality (e.g. flowcharts) is also utilised as another kind of scaffolding.

6.8 Findings and Discussion

6.8.1 Delivery of the Materials – Integration of Content and Language Teaching

This section reports how the teachers used the materials in lessons, particularly some useful strategies that they adopted to incorporate language teaching into their science lessons. When designing the materials, the university team paid special attention neither to 'interrupt' science teachers' inquiry-based approach nor to 'foreground' language teaching in science lessons. Hence, although the materials still aimed at integrating content and language teaching, the university team treated the materials as consolidating tasks, helping students to consolidate the science concepts learned while developing science literacy. The university team expected the teachers to use the materials after they finished teaching the concepts (i.e. the whole breathing mechanism). However, when the teachers tried out the materials, they skilfully incorporated the materials into their content teaching, thereby integrating content and language teaching rather effectively and successfully. For example, when we observed her lesson, T1 just started the topic of breathing mechanism. She first introduced the various parts of body involved in the breathing processes and illustrated the breathing processes with the Powerpoint slides and video clips prepared by the textbook publisher. She then asked her students to complete Activity 1 of the materials to consolidate what she had talked about, particularly the language involved (i.e. the collocation between different body parts and verbs/actions). Excerpt 1 briefly demonstrates how T1 shifted between content and language when she introduced the body parts and actions involved (see the Appendix for transcription conventions).

Excerpt 1: T1 – Body parts and the corresponding actions [22:13 – 22:38]

At the beginning of the lesson, T1 briefly introduced the different body parts involved in the breathing mechanism. Then, she asked all the students to stand up and feel the movements of their ribs when breathing in and out. She further asked the students why it was more difficult to hold the action of 'breathing in'. The following shows how she addressed this question.

> T1: … The idea is that because the intercostal muscles contract. So first of all, T1 [*T1 addressing herself*] would like to talk about the muscles, the muscles, the muscles can have two conditions. One is contract [*T1 holding her fist tightly*], and one is relax [*T1 softening her voice*]. So, in your test or exam, don't write the muscle 'extend' and 'reduce'. I don't know what is it about. So you should use correct verb, that is 'contract' and 'relax'. Is that OK? So, when we breathe in, the intercostal muscles, muscles [*T1 emphasising plural 's'*], because there are many muscles between the ribs, so muscles contract. And when the muscles contract, it brings about, so, our ribs move upward and outward. So can you feel that, just now, our ribs, OK, move out, the rib cage, the whole rib cage move outward and upward? So in this case, why it is difficult to hold? Because the muscles contract, contract, it cannot contract for a long time, so this is the reason ….

In this excerpt, T1 attempted to explain why it is more difficult to hold the action of breathing in, but she drew students' attention to the language used in

Lines 2–7, when she reminded students that they '*should use correct verb, that is "contract" and "relax"*' (Lines 4–5) and that '*muscles*' should be in the plural form '*because there are many muscles between the ribs*' (Lines 6–7). Such explicit instruction of the collocation between the body parts and action verbs was probably due to Activity 1 of the designed materials, which reminded the teachers to provide some scaffolding for the students at the vocabulary/phrase level.

On the other hand, T2's class had already completed Activity 1 and 2 in previous lessons. T2 had asked the students to attempt Activity 3 (the jumbled sentence task) as homework. At the beginning of the lesson we observed, T2 first talked about the 'bell jar model', which resembles the breathing mechanism. Through this, T2 revisited the key words and major steps involved in the breathing mechanism. Then, he checked the answers to Activity 3 with the students, before they proceeded to Activity 4, the individual writing task. Excerpts 2 and 3 present two episodes about how T2 guided students to write a coherent text, either implicitly or explicitly.

Excerpt 2: T2 – Implicit highlighting of connectives [08:41 – 09:56]

This episode took place at the beginning of the lesson. T2 revisited the breathing mechanism with the students using the 'bell jar model'. After illustrating the bell jar model, T2 discussed with the students how the model represented the breathing mechanism with some PowerPoint slides.

1 T2: OK, the balloon becomes bigger. Here, again, this is the part [*T2 pointing at the word*
2 '*balloon' on the PowerPoint slide*], in blue, even though the colour is not very sharp,
3 blue. Red is the action [*T2 pointing at the word 'bigger'*], just like Activity 2. OK, then
4 what happens? What happens then?

5 [*pause; T showed the next PowerPoint slide; Ss laughed, as the answers were already*
6 *shown*]

7 T2: OK, when the rubber sheet is pulled downward, done by me, OK, pulled it down,
8 then what happens to the volume of the bell jar? The volume, the volume, when it was
9 pulled downward, the volume become _____? Larger, or say increases of course. So, the
10 volume increases. Then what happens to the gas pressure? When the volume becomes lar-
11 ger, the gas pressure becomes _____? Becomes _____?

12 Ss: Lower

13 T2: Lower. Look at these [*T2 pointing at the coloured words in the PowerPoint slide*].
14 Here, you can see different parts [*T2 pointing at the phrase 'the rubber sheet'*], action
15 [*T2 pointing at the phrase 'is pulled down'*], volume of the bell jar, different parts, action
16 [*T2 pointing at the word 'increases'*], and the gas pressure, action [*T2 pointing at the*
17 *phrase 'becomes lower'*], right? Just like what we did before. How about this here
18 [*T2 pointing at the word 'when'*]? What is this? 'When'. Think about that, OK? You will
19 see other green words later ….

In Excerpt 2, T2 revisited the relationship between the action of the diaphragm, the volume of the chest cavity and the gas pressure inside the chest cavity, using the corresponding parts of the bell jar model (Lines 6–10). During the revision of concepts, he slightly modified the PowerPoint slides provided by the textbook publisher by changing the colour of some of the words to highlight the different body parts (in blue), actions (in red) and conjunctions (in green) (as mentioned in

Lines 12–17 of Excerpt 2). In addition to the different parts and actions, towards the end of the episode, T2 further prompted students to think about what the green word '*when*' represented (Lines 16–17). By highlighting the use of connectives implicitly with such a strategy, T2 was actually paving the way for the final task (Activity 4), as Excerpt 3 shows.

Excerpt 3: T2 – How to write a sequential explanation [17:01 – 19:18]

After checking the answers to Activity 3 (the jumbled sentence task), T2 asked the students to put Activity 3 into their textbook and just leave Activity 1 and 2 (i.e. the exercise on collocations of body parts and verbs, as well as the flowcharts) on the table. T2 then distributed the final task (sequential explanation writing) to the students and gave the following instructions.

1 T2: OK, what is the last shot? The last shot is [*pause; T2 showing the PowerPoint slide*]
2 try to write a sequential explanation. Now I know that, I know that if you have Activity 3
3 in your hand, you will just copy and paste, right? So, I'll try to ask you to do it just
4 referring to Activity 1 and 2. Now remember, I'm not asking you to dictate, have dicta-
5 tion, OK? I just ask you to figure out how all these actions, all these actions, bring about
6 the air movement in and out of your lungs. I'll just ask you to do breathing in only. OK?
7 Now maybe you think it's a little bit difficult, but you can see here again [*T2 pointing at*
8 *the PowerPoint slide*]. It gives you some hints. When breathing in, the intercostal muscles
9 contract and _____? [*T2 reading out the first sentence provided in the worksheet*] So
10 refer to the Activity 2, you have this, right? [*T2 referring to the flowchart*] If we start
11 from here, then what happens? What is the event that happens afterwards? (pause) Will be
12 the _____?

13 S1: (…)

14 T2: Yeah, OK? Hey, wait, wait, wait, wait, wait. Not just, not just, er, write what I say.
15 You try to think about it. If these happen, if you mention these, OK, then can you just go
16 down? No, because?

17 Ss: At the same time.

18 T2: Yeah, this is 'at the same time', right? You (…). So these two events happen at the
19 same time. Just like this. You try to figure out the whole paragraph. OK? …

In Excerpt 3, T2 was guiding students to write a sequential explanation text with the help of the key words in Activity 1 and the flowcharts in Activity 2, instead of having a '*dictation*' (Line 4). This is actually the independent construction stage. T2 showed how students could start by referring to the flowcharts (Lines 6–17). Later in that episode, T2 also reminded the students not to forget '*the green words*', which were the conjunctions he had mentioned before. Students were then given 10 minutes to complete the task on their own, and as the sample work in the next section shows, most students could construct the short texts rather successfully.

One key issue concerning CLIL is how teachers, especially content subject teachers, actually integrate content and language teaching (Davison & Williams, 2001), since they may lack knowledge of language teaching pedagogy, given their training as subject specialists (Koopman et al., 2014). For science teachers in particular, there may be tension between inquiry-based pedagogy and explicit language scaffolding, with the fear of 'foregrounding' language teaching

(Weinburgh et al., 2014). The lesson excerpts presented in this section may, to a certain extent, provide some insights into this issue. Guided by genre-based pedagogy, science teachers could weave language teaching into the process of knowledge construction, with strategies such as occasional language-focused instruction (T1 in Excerpt 1), visualising/highlighting the use of connectives (T2 in Excerpt 2), and explicit reminders of language usage (T2 in Excerpt 3). In this way, the process of knowledge building will not be seriously disrupted, and students can develop their academic literacy in a contextualised way. All these strategies and examples demonstrate how genre-based pedagogy can be integrated into science lessons to assist students in grasping content and language simultaneously. However, it may be noticed that teacher–student interaction was rather limited in the lesson excerpts shown above, which may raise the question whether the principle of 'guidance through interaction' was observed. We would argue that the above episodes mainly show how the teachers drew students' attention to the language features of the genre and prepared students for the independent construction task. Therefore, the teachers dominated the classroom talk, and students mainly responded by doing the tasks. More teacher–student interaction was observed later when the teachers checked the answers to the tasks with the students, during which the language features of the genre were reinforced when the teachers commented or elaborated on students' responses. In addition, when the teachers asked the students to complete the tasks, some of them assigned students to do so in groups (e.g. T1 made Activity 2 a group task and asked students to complete the enlarged flow charts in groups of 5–6). This in turn promoted more peer interaction.

6.9 Teachers' Reflection and Students' Work

This section presents the teachers' views on the effectiveness of the materials designed for this project, together with some evidence gathered from students' work.

When being asked their general impression of the materials after the try-out, all the three teachers showed very positive responses and found the materials useful and '*user-friendly*' (T1). In particular, all the three teachers found the materials effective in drawing students' attention to language while teaching scientific concepts. 'Language' here does not only refer to subject-specific vocabulary, but also words like articles and connectives, which are essential for students to express their ideas in complete sentences and coherent texts. As T3 commented, '*I think the jumbled sentence (Activity 3) is pretty good. It makes students pay attention to and clearly understand the meaning of those vocabulary items. … Very often they (students) don't care about those articles and connectives. But the jumbled sentence (activity) forced them to put those things back. They cannot avoid them. But during the process of organising, when we checked the answers, they realised they put (the words) in the wrong places. They would then pay more attention*

4. movements / of / cavity / these / chest / the / volume / increase / the / and / gas / the / cavity / inside / the / decreases / chest / pressure / therefore

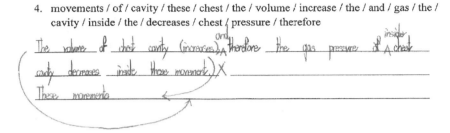

Extract 6.1 A student's classwork in Task 3 (from T3's class)

next time'. Extract 6.1, which was taken from a student's classwork, illustrates what T3 mentioned.

This student actually got most of the concepts correct (e.g. the volume of the chest cavity increases; the gas pressure inside the chest cavity decreases), but he did not know where to put the phrase '*these movements*', which refers to the preceding movements of the ribs and diaphragm. The jumbled sentence activity would then help the students to engage in deeper processing of the language and raise their language awareness.

Similarly, T2 realised the importance of making students aware of the usage of '*logical link*' (i.e. connectives). As he reflected, '*The flowchart mentions "at the same time" and "these movements". In fact, these are very important link. Even if (students) can combine those actions and structures, or the relationship between structures and actions, whether they can link all these up depends on those links*'. Such awareness probably explains why T2 highlighted such connectives as '*when*' and '*therefore*' in his lesson, as demonstrated in Excerpts 2 and 3 above. After the try-out, he found that '*the effectiveness is pretty good, at least 80% or above (could complete Activity 4). The students in my class are relatively weak, but what they came up with was very close (to the target output). This is better than before*'.

The usefulness of the materials could be demonstrated with some students' sample work in Activity 4. Extract 6.2 was written by a student in T2's class, whom he commented on as '*relatively weak*'. That may explain why some grammatical errors (e.g. subject-verb agreement, sentence formation) could still be found, but in general, the concepts were accurate and the text was rather coherent, as the student did try to use such connectives as '*when*' and '*so*' to link up the ideas. Extract 6.3 was written by a student in T3's class. His output was even closer to the target text, except that some 'logical links' (e.g. '*at the same time*', '*these movements*', '*therefore*') are missing.

Even more encouragingly, the effectiveness of the materials in promoting academic literacy for science can be further reflected by students' performance in a formal test conducted by the school later. In that test, students were required to write a sequential explanation that explains how the breathing in mechanism takes

A. BREATHING IN

When breathing in, the intercostal muscles contract and _diaphragm will contract then the ribs become upwards o~~ff~~ and outward and the diaphragm become flattened, it ca~~u~~sing the volume of chest cavity increase therefore the gas pressure of chest cavity ~~increase~~ decrease, when the gas pressure of chest cavity become lower than atmosphere pressure so air drawn in lungs._

Extract 6.2 A student's classwork in Activity 4 (from T2's class)

<u>**Breathing in**</u>

When breathing in, the intercostal muscles _contract. The ribs move upwards and outwards. At the same time, the diaphragm contracts and becomes flattened. These movements causes the volume of chest cavity increases. Therefore, the gas pressure inside the chest cavity decreases. When gas pressure of chest cavity becomes lower than the atmospheric pressure, air is drawn into the lungs._

Extract 6.3 A student's classwork in Activity 4 (from T3's class)

place. Extract 6.4 shows how a student performed on that question. The student clearly demonstrated his understanding of the breathing mechanism with a well-constructed sequential explanation text. In his answer, phrases like '*at the same time*' and '*so*' were used to connect different ideas.

Hence, from the teachers' reflection and students' work, it seems that the materials, designed based on genre-based pedagogy, were effective in helping students to produce the target scientific text, thereby developing their science literacy. What the teachers highlighted as particularly useful is the bridging between vocabulary and text level with more explicit instruction and language-focused practices, something which is often ignored by science teachers, who tend to put more emphasis on the teaching of concepts (Lo, 2014; Tan, 2011). As highlighted in the literature review, academic literacy is considerably different from everyday language and it is not sufficient to simply adopt the 'language bath' approach, hoping

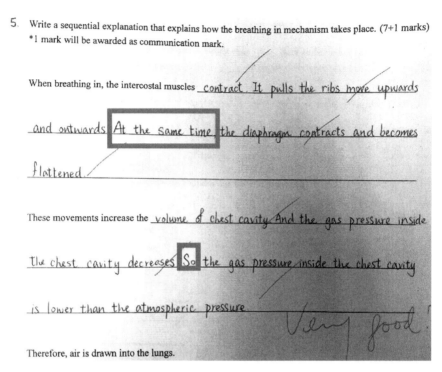

Extract 6.4 A student's performance in a formal test conducted by the school (from T3's class)

that students would incidentally acquire academic literacy (Morton, 2010). This is especially true for EFL students, who often lack regular contact with the target language outside classrooms. Therefore, extra support in the form of more explicit instruction and scaffolding is necessary (Llinares et al., 2012).

We acknowledge that from the lessons observed and the sample work collected, we could not say much about students' learning of the scientific concepts. However, we would argue that the aim of developing the set of materials based on genre-based pedagogy was to 'counterbalance' (Lyster, 2007) the content-oriented lessons with more focus on explicit support for language. Hence, we tended to pay more attention to students' academic literacy development. Yet, we believe that the set of materials could reinforce students' conceptual understanding, since they were guided to describe the processes through appropriate and precise academic language. In many CLIL classrooms in Hong Kong, it is likely that the teacher would use some L1 to explain the concepts thoroughly and/or to check students' conceptual understanding, however, without also helping students to express it in L2 (Lin & Lo, 2017). Our scaffolding in this study provides one way of helping teachers to ensure that students also learn how to express concepts in L2.

Although it seems that all the teachers involved agreed with the usefulness of the materials and language scaffolding in their lessons, they did reflect on some issues when using the materials. The first thing mentioned concerns how to integrate content and language teaching in science lessons. T1 said, '*I think before using this set of materials, we may need more time to prepare, or to see how to integrate (it) into the lessons. The original plan was perhaps we taught (the content) first, and then used the activities. But I think eventually what we did and the outcomes were different. We all taught some (content) and then did some (activities)*'. Such reflection echoes what we illustrated above when discussing the teachers' practices of integrating language scaffolding and science pedagogy. In collaboration with the university team, the teachers were provided with a set of materials, but how they actually incorporated that into their lessons to support students' learning also depends on teachers' own experience, pedagogical awareness and understanding of their students.

Another issue related to the integration of content and language teaching concerns the practical issue of 'time'. It is inevitable that some extra time is needed to provide more explicit scaffolding for academic literacy, as T3 commented that he had overestimated his students' ability and it actually required plenty of time to help students proceed from understanding the concepts to expressing their understanding through appropriate academic language. T2 expressed similar concerns and he thought it was impossible to provide such '*intensive*' language support for each topic. This is in line with what previous studies found – the scramble for time to cover the content in the syllabus has been identified as one hindrance to CLIL, especially in examination-oriented contexts where content subject teachers bear the pressure to prepare students for high-stakes examinations (Lo, 2014; Tan, 2011). However, T2 believed that '*if we could choose one to two topics and let students experience (academic language learning) or develop their language awareness*', it would be worth the extra time spent. Such a comment may lead us to think about the solution to the practical constraint of time in terms of the overall curriculum design and perhaps language across the curriculum. It may be the case that content subject teachers cannot afford too much time to focus intensively on language scaffolding in their lessons, but it is still possible to incorporate language teaching in some of their lessons if the curriculum is planned in such a way that different subjects target at similar genres for recycling or reinforcement (i.e. horizontal curriculum mapping), or that important language features of a particular content subject (e.g. reports and explanation texts in science) are introduced systematically across grade levels (i.e. vertical curriculum mapping) (Lin, 2016).

6.10 Conclusion

This chapter reports the collaboration between university language specialists and science teachers in a secondary school on implementing genre-based pedagogy in CLIL science lessons. Based on the multiple sources of data collected, both the university team and the science teachers believed that the materials designed based on the principles of genre-based pedagogy were effective in helping the students to achieve the target language objective – being able to write sequential explanation texts. We would then argue that this demonstrates the effectiveness of applying genre-based pedagogy in facilitating students' academic literacy development, which corroborates Morton's (2010) proposal that the genre-based approach is a systematic way of incorporating language scaffolding into content teaching.

Despite our successful experience of implementing the genre-based pedagogy in CLIL, the teachers in this project did reflect on some issues, including the practical constraint of lesson time. We would like to propose language across the curriculum as a potential solution, with teachers of different subjects collaborating with each other to plan a more integrated curriculum which focuses on different academic text types in different subjects and at different grade levels (Lorenzo, 2013). Apart from this, we also acknowledge other limitations of this small-scale project. In particular, the teachers involved in our project were experienced and dedicated and their students were 'Band 1' students, whose academic ability and English proficiency level were above average. We could perceive more challenges in helping students with lower English proficiency to develop science literacy, which would warrant further research.

6.11 Appendix

6.11.1 *Transcription Conventions*

T1, T2, etc. = Teacher 1, Teacher 2, etc.
S1, S2, etc. = single student
Ss = more than one student
(…) = inaudible utterances
(italics) = words added to make the utterances comprehensible
[ ] = nonverbal actions or author's comments
___ (at the end of questions) = short pauses indicating blank filling questions

References

Coyle, D., Hood, P., & Marsh, D. (2010). *CLIL: Content and language integrated learning.* Cambridge: Cambridge University Press.

Dalton-Puffer, C., Nikula, T., & Smit, U. (2010). Language use and language learning in CLIL: Current findings and contentious issues. In C. Dalton-Puffer, T. Nikula & U. Smit (Eds.), *Language use and language learning in CLIL classrooms* (pp. 279–291). Philadelphia: John Benjamins.

Davison, C., & Williams, A. (2001). Integrating language and content: Unresolved issues. In B. Mohan, C. Leung & C. Davison (Eds.), *English as a second language in the mainstream* (pp. 51–70). Harlow: Pearson Education Ltd.

de Oliveira, L. C. (2010). Enhancing content instruction for ELLs: Learning about language in science. In D. Sunal, C. Sunal, M. Mantero & E. Wright (Eds.), *Teaching science with Hispanic ELLs in K-16 classrooms* (pp. 135–150). Charlotte: Information Age.

de Oliveira, L. C., & Iddings, J. (Eds.) (2014). *Genre pedagogy across the curriculum: Theory and application in U.S. classrooms and contexts.* London: Equinox Publishing.

Derewianka, B. (2003). Trends and issues in genre-based approaches. *RELC Journal, 34*(2), 133–154.

Echevarria, J., Vogt, M. E., & Short, D. J. (2010). *Making content comprehensible for secondary English learners: The SIOP model.* Boston: Allyn & Bacon.

Fan, C. C., & Lo, Y. Y. (2016). Interdisciplinary collaboration to promote L2 Science literacy in Hong Kong. In A. Tajino, T. Stewart & D. Dalsky (Eds.), *Team teaching and team learning in the language classroom: Collaboration for innovation in ELT* (pp. 94–112). New York: Routledge.

Fang, Z., Lamme, L., & Pringle, R. (2010). *Language and literacy in inquiry-based science classrooms, grades 3-8.* Thousand Oaks: Corwin Press and Arlington: National Science Teachers Association.

Gibbons, P. (2009). *English learners, academic literacy, and thinking: Learning in the challenge zone.* Portsmouth: Heinemann.

Hand, B. M., Alvermann, D. E., Gee, J., Guzzetti, B. J., Norris, S. P., Phillips, L. M., et al. (2003). Guest editorial: Message from the "Island Group": What is literacy in science literacy? *Journal of Research in Science Teaching, 40*(7), 607–615.

Hyland, K. (2007). Genre pedagogy: Language, literacy and L2 writing instruction. *Journal of Second Language Writing, 16*(3), 148–164.

Koopman, G. J., Skeet, J., & de Graaff, R. (2014). Exploring content teachers' knowledge of language pedagogy: A report on a small-scale research project in a Dutch CLIL context. *Language Learning Journal, 42*(2), 123–136.

Lin, A. M. Y. (2016). *Language across the curriculum: Theory and practice.* Dordrecht: Springer.

Lin, A. M. Y., & Lo, Y. Y. (2017). Trans/languaging and the triadic dialogue in Content and Language Integrated Learning (CLIL) classrooms. *Language and Education, 31*(1), 26–45.

Llinares, A., Morton, T., & Whittaker, R. (2012). *The roles of language in CLIL.* Cambridge: Cambridge University Press.

Lo, Y. Y. (2014). Collaboration between L2 and content subject teachers in CBI: Contrasting beliefs and attitudes. *RELC Journal, 45*(2), 181–196.

Lorenzo, F. (2013). Genre-based curricula: Multilingual academic literacy in content and language integrated learning. *International Journal of Bilingual Education and Bilingualism, 16*(3), 375–388.

Luke, A., Freebody, P., & Land, R. (2000). *Literate futures: Review of literacy education.* Brisbane: Education Queensland.

Lyster, R. (2007). *Learning and teaching languages through content: A counterbalanced approach.* Amsterdam: John Benjamins.

Morton, T. (2010). Using a genre-based approach to integrating content and language in CLIL. In C. Dalton-Puffer, T. Nikula & U. Smit (Eds.), *Language use and language learning in CLIL classrooms* (pp. 81–104). Amsterdam: John Benjamins.

Norris, S. P., & Phillips, L. M. (2003). How literacy in its fundamental sense is central to scientific literacy. *Science Education, 87*(2), 224–240.

Poon, A. Y. K. (2010). Language use, language policy and planning in Hong Kong. *Current Issues in Language Planning, 11*(1), 1–66.

Rose, D., & Martin, J. R. (2012). *Learning to write/reading to learn: Genre, knowledge and pedagogy in the Sydney school.* London: Equinox.

Schleppegrell, M. (2004). *The language of schooling: A functional linguistics perspective.* New York: Routledge.

Tan, M. (2011). Mathematics and science teachers' beliefs and practices regarding the teaching of language in content learning. *Language Teaching Research, 15*(3), 325–342.

Vygotsky, L. S. (1978). *Mind in society: The development of higher psychological processes.* Cambridge: Harvard University Press.

Weinburgh, M., Silva, C., Smith, K. H., Groulx, J., & Nettles, J. (2014). The intersection of inquiry-based science and language: Preparing teachers for ELL classrooms. *Journal of Science Teacher Education, 25*(5), 519–541.

Chapter 7
Language, Literacy and Science Learning for English Language Learners: Teacher Meta Talk Vignettes from a South African Science Classroom

Audrey Msimanga and Sibel Erduran

Abstract There is considerable research on the role of language in science teaching and learning from contexts in which non-native speakers of English are taught science by teachers who are either native speakers of or proficient in English as the language of learning and teaching (LOLT). In many South African classrooms, students who are non-native speakers of English are taught in English, by teachers who are also non-native speakers. Furthermore, many students and teachers in these classrooms speak more than one local language, such that English is their third or subsequent language. Thus, most South African students are English additional language learners (EALs) as opposed to being English second language learners (ESLs). Many South African classrooms are therefore multilingual by nature. Yet, very little is known about the pedagogic demands of teaching science to EALs. Research on EALs experiences in science lessons in the South African context has potential not only to contribute to the broader literature on EAL teaching but also to inform teacher education on how to prepare teachers for these environments. Meta talk is one of the teaching strategies that may help facilitate EALs involvement in the lesson as well as promote deep engagement with content. The chapter discusses some vignettes from whole class discussions observed in a multilingual science classroom showing how one South African teacher uses meta talk to get his learners to engage with science concepts.

Keywords Meta talk · pedagogy · language · ESLs · multilingualism

---

A. Msimanga (✉)
University of the Witwatersrand, Johannesburg, South Africa
e-mail: audrey.msimanga@wits.ac.za

S. Erduran
Department of Education, University of Oxford, Oxford, United Kingdom

National Taiwan Normal University, Taipei, Taiwan
e-mail: sibel.erduran@education.ox.ac.uk

7.1 Introduction

There is considerable research on the role of language in science teaching in which non-native speakers of the language of learning and teaching (LOLT) are taught science by teachers who are native speakers of the LOLT. Examples include migrant students in the USA and Spain, taught by native speakers of English or Spanish. The situation is different for most South African students. Although the Language-in-Education-Policy (Department of Education, 1997) allows schools to choose any of the nine official local languages as the LOLT, most schools choose English, even in situations where neither the students nor the teachers are proficient in English (Probyn, 2006; Setati & Adler, 2000). Thus, most South African students who are not proficient in English are taught in English by teachers who are themselves not first speakers of the English language. The situation is complicated particularly in the Gauteng province of South Africa. Gauteng is the most highly urbanised province experiencing high levels of immigration both from other provinces and from outside South Africa. Thus, many students and teachers speak more than one local language, such that English is not their second language (as would be the case for their counterparts in some international studies), but it is their third or subsequent language. In other words, many South African students are not English second language learners, ESLs but English Additional Language Learners or EALs (Department of Basic Education, 2011; Janks, 2010). Thus, while there are common concerns about language in science teaching and learning in South Africa as in other countries, there are some contextual complexities that play out in a unique way in South African science classrooms. In addition to language, an important factor is the differential access to science that persists as a legacy of the apartheid era. More than 20 years after the establishment of democracy, performance in science and mathematics continues to follow racial lines (see e.g. Carnoy, Chilisa, & Chisholm, 2012; Howie, Scherman, & Venter, 2008; Mji & Makgato, 2006). Most EALs, in the poorer township and rural schools continue to underachieve in mathematics and science. They are, therefore, still excluded from entry into science programmes in tertiary education and from taking up science-related careers. Despite this being the case, only a few South African teacher education programmes specifically prepare teachers to teach science to EALs.

This chapter reports on findings from a project that worked with both practising and new teachers to identify teaching strategies which create opportunities for learning science for EALs. We focus on one such strategy, 'teacher meta talk'. We consider teacher meta talk as a teaching strategy with potential to facilitate student involvement in the lesson as well as promote deep engagement with science concepts. The broader data set that form the basis of the analysis came from three teachers' classrooms, but in this chapter we use excerpts from one teacher's lessons. We show how he used meta talk to mediate student participation in class activities and to foster meaning-making during whole class discussions. The results are discussed within the context of learner involvement, conceptual engagement and understanding of science concepts. We discuss the implications for teacher education in South Africa as well as teacher professional development in general.

7.2 Theoretical Framework

We take a socio-cultural approach in framing teaching and learning of science in the classroom. The role of social interaction, particularly language and talk, in socio-cultural theories of learning has taken centre stage in research on science classroom discourse globally over the past few decades (see e.g. Lemke, 1990; Mercer, Wegerif, & Dawes, 1999; Vygotsky, 1978). Recognition of the role of social mediation, particularly in the learning of science is important in shifting views and understandings of the nature of science, not as a ready-made body of knowledge for transmission to learners but as a messy, socially constructed, value-laden enterprise which must be engaged with in order to be understood (Erduran & Dagher, 2014). Talk is an important cultural and psychological tool in the mediation of this construction of both shared and individual understandings of science knowledge. Teachers who take this view of learning then adopt socio-cultural approaches to pedagogy (Gibbons, 2007). For EALs such approaches can serve dual purposes: fostering understanding of science knowledge while facilitating language development. As Gibbons says, there are implications for classroom practice. Teachers have to be able to orchestrate classroom interaction in specific ways that achieve these dual goals of classroom discourse.

This chapter reports on how one teacher used meta talk as a tool in the social space of classroom discussion to mediate meaning-making for learners engaging in a language in which they are not proficient. We consider teacher meta talk as a teaching strategy with potential to facilitate EAL student involvement in the lesson as well as to promote engagement with science concepts.

Meta talk seems to have its origins in language education where it was initially defined within the broad frame of meta language. For instance, Schiffrin (1980) made the following statement about meta language:

> Because human language can be used to talk about virtually anything, individuals conversing with one another have available an infinite number of topics about which they can talk. One topic is talk itself ... many conversations allow talk to emerge as a subtopic within ongoing talk about something else. (p. 199)

According to Schiffrin, therefore, meta language is the use of language to focus on talk as a topic. As she later put it, 'Language can be used to talk about itself, that is, it can serve as its own meta-language' in 'talk about talk' (p. 200). Schiffrin's proposed framework for analysing conversations placed focus of the analysis at three different aspects of talk. The focus could be on the code or how the interlocutors talk about language. Focus could also be placed on the speakers themselves and how they talk about their own and the other speakers talk. Finally, analysis could focus on the messages that are being relayed about language or talk. She noted, however, that oral messages seldom serve a single purpose, but that they are multi-functional. They can serve a meta lingual function as well as a communication function, to prevent a breakdown in communication or to enlist the other's participation. The latter assertion is important for analysis of classroom talk as we propose to use this framework in this chapter. In the complex interaction between a science teacher and his students the main focus (at least consciously) is the science content under consideration. The tendency, therefore, is for both teacher and students to focus on the science content and not the ways in which they talk about their own talk or their own thinking unless probed to do so. Thus, it is important to make teachers aware of the ways to make language visible in the science classroom.

Reporting on work in a different context, Faerch (1985, p. 185) used the concept of meta language to analyse communication in foreign language (FL) classrooms. He defined, 'those portions of … lessons in which teacher and student focus on the linguistic code rather than on content. I refer to such phases of discourse as *meta talk*'. Faerch distinguished between meta transactions, the parts of discussion that focused on the FL code, and content transactions, which focused on non-linguistic content. He argued that meta talk can take place in both types of transactions. Meta talk in a science lesson may be expected to take one or both of these two forms: talk focusing on the language of science as opposed to talk about the English language as the LOLT or the language in which the discussion happens. The language of science is not immediately explicit, even to native speakers of English, as the LOLT. And yet most science classroom discourse in South Africa and the rest of the world tends to be dominated by content transactions and hardly any meta transactions. Content transactions, in this case, would be discourses in which the teacher's focus is on the science content. Meta transactions are more debateable in a science classroom. They could include any parts of the discourse in which the teacher focuses student attention on the 'code' or the language, whether the language of science or the language in which science is being discussed, the LOLT. However, these are rare in South African science classrooms where teacher talk tends to dominate and interaction is more authoritative focussing on disciplinary knowledge (Webb, 2010). As Faerch (1985) argues, meta talk can be a good indicator of teaching methodology and of the views that teachers hold about learning. Teachers could adopt one of two possible positions:

> The ideal of this natural approach is that meta talk in the classroom should be reduced to a minimum and that grammar rules belong to grammar books, to be studied out of (*science*) class. A diametrically opposite view of the role of meta talk is that it constitutes an essential part of FL classroom discourse …. Teachers of this opinion, or teaching Within a context where they need to acknowledge this, will typically either reserve part of the lesson for meta talk and practice, or they will introduce meta talk whenever content provides a clue to do so. (Faerch, 1985, p. 185)

Of course the quotation above refers to language teaching where, as Faerch observes, 'metalinguistic and metacommunicative knowledge is one of the explicitly formulated goals' (p. 200). However, the same could be argued of science classroom discourse. The focus of any science classroom discussion is to afford access to science concepts, the language of science and the requisite scientific communication skills, all of which can happen together or at different times in the same lesson or across a range of lessons. Where this discussion takes place in a language in which the student is not proficient, opportunities must be created for access to the LOLT at the same time. Thus, for EALs meta talk would have even greater potential to support learning. This places an added burden on the teacher who must, therefore, be aware of its potential and be able to model it and generate awareness among his/her students. Moate (2010) argues that meta talk is the explicit awareness of talk as a tool and that as a skill such talk can be practiced and honed. It is, therefore, important for teacher education and teacher professional development to understand teacher meta talk as an important pedagogical tool in both content and meta transactions in EALs science classrooms. For our data analysis we adopted Faerch's approach, identifying meta talk as any part of the discussion that focuses on language. We looked at meta talk about the technical language of science as well as about the LOLT.

We analysed transcripts of some classroom discussions to determine how teacher communication in the form of meta talk mediates student engagement with science. First, we identified incidences of teacher meta talk and then determine its explicit or implicit role in shaping, selecting or marking student ideas. In addition to this focus on engagement we also identified other functions of meta talk such as classroom management and affective functions as identified by Moate (2011) with Finnish teachers and students.

Our interest was in how the teacher talked about how he was talking about science and the activities under discussion as well as in how he engaged in talking about their ideas and their thinking about the content under discussion. In order to understand the nature and role of teacher meta talk in our sample, we drew on a number of sources both from language education and science education. We refer to the work of Yore and Treagust (2006) as well as Rincke (Rincke, 2011) on language in the science classroom. Rincke proposes three languages of science classrooms: the home language, the language of instruction and the language of science, and argues that research on second language learning is useful in understanding how ESLs or EALs might transition between the languages of the science classroom. According to Rincke, learning specific phrases is more beneficial for ESLs in acquisition of the target language than memorisation of single words. She argues that these specific or 'automated phrases' have linguistic environment. For example, it is useful for ESLs (and EALs) when learning the words 'decision' and 'conclusion' to understand that one 'takes or makes' a decision and 'arrives' at a conclusion. Teachers, therefore, need to *'model scientific language by explaining to students how they themselves are combining terms together in sentences'* (Rincke, 2011, p. 235). Teachers are not always given the training and preparation required to be able to do what Rincke suggests for ESLs, much less so South African science teachers. In fact many science teachers would be trained to focus on the subject and its teaching, rather than issues related to language. Yet for ESLs and EALs, the subject and the language that is conveying the subject cannot be separated. Meta talk has the potential to address the gap in teachers' understanding of language issues related to EALs. Meta talk can be used to mediate learning of content as well as to develop the linguistic proficiency that students require in order to engage effectively in classroom discussion and/or science communication outside of the classroom. Also, meta talk provides a lens for understanding how teachers who are themselves non-native speakers of English consciously or unconsciously mediate engagement with science concepts for EALs.

7.3 Methodology

As part of a bigger intervention at high school level in the Gauteng province of South Africa, we observed interactions between teachers and their students in Grades 10–12 (age 16–18) science classrooms. The data reported on in this chapter were collected 3 years into the project, following teacher participation in several workshops and group discussions at which they were exposed to constructivist approaches to pedagogy. The teachers were recruited on a voluntary basis. Their students were invited to participate and then presented with information sheets about the project after which their informed consent was obtained. Since some of them were minors (below 18) their parents/guardians' consent was also sought. For this project all the students consented to audio and video recording.

Argumentation was selected as the key teaching strategy for the project and the teachers were introduced to the strategy during the workshops. Argumentation involves the justification of claims with evidence. It is a process that underpins scientific reasoning because of the centrality of evidence to science. A great deal of research has been carried out about the role of argumentation in science education in recent years (Erduran, Ozdem, & Park, 2015). The researchers co-taught some lessons in the first year of the project to demonstrate the use of argumentation in whole class teaching and for small group discussions. This was in line with research findings that teachers do need to be taught argumentation skills before they can confidently use the strategy in their own teaching (see e.g. Erduran & Jimenez-Aleixandre, 2007; Simon, Erduran, & Osborne, 2006). In order to contextualise argumentation and its potential for promoting student involvement and inquiry, teachers were made aware of the various discourse types that are possible in a science classroom. Types of variations involve different combinations of talk based on initiation, response, evaluation which can appear in different permutations. In a science lesson, different variations can be used depending on the purpose of the lesson and goals that the teacher has at a particular point in the lesson. We discussed the more traditional Initiation-Response-Evaluation/Feedback (IRE/F) triads (Mehan, 1979) and explained how most teaching in South African classrooms still tends to follow this teacher-centred closed form in which the teacher initiates interaction and a learner responds, to which the teacher provides some feedback before proceeding to initiate the next interaction triad. We then introduced the teachers to the more dialogic discourses such as Mortimer and Scotts' (2003) extended IRPRPE and IRPRRR chains (where P denotes prompting) which open up for more elaborate student contributions. In the extended chains, the teacher initiates the interaction but allows students to interact with each other by questioning or supporting or extending and explaining their peers' contributions, thus creating chains of interaction. The role of teacher questioning techniques in shaping these forms of engagement was discussed. Some of these techniques were modelled in the participating teachers' classrooms during co-teaching sessions. Teacher-researcher reflection sessions were conducted after selected lessons.

Each of the teachers worked differently with the strategies that were introduced during the intervention and produced various hybrid forms of argumentation-based teaching. They applied different argumentation skills into their classroom discussion as well as their questioning techniques, such as evaluating evidence and drawing conclusions from evidence. In the case of Mr McFar's lessons, the interaction manifested as meta talk and we wanted to understand the nature and role of meta talk in his lessons.

Data were collected in the form of audio and video recordings of the lessons. The lessons were a mix of teacher-centred exposition and learner discussion, and engagement with definitions of elements, compounds and mixtures. In other words, sometimes the conceptual knowledge was being developed through discussion themselves while at other times, the teacher mediated the introduction of the concepts'. The recordings were transcribed and the transcripts analysed for teacher meta talk. In this chapter we discuss a form of teacher–student interaction that emerged in one of the teachers' lessons. We have changed his name to Mr McFar. All the names of the student participants in the excerpts have also been changed and pseudonyms are used.

7.4 Results

Mr McFar used meta talk quite frequently in his lessons. He used the strategy to mediate student meaning-making during whole class discussion. He explained what he was doing or saying and what he was asking his learners to do or say and why. He used meta talk to make explicit the connections between concepts and ideas. He did this by asking probing and open-ended questions about what the students were saying and why they said it. We did not see this use of meta talk in the other teachers' classrooms in the project.

The excerpt below illustrates how Mr McFar showed links between concepts through meta talk. In this episode the class was trying to determine whether momentum was a scalar or vector quantity:

Teacher: according to eh Kelvin's definition momentum, that it can be regarded as a measure of the product of the mass and the velocity. Now think about mass in terms of the quantity … can we regard mass as a vector quantity or is it a scalar quantity?

Students: vector … scalar … scalar … vector

Teacher: now **I will say that again** think about it carefully

Students: (*all talking at the same time*)

Teacher: think about mass how do we regard mass **because he has used the words mass and velocity**

Len: scalar

Teacher: why?

Len: because yah the mass is got size

Langa: yah it does ...

Melo: mass is got size

Teacher: **so why am I asking this**? Because one of the biggest problems that we experience is that most of us cannot distinguish between this and that (*underlining vector and scalar*). So let's just refresh quickly. Len you said this is scalar and why are you saying this is scalar?

Sindi: because it has size ...

Teacher: thank you very much so we only have size **which can be also be referred to as ...**?

Kelvin: magnitude

Teacher: magnitude. So here (*pointing at the word 'mass'*) we have size or magnitude

Sipho: **no direction**.

In the opening turn the teacher reiterated Kelvin's definition of momentum and then to provide the class with a cue to think through the question he asked them to 'think about mass in terms of the quantity can we regard mass as a vector quantity ...?' The class chorused variable answers to which he told the learners to 'think about it carefully'. When Len says that mass is a scalar quantity (Line 10) the teacher asks 'Why?' thus creating an opportunity not only for Len to justify his answer, but also for the rest of the class to think through the various answers they had chorused earlier in Line 5. In response, Len provides justification for his answer drawing from the scientific evidence that mass has size. In Line 15 the teacher then engages in meta talk about his reasons for the probing, 'so why am I asking this?' In so doing the teacher helps learners recall information from past lessons and from earlier grades that relates to the current lesson. The teacher's question at the close of the turn (Lines 18–19) and what the teacher did with the answer in Lines 21–22 indicates a shift in teaching purpose as he cues the learners to introduce a more 'scientific' term for size (magnitude). The episode closes with an unsolicited contribution from Sipho in Line 26, providing support for the argument that mass is a scalar quantity since it only has magnitude and 'no direction'. This could be construed as an indication that Sipho has reached an understanding of the difference between vector and scalar quantities. For a non-speaker of English, it is important to point out the substitution of magnitude for size as the similarity in meaning of the two words may not be obvious and learners need to be made awareness of the specific use of one (and not the other) in science.

The next excerpt comes from a chemistry lesson, an introduction to properties of compounds. Here again, we see Mr McFar making the links between current and previous coverage of similar content explicit:

Teacher: They are elements what else? They are elements, they are gases what else do you know about hydrogen and oxygen? So we are using the terms now elements we are using the term gases we are using what else? If you think about hydrogen ... Tebogo?

Tebogo: I would say Sir, compound

Teacher: **Now you are using another word**. We are saying we know what is an element (*teacher writing words on the board*) we know what is a compound a chemical compound.... Now just briefly is there anyone in this class who can refresh our memory with regard to an element? What do you understand by the word element? Simphiwe?

Simphiwe: A substance that cannot be broken up into smaller particles

Teacher: He says a substance that cannot be broken up into smaller particles. Remember that Monday we struggled to find this. Any other person that can tell us the definition of compound?

In the episode above the teacher helps his learners to recall word definitions and to link to work done in a previous lesson. He draws learners' attention to related concepts and principles within and between lessons. By talking explicitly about how he and the learners were talking about the words and concepts, he makes clear the progression from simpler to more complex concepts:

Teacher: Right while you are still thinking about a gas I am looking at hydrogen and I am looking at oxygen can we classify hydrogen or oxygen as a mixture? **So I am asking what is a mixture** (*writes 'mixture'*)

Sebuka: (*shouts*) adding ... adding

Teacher: Sebuka? (*addressing learner who is shouting*)

Sebuka: adding different things

Teacher: **different chemical substances put together**

Class: Yes

Teacher: That is a mixture. So how does a mixture differ from a compound? **If there is a difference**. Siviwe?

Siviwe: In a compound elements are in fixed proportions while a mixture is different ...

Teacher: **Please listen to that**. A compound (*writing on the board*) has fixed and **the word he uses is** pro ...?

Siviwe: [... portions]

Teacher: [... portions] And a mixture? It doesn't have a specific ratio isn't it so? **Now why am I doing this again? because I'm gonna come back to a molecule** but our aim for and our objective for today is to look at how Magnesium will react with ... air. What chemical in air?

In turn 59 at the beginning of the episode the teacher uses meta talk to explain a question. After asking the learners if hydrogen or oxygen gases could be regarded as mixtures, he elaborates 'So I am asking what is a mixture?' The previous episode closed with a reminder about Monday's struggle to define an element. The teacher here brings to contrast an element and a mixture by referring to oxygen and hydrogen and then in meta talk explains what the actual question is.

Later in this lesson, Mr McFar used a practical demonstration to show the reaction of magnesium with oxygen. He had a few learners upfront to help ignite the magnesium strip and this led to the following discussion:

Class: Oh oohh aah (*as magnesium ribbon ignites*)

Teacher: Right there we are. Please don't shake it just turn your hand around a little bit. I want you all to have a look at it. **This is ... I don't want to use the word**. But this is something that was formed after magnesium and oxygen have reacted (*laugh*) **I can't use the other word for now**. So please look at this colour and write it down as well. So this becomes **that word that I want**, a (*pause*) of magnesium and oxygen

Smitherson: what is that word? mixture of ingredients?

Teacher: Ok, I am not gonna waste time I am gonna tell you. It is the product of, product of. So now I already have this part of the word. When magnesium and oxygen react we get a product. So the right hand side of the chemical equation is called the product or products. Now what do we call the left hand side? If that (*pointing to right side of equation*) gives us the product what will cause us to have a product?

Smitherson: mixture

Neliswa: ingredients

Dana: (*inaudible*)

Teacher: **he says a ...? say that again** Dana

Dana: reactant

Teacher: Right. So we are saying these are the substances that reacted to form a product. **Reactants plural form**.

Amidst the learners oohs and aahs at the bright flame of the burning magnesium ribbon, Mr McFar shifted the discourse to commence development of the scientific story linking what they had just seen with the previous discussion and moving on to the focus of the day's lesson, properties of compounds. The episode starts with a teacher utterance in which he uses meta talk several times. He was targeting a specific word to describe the outcome of the chemical reaction, '... *This is ... I don't want to use the word ... I can't use another word for now. ... So this becomes that word that I want of magnesium and oxygen*'. He wanted to get the learners to come up with the word 'product' by getting them to think of a word that describes the outcome of the reaction. When the learners failed to come up with the word, he volunteered the answer so as not to 'waste time'. This was a common statement in Mr McFar's lessons. He would give the learners a chance to come up with the answer, always providing clues. Occasionally he would cut the interaction short and provide the answer so as 'not to waste time'. When asked about this in the reflection discussions he explained that while he wants to help his learners think for themselves and practice talking in class, he was also mindful of the curricular requirements and the strict timelines in which he had to cover the prescribed content. This is a common tension for many South African teachers. Recent research findings show that while teachers are eager to take up and use teaching strategies that have been shown to promote science learning for EALs, they often have to make pragmatic decisions in the interest of time and meeting deadlines (Mji & Makgato, 2006; Msimanga & Lelliott, 2012; Probyn, 2009).

The next excerpt is from a discussion towards the end of the lesson. This excerpt illustrates how Mr McFar again used meta talk to mediate different forms of link-making. Scott and colleagues recognised three ways in which a teacher can help learners make connections between concepts and across topics or lessons. They identified link-making for knowledge building, in which a teacher makes explicit the connections between concepts; link-making for continuity, which shows how different parts of the lesson or different lessons build on each to develop concepts and ideas; as well as link-making for emotional engagement, which is a way in which the teacher can appeal to student interest and emotional engagement to facilitate science learning (Scott, Mortimer, & Ametller, 2011). Mr McFar seemed to draw on all three forms of link-making.

Earlier in this lesson as he took the class through recall of definitions of the terms molecule, element, mixture and compound, he asked 'Why am I doing this again? Because I'm gonna come back to the molecule later. ...' In the next excerpt we show how he makes the links to that earlier conversation. The class was now discussing an equation, $Mg + O_2 \rightarrow MgO_2$, which Kgotso and Zama's group had put up on the board for the combustion of magnesium in air:

Class: (*chorus answer*) mg mg mg

Teacher: Right so the symbol for magnesium is Mg (*points to the expression*)

Thabisile: plus O-2

Teacher: **what are you saying** Thabisile? **Say that again**

Thabisile: (*silent*)

Teacher: hah? **Why are we writing it as O-2 and not as O? Why are we** writing it as O-2 and not as O?

Kgotso: oxygen is diatomic

Teacher: **Thank you very much**.... So oxygen is a diatomic molecule. Is there anyone else that will explain to us **what we mean by the term** diatomic? Remember I said when I started that I will come back to the specific definition a specific term. What does di-mean? Kabelo? Lerato? Melo? Bongani?

Bongani: Two

Teacher: **Thank you**. So, **Kgotso wrote this** (*expression*) **down** for Magnesium.

The episode above illustrates the teacher's use of meta talk to link the current discussion to earlier ones in the same lesson. Again, the teacher used probing questioning to open up the discussion, resulting in an interesting discourse type in the form of Mortimer and Scott's (I)RRPRPRE closed chains. Mr McFar and the class were now reviewing the equation that Kgotso and Zama had put up on the board. The teacher had initiated the conversation by asking what the symbol for magnesium was, to which the class now responded in chorus 'mg mg'. However, before he could continue Thabisile made a (unsolicited) contribution, calling out 'Plus O-2'. This was an unsolicited learner contribution but appropriately presented as the next term in the equation that they were constructing. The teacher followed this up with a meta talk utterance that seems to have two of Schiffrin's (1980) focus points. There appears to be a focus on the message of Thabisile's utterance of turn 179 as the teacher asks what seems to be a clarification question, 'what are you saying Thabisile? Say that again'. However, the rephrased question in turn 182 shows that he was actually focusing on the content of the utterance, probing Thabisile's reasoning. Thabisile's reaction, in turn 181 is interesting.

Her reluctance to repeat her answer seems to suggest that she may have interpreted the teacher's response in turn 180 as negative evaluative feedback, that her answer was incorrect. However, from the teacher's follow up in turn 182 it becomes clear that he was genuinely probing for Thabisile's understanding of 'O-2' as she put it. Since the discussion was about molecules, the teacher may have wanted to confirm that she and her peers understood the difference between the molecule of oxygen, O_2 and its atom, O. This seems to be confirmed in the way the teacher responds to Kgotso's answer that oxygen is diatomic (turn 183). The teacher thanks Kgotso (turn 184) and repeats his answer, thereby affirming his contribution and marking his answer as correct and probably important. Thanking Kgotso for clarifying Thabisile's answer could be seen as creating a safe space for the two learners, link-making for emotional engagement according to Scott et al. (2011). We see this again in turn 186.

In turn 184, the teacher then uses meta talk to link this episode to an earlier one where he had said that he would come back to the definitions later in the lesson, 'Remember I said when I started that I will come back to the specific definition a specific term' (turn 184). This was link-making for continuity, showing the learners how this part of the lesson links to the beginning. Meanwhile he was also mediating for the learners another way of making connections, link-making for knowledge building (Scott et al., 2011). He was demonstrating how the concepts are linked. He had raised the questions in turns 180, 'What are you saying Thabisile?' and 182, 'Why are we writing it as O-2 and not as O?' to check understanding of the concept of molecules (O_2 vs. O). In the rest of the episode he questioned, probed and provided cues for learners to be able to distinguish between a diatomic molecule and its atomic (element) form.

Often in reflection meetings Mr McFar would refer to the importance of making links between lesson episodes and between lessons explicit, especially for ESLs. He also sometimes told his learners this. For instance, at the beginning of this lesson on momentum, he said to the class 'now I said I said to them (referring to the project team) that my introduction will be on types of collisions but when I got here I realised something else. So I just changed my introduction to check if people still remember these concepts and to show you how these things are all related from the last lesson to today's lesson. Now I need to go back to my plan and that's where we are starting and we go forward'. This was common in Mr McFar's interaction with his class, explaining why he was doing or saying something at the time.

7.5 Discussion

The potential of meta talk as a teaching strategy for EALs seems to be threefold: making explicit the purpose of various forms of classroom talk; making connections between different sections of the same lessons and/or previous lessons or grades; and explaining or calling attention to important terms, thus, mediating conceptual engagement. In Mr McFar's case meta talk sometimes took the form of questions such as, '*So why am I asking this?*' or '*So why am I doing this?*' and thus foregrounding important content or skills required of students. To mediate link-making or to make connections he used statements like, '*I will come back to it*' and when he did, '*I said I will come back to it* …' Finally, Mr McFar used meta talk to alert students to important terms, '*So we are using the term …, we are using the term …*' or '*Now you are using another word …*'. Meta talk sometimes works, sometimes not. At times Mr McFar was able to establish a rich dialogic discourse, while at others talk tended to remain closed and teacher centred in spite of the use of meta talk to help open up talk. However, overall the use of meta talk may still be helpful in addressing the language problem for those teachers who do not believe in use of home language, but are happy to scaffold student struggles with learning in a language they are not proficient in.

Learning and teaching science through a non-native language is a challenge for teachers and learners. It places demands on teachers to think about language. Meta talk can help teachers address EALs difficulties in science classrooms. While we acknowledge that meta talk as a strategy would benefit all science learners, whether native speakers of English or EALs, we argue that as teachers' skills in managing meta talk in the lessons improve, they become better equipped in dealing with EAL-related difficulties as well.

In this chapter we have demonstrated teacher use of meta talk in a South African high school science classroom for purposes other than classroom management. Mr McFar used meta talk to show conceptual connections and to point learners to continuities between concepts and within the lesson. He successfully used meta talk to mediate conceptual understanding for his learners. We illustrated meta talk within Faerch's (1985) teacher-student content transactions and demonstrated its potential as a teaching strategy to enhance learning of science by EALs. Teacher education can draw on these findings to plan programmes that can empower teachers for effective science teaching to EALs in South Africa and elsewhere. Further research is also required to provide empirical evidence of the effectiveness of the strategy and to demonstrate the learning gains for the EALs.

Acknowledgements We are indebted to Mr. McFar and his class. The funding for the research from which the data is drawn was provided by DIFID, the University of the Witwatersrand in Johannesburg. Attendance at the ESERA conference where the first version of this chapter was presented came from an NRF KIC2015 grant to the first author.

References

Braund, M., Lubben, F., Scholtz, Z., Sadeck, M., & Hodges, M. (2007). Comparing the effect of scientific and socio-scientific argumentation tasks: Lessons from South Africa. *School Science Review, 88*(324), 67–76.

Carnoy, M., Chilisa, B., & Chisholm, L. (2012). *The low achievement trap: Comparing schooling in Botswana and South Africa*. Pretoria: HSRC.

Department of Basic Education. (2011). *Curriculum and assessment policy statement grades 1–3: English first additional language*. Pretoria: Department of Basic Education.

Department of Education. (1997). *Language-in-education policy*. Pretoria: Department of Education.
Department of Education. (2011). *National curriculum statement grades R-12: Curriculum and assessment policy (CAPS) physical sciences*. Pretoria: Department of Education.
Erduran, S., & Dagher, Z. (2014). *Reconceptualizing the nature of science: Scientific knowledge, practices and other family categories*. Dordrecht: Springer.
Erduran, S., & Jimenez-Aleixandre, M. P. (Eds.). (2007). *Research in argumentation in science education: Perspectives from classroom-based research*. Dordrecht: Springer.
Erduran, S., Ozdem, Y., & Park, J. Y. (2015). Research trends on argumentation in science education: A journal content analysis from 1998-2014. *International Journal of STEM Education 2015, 2*, 5–12. https://doi.org/10.1186/s40594-015-0020-1
Faerch, C. (1985). Meta talk in FL classroom discourse. *Studies in Second Language Acquisition, 7*(2), 184–199.
Gibbons, P. (2007). Mediating academic language learning through classroom discourse. In J. Cummins & C. Davison (Eds.), *International handbook of English language teaching* (pp. 701–718). New York: Springer.
Howie, S., Scherman, V., & Venter, E. (2008). The gap between advantaged and disadvantaged students in science achievement in South African secondary schools. *Educational Research and Evaluation: An International Journal on Theory and Practice, 14*(1), 29–46. https://doi.org/10.1080/13803610801896380
Janks, H. (2010). Language, power and pedagogies. In N. H. Hornberger & S. L. McKay (Eds.), *Sociolinguistics and language education* (pp. 40–61). Bristol: Multilingual matters.
Lemke, J. L. (1990). *Talking science: Language, learning and values*. Norwood: Ablex Publishing.
Mehan, H. (1979). *Learning lessons: Social organisation in the classroom*. Cambridge: Harvard University Press.
Mercer, N., Wegerif, R., & Dawes, L. (1999). Children's talk and the development of reasoning in the classroom. *British Educational Research Journal, 25*(1), 95–117.
Mji, A., & Makgato, M. (2006). Factors associated with high school learners' poor performance: A spotlight on mathematics and physical science. *South African Journal of Education, 26*(2), 253–266.
Moate, J. (2010). The integrated nature of CLIL: A sociocultural perspective. *International CLIL Research Journal, 1*(3), 38–45.
Moate, J. (2011). Reconceptualising the role of talk in CLIL. *Journal of Applied Language Studies, 5*(2), 17–35.
Mortimer, E., & Scott, P. (2003). *Meaning making in secondary science classrooms*. Maidenhead: Open University Press.
Msimanga, A. (2013). *Talking science in South Africa: Case studies of grade 10-12 science classrooms in Soweto*. Johannesburg. University of the Witwatersrand.
Msimanga, A., & Lelliott, A. (2012). Making sense of science: Argumentation for meaning-making in a teacher-led whole class discussion. *African Journal of Research in Mathematics Science and Technology Education, 16*(2), 192–206.
Probyn, M. (2006). *Language and learning science in South Africa*. Paper presented at the Annual conference of the Southern African Association for Research in Mathematics, Science and Technology Education (SAARMSTE2006), Pretoria.
Probyn, M. (2009). Smuggling the vernacular into the classroom': Conflicts and tensions in classroom codeswitching in township/rural schools in South Africa. *International Journal of Bilingual Education and Bilingualism, 12*(2), 123–136.
Rincke, K. (2011). It's rather like learning a language: Development of talk and conceptual understanding in mechanics lessons. *International Journal of Science Education, 33*(2), 229–258.
Schiffrin, D. (1980). Meta-talk: Organizational and evaluative brackets in discourse. *Sociological Inquiry, 50*(3–4), 199–236.

Scott, P., Mortimer, E. F., & Ametller, J. (2011). Pedagogical link-making: A fundamental aspect of teaching and learning scientific conceptual knowledge. *Studies in Science Education, 47*(1), 3–36.

Setati, M., & Adler, J. (2000). Between languages and discourses: Language practices in primary multilingual mathematics classrooms in South Africa. *Educational Studies in Mathematics, 43*(3), 243–269.

Simon, S., Erduran, S., & Osborne, J. (2006). Learning to teach argumentation: Research and development in the science classroom. *International Journal of Science Education, 28*(2–3), 235–260.

Vygotsky, L. S. (1978). *Mind in society: The development of higher psychological processes.* Cambridge: Harvard University Press.

Webb, P. (2010). Science education and literacy: Imperatives for the developed and developing world. *Science, 328*(5977), 448–450.

Yore, L. D., & Treagust, D. F. (2006). Current realities and future possibilities: Language and science literacy—empowering research and informing instruction. Special issue: Natural science, cognitive science and pedagogical influences on science literacy: Empowering research and informing instruction. *International Journal of Science Education, 28*(2–3), 291–314.

Chapter 8
The Content-Language Tension for English Language Learners in Two Secondary Science Classrooms

Jason S. Wu, Felicia Moore Mensah and Kok-Sing Tang

Abstract Investigating the use of native languages (L1) in secondary science remains an unaddressed need in global scientific literacy. While past research in this area has largely focused on primary school students, more clarity is needed on the role of secondary school students' L1 use in the classroom as the language of science becomes more specialized at a higher level. This chapter details two studies investigating L1 use in secondary science classrooms in New York and Singapore. The study employs qualitative and quantitative methods, including surveys, interviews, observation, and audio recording of student discourse. We find that the L1 can be used for learning scientific content, but is seen by some students as a hindrance to the acquisition of the majority language. This is seen when comparing in-class native language use and data from surveys and interviews. We propose that this reflects a *content-language tension* that exists in many linguistically diverse science classrooms. This tension highlights competing goals of content learning and acquisition of the majority language. We conclude with a discussion of implications for addressing scientific literacy on a global scale.

Keywords English learners · English language learners · translanguaging · content-language tension · content-based language instruction · sheltered instruction · secondary science

---

J.S. Wu (✉) · F.M. Mensah
Department of Mathematics, Science & Technology, Teachers College, Columbia University, New York, USA
e-mail: jasonwu@columbia.edu

F.M. Mensah
e-mail: moorefe@tc.columbia.edu

K.-S. Tang
School of Education, Curtin University, Perth, Australia
e-mail: kok-sing.tang@curtin.edu.au

© Springer International Publishing AG 2018
K.-S. Tang, K. Danielsson (eds.), *Global Developments in Literacy Research for Science Education*, https://doi.org/10.1007/978-3-319-69197-8_8

8.1 Introduction

Addressing global needs in scientific literacy would not be complete without considering the rapidly growing population of second language learners, in this case English language learners (ELLs), in science classrooms. Such students often fall behind when facing the dual burdens of content learning and language acquisition. Improving scientific literacy for ELLs is thus imperative for both equitable science education and ultimately for future economic and technological success and personal needs. For example, one area that remains unaddressed is students' use of their native languages (L1) in secondary science learning. Although secondary students often use their L1 in the classroom, there has been little characterization of it and little consensus on how it influences their learning of science. In this chapter, we provide a review of literature and theoretical framework before presenting findings from two studies conducted in New York City and Singapore on ELLs learning science. We highlight key findings before discussing the implications of the content-language tension in the ELL science classrooms. In light of our discussion, we conclude with recommendations for supporting the development of scientific literacy amongst second language learners in science classrooms.

8.2 Literature Review

Large-scale reviews on the science education for ELLs highlight a need for further research on student L1 use in secondary science (Janzen, 2008; Lee, 2005; Rollnick, 2000). These reviews generally conclude that the L1 can be used to support learning conceptual information in the science classroom. However, we find that empirical research specifically investigating this is generally lacking. Unfortunately, most research in bilingual education has focused on elementary students, largely ignoring the experiences of older bilinguals in secondary science (August & Hakuta, 1997; Christian, 2001; Janzen, 2008; Ruiz-de-Velasco & Fix, 2000). The research that does exist comes from a variety of fields, owing to the interdisciplinary nature of the issue.

A consistent finding throughout the literature is the need to provide spaces in which different discourses could come together in the science classroom (Janzen, 2008; Lee, 2005; Rosebery, Warren, & Conant, 1992). For example, Moje, Collazo, Carrillo, and Marx (2001) find that students' various discourses, that of the home and the science classroom, compete and conflict with one another in the classroom. They argue for *third spaces* which allow for the synthesis of students' various discourses. One form of this is known as *translanguaging*, which is the fluid use of multiple linguistic resources in the classroom (Creese & Blackledge, 2010; García & Wei, 2014). Translanguaging as a paradigm eschews language separation and embraces the fluid use of all available linguistic resources for learning. This is based on the premise of knowledge and skill transfer between the first and second language (Cummins, 1979). Using analysis of student–teacher interaction, Lin and Wu (2014) suggest that translanguaging can be an essential tool to support learning in the science classroom. Other researchers offer home language support as a strategy for the integration of science and English proficiency, and there is large support for promoting language development through inquiry-based learning and enacting scientific practice (Lee & Buxton, 2013; Lee, Quinn, & Valdes, 2013). A central premise to this approach is that students can develop language through use in authentic scientific contexts provided by inquiry instruction, rather than rote instruction.

An alternative approach that has been advocated to address the disconnect between the home language and the language of science is the explicit instruction of scientific English (Snow, 2008). This approach has likely been bolstered by research characterizing scientific English (Snow, 2010; Wong Fillmore & Snow, 2000) and studies finding English proficiency to be well-correlated with science content achievement (Lee, Penfield, & Buxton, 2011; Maerten-Rivera, Myers, Lee, & Penfield, 2010; Torres & Zeidler, 2002). However, standardized science exams have intrinsic linguistic demands that reflect more than just content understanding (Butler, Stevens, & Castellon, 2007; Kieffer, Lesaux, Rivera, & Francis, 2009; Noble et al., 2012). Correlations between English proficiency and exam performance support the conclusion that language is a primary barrier in assessing content knowledge, but not necessarily in learning content knowledge. Anstrom et al. (2010) note that researchers are still divided on whether using scientific English is a prerequisite for learning scientific content.

Lee, Quinn, and Valdes (2013) suggest that a focus on language skills originated from *content-based language instruction* (CBLI), where academic subject matter such as history or science is used as a medium for language instruction (Met, 1991; Scarcella, 2003). CBLI later evolved into *sheltered instruction* (SI) models, which sought to equip content-area teachers with language acquisition strategies to simultaneously target language and content-area objectives (Echevarria, Short, & Powers, 2006). It is important to note that while SI focuses on making content comprehensible, L1 use is usually discouraged in program design, except in programs for ELLs with little to no English proficiency (Genesee, 1999). This likely reflects the desire to facilitate transition to English and reduce dependence on L1 use. There have been some instances where such practices have led to an overemphasis in language instruction (Bruna, Vann, & Escudero, 2007). Lee and colleagues describe this as placing emphasis on "the study and practice of language elements rather than on immersion in rich environments that use language for sense making" (2013, p. 231).

Moreover, there is evidence suggesting that a strong emphasis on language instruction may at times hinder learning scientific content. Take for example a sizeable study of 440 students conducted by Echevarria and colleagues (2006), which evaluated the impact of an SI model on student outcomes as measured by an expository academic writing task. While measures of academic writing ability improved, the domain relying on content mastery (support or elaboration) showed no significant improvement when compared with control (Echevarria et al., 2006). These results suggest that students receiving SI may benefit in the language domain but miss out on deeper conceptual understandings. The relationship between the content and language domains, then, becomes a key issue for second language learners. In most instances, acquiring the language of science in the majority language is seen as a primary goal. Although use of the L1 may support content learning, it may be seen as hindering acquisition of the majority language.

There is some research which supports that learning scientific content provides a conceptual basis for the acquisition of scientific English. For instance, Brown and Ryoo (2008) found that using vernacular language to develop everyday understandings of scientific content resulted in a significant improvement on multiple choice and open-ended response questions. Other empirical research disaggregating content and language is scarce. Some studies have attempted to investigate the effects of L1 use in the classroom either amongst students or during instruction (Kearsey & Turner, 1999; Reinhard, 1996; Tobin & McRobbie, 1996), but these did not effectively isolate the interactions of native language (L1) use and content understanding. Overall, more empirical research is needed, prompting the motivation for this study.

8.3 Theoretical Framework

The theoretical framework for this research study is largely attributed to Cummins (1980), who posited a *common underlying proficiency* for bilingual students. This assumes that first and second languages are interdependent, and that academic skills and knowledge are transferrable across those languages. Cummins (1979) was first to draw on the distinction between *cognitive academic language proficiency* (CALP) and *basic interpersonal communicative skills* (BICS). He observed that although second language students could develop oral communicative fluency relatively quickly, many still struggled with the linguistic demands of academic content areas. This contrast was used to highlight the need for native language instruction while transitioning students develop their L2 CALP (Collier, 1987; Cummins, 1979, 1981; Hakuta, Butler, & Witt, 2000). Assumptions about language transfer, interdependence, and CALP provide the starting point for the analysis of relationships between language use, science achievement, and the development of content understanding and language acquisition.

8.4 Methodology

We conducted studies examining L1 use in science classrooms in New York City and Singapore. Although distinct in many ways, the two settings share characteristics that are common to many science classrooms with ELLs. Both countries have a heterogeneous population and adopt English as the official language and the language of instruction. At both school sites, there are large immigrant communities that speak a common L1 and acquiring English is a primary goal among the students. These similarities allow us to identify common trends that may be found in a wide spectrum of second language science classrooms. The primary research question guiding both studies was "How is the L1 used in the secondary science classroom?"

8.4.1 New York City

In New York City, data collection was performed in a public high school designed for students who have recently arrived to the United States and have low English proficiency. The science class under investigation was comprised of 19 students, ranging from 14 to 18 years of age, five of whom were female. Spanish was the dominant L1, although Arabic and French were also spoken amongst a few of the students.

The class was equivalent to a 9th grade introductory biology course. Employing a sheltered approach, instruction was primarily in English, but supported by strategies such as translation and the use of cognates. The instructor was a native Anglophone with a basic knowledge of Spanish. The course culminated in a standardized exam given in English that is required for graduation. Of the students selected as interview participants, length of residency in the United States ranged approximately from 3 to 18 months. Many of the students immigrated from the Dominican Republic, generally with relatively low levels of English proficiency and science content knowledge.

8.4.2 Singapore

In Singapore, our study took place at a private international school that caters to international students from various East and Southeast Asian countries who have moved to Singapore for secondary schooling in order to acquire English and ultimately pursue tertiary education at English-speaking universities. Many of these students and their parents chose Singapore because of its large Asian population (predominantly Chinese) and a reputable English-speaking education system. The classrooms under investigation were three Grade 7 general science classes with a total of 70 students. Almost all instructions were in English, with Chinese occasionally (less than once per class period) used for clarifying instructions or concepts. The instructor was a native Chinese-Singaporean with only a basic level of Mandarin Chinese. The predominant L1 was Mandarin Chinese (61%), followed by Thai (17%). One class of 24 students was selected for qualitative data collection. The average age was 14, and ages ranged from 13 to 17. The average length of residency for these students was 12–24 months and most arrived with relatively low English proficiency. In all, 89 students from the two settings participated in the study.

8.4.3 Data Collection and Analysis

Data were collected using focus group interviews, audio recording of student discourse, and classroom observation over a 6-week period in both locations. Studies at both sites received Institutional Review Board approval. Participation was voluntary and required students' informed consent. Identifying information and audio recording data was kept confidential, and names in excerpted data were replaced with pseudonyms. Survey and interview questions were translated and explained for clarity, and care was taken to ask questions in non-leading ways. Teacher and student interviews from New York City and Singapore were conducted using structured interview protocols with questions focusing on L1 use in the science classroom and the use and acquisition of English. In total, we performed 11 student focus group interviews and four teacher interviews across both sites. During class, students were purposefully selected based on their in-class use of Spanish for recording of peer-to-peer discussion during group work (Miles, Huberman, & Saldaña, 2014).

In Singapore, we also collected achievement and survey data. All 70 students completed the in-class survey; however, achievement data were available for only 60 students. Mathematics and the English language achievement data were obtained from entrance exams which students took 1–2 years earlier upon admission to the school. Science achievement data were based on in-class performance over the course of the previous semester, including periodic tests and a mid-year exam. The survey instrument included self-reported measures of language proficiency and comfort level with English, as well as questions about whether or not using another language besides English helps students' scientific English or scientific concepts. These were reported as Likert scores based on a 5-point scale.

We analyzed the data collected from New York City and Singapore using standard qualitative methodology, including iterative coding, triangulation, and member checking to identify important themes (Boeije, 2010). Our approach adopted a phenomenological framework, which allows for key findings to emerge from a focus on the essence of an observed, lived phenomenon (Creswell, 2013; Wojnar & Swanson, 2007). Three stages of open, axial, and selective coding generated salient themes in the data (Boeije, 2010). Codes were assigned in a directed manner (Hsieh, 2005), drawing from the extant literature base and theoretical framework identified earlier. Special attention was paid to addressing potential biases during all stages of research, including observation, coding, analysis, and presentation of findings (Armour, Rivaux, & Bell, 2009; Wojnar & Swanson, 2007). Peer debriefing and conversations amongst the authors assisted in identifying bias and making stronger interpretations from the data and analysis leading to consensus of findings from the study. The findings of the study are presented in the next section as representative excerpts to highlight language use and the *content-language tension* that exists in many linguistically diverse science classrooms.

8.5 Results

The principal research question of the study was *how do ELLs use their L1 in a secondary science classroom*? This question was answered by focusing on essential characteristics of L1 use during qualitative analysis. Thus, to address this question, the findings are presented as two separate cases, starting with New York City and followed by Singapore. At the end, both are discussed together as the findings pertain to understanding L1 use in science classrooms.

8.5.1 New York City

We found that the L1 was frequently used for direct translation of vocabulary and content explanations. This is clearly shown in Excerpts 1 and 2, where students used their L1 in class to discuss photoreceptors and the human body. At the same time, focus group interviews revealed that these same students had divided opinions regarding L1 use. Some students found it helpful in learning content, whereas others considered it a hindrance in learning English (see Excerpt 3). In their disagreement, we see competing goals of content and language learning at play in the classroom. This is evidenced by the excerpts below.

8.5.1.1 Direct Translation of Vocabulary and Content

Spanish was frequently used during small-group discussion for translation of key vocabulary and content. In the example below, students are discussing the photoreceptors of the eye when reviewing a question packet.

Excerpt 1

Miguel:	No entendi, que es lo que tengo que hacer aqui? Tengo que comparar?
	I don't understand, what do I have to do here? Do I have to compare?
Juan:	Nooo [Emphatic]. ¿Qué es un fotoreceptor? Lo que envía la información y las olas senal al cerebro. Eso es un receptor. Receptor [Emphasizes pronunciation]
	No, "What is a photoreceptor?" It's what sends the information and wave signals to the brain. That's a receptor.
Miguel:	Oh, un receptor. Por ejemplo … cuando te expliqué lo que era la vista.
	(continued)

(continued)

	Oh, a receptor. For example, when I was explaining what vision was.
Leon:	Sí. *Yeah.*
Miguel:	El fotoreceptor es era lo que manda el senal al cerebro, cuando la luz toca el sensor. *The photoreceptor is what sends the signal to the brain, when the light touches the sensor.*
Juan:	*[Emphasizes pronunciation]* Fo-to-re-cep-tor. Pero ahora estamos hablando del receptor. Esdiferente. *Pho-to-re-cep-tor. But right now we're talking about the receptor. It's different.*
Leon:	Es lo mismo. *It's the same.*
Juan:	Es lo mismo, hace lo mismo. *It's the same. It does the same thing.*

Here, Juan clarifies the instructions and gives an explanation at the same time. It is seen in this first excerpt the L1 is employed to label and explain the scientific concept of *photoreceptor*, a direct cognate. Miguel then repeats the explanation reiterated by Juan. Finally, Juan's claim that *receptor* and *photoreceptor* are different is discussed and negotiated amongst the group of three students. This type of discussion, where students mutually discuss the validity of a scientific idea, has been described as a central feature of socially constructed scientific knowledge (Driver, Asoko, Leach, Scott, & Mortimer, 1994). All of this is occurring as the L1 is being used as the primary language for conceptual understanding.

Excerpt 2

Freyan:	Ven vamos a hablar del cuerpo human, porque para aprender. Tú entiendes lo que es el cuerpo humano? *Come on, we're going to talk about the human body in order to learn. Do you understand what the human body is?*
Julia:	Bueno cuerpo humano, es el cuerpo … *Ok, human body, it's the body …*
Freyan:	Como está lleno de células, de nervios, de venas. *It's filled with cells, nerves, veins.*
Valerie:	Sangre. *Blood.*
Julia:	Sangre. *Blood.*
Freyan:	¿Qué mas tú entiendes? *What else do you know?*
Valerie:	¿Qué tu entiendes? *What do you know?*
Freyan:	Tiene cerebro.

(continued)

(continued)

	It has the brain.
Valerie:	Cerebro.
	Brain.
Freyan:	Tiene los cinco sentidos.
	It has the five senses.
Valerie:	Tiene los cinco sentidos, hueso.
	It has the five senses, bone.

Here, a group of three students are reviewing the human body. Freyan initiates the conversation, using the L1 to ask questions to elicit understanding. Vocabulary terms related to the human body are recalled, though with limited elaboration. Later in this conversation, there is some mention of pertinent scientific vocabulary in English, but the vast majority of the discussion and questioning occurred in the L1. Overall, Excerpts 1 and 2 show L1 use for elaboration of conceptual knowledge and key vocabulary terms. According to Cummins' language interdependence, such information should eventually transfer to English (e.g., during oral recall or assessment).

8.5.1.2 Transitioning to English

Although there is evidence of L1 use for content learning, interview data revealed some students' reservations about using the L1 when transitioning to English. In Excerpt 3, Elias and Freyan immediately respond to oppose L1 usage because it hinders English acquisition.

Excerpt 3

Researcher:	When you get things in two languages does that help?
Miguel:	Entendemos mejor.
	We understand better.
Freyan:	No, sometimes. Sometimes [but] it doesn't help them.
Researcher:	Leon, do you feel that using two languages, that translating helps?
Leon:	En mi lengua todo es más fácil.
	In my language, everything is easier.
Freyan:	Not helps much because you will not learn in English. You will just learn about the term we're talking, not the language.
Elias:	To be honest, I really don't like translation. English is a language. Spanish is another language. You learn English, not English translated Spanish, right? So it's not the same, so I get mad when people translate the word, that blocks me [from learning] English.
Freyan:	I agree because ….

(continued)

(continued)

Elias:	Because I want to know the English, not the translation.
Freyan:	When everything is in English, or another language that you don't know, without translator, you will learn because you will learn.
Freyan:	Because with a translator, te va atrás un poco más. *You lag behind a little more.*

Here, students express disinterest with using the L1 in class. They would rather learn English and not a "translated" English. In spite of this fact, the L1 is still used because it is seen as necessary for the low proficiency students. To paraphrase a famous theologian, "What they do not want to do, they do. And what they do want to do, they do not do." This internal schism, within Elias and Freyan as individuals, and within the classroom as a social community, is a visible expression of the content-language tension. While six of the eight interviewed students supported using the L1 for translation, two students did not. Freyan and Elias preferred an English-only environment without translation. They felt L1 translation slowed the class down and slowed down the process of language acquisition. Others in the class felt L1 use helped them transition and made the class comprehensible. This has been previously described as *comprehensible input*, an important condition for second language acquisition (Krashen, 1981). The students stated that without translation, they would not understand anything.

Moreover, it is important to point out that Mr. Williams, the classroom teacher, gives a very similar account in his individual interview. In Excerpt 4, he states that if the class were entirely in the L1, the students would "learn a lot more," but that this competes with the need to acquire English quickly.

Excerpt 4

Researcher:	How do you see them using their native language, and what role do you think it plays?
Mr. Williams:	I think that currently they are primarily using their native language. For most students, they are getting the knowledge in English, translating it into their native language, and then understanding it in their native language. There are some students who are learning in English and thinking in English. And so they don't need to take the detour through their native language.
Researcher:	Do you see the English-only approach as effective?
Mr. Williams:	I think it's effective in teaching English, which is why we do it. Obviously, if I did the entire class in Spanish, the Spanish students would learn a lot more. This school is not focused on getting them to learn science; it's focused on getting them to learn science and English. […] I think that ultimately for the students, it helps them more in their life to learn the English, than to learn the content. They're learning the content partly so they can pass the test. And the tests are just a bunch of facts that don't really help them in their life.

Mr. Williams described how most students in his class, who are first or second year ELLs, rely on their L1 for learning science, but mostly to learn English. This is likely through translation and comprehensible input (Krashen, 1981). Although the L1 was seen as effective in teaching content, this goal was secondary to the practical goal of transitioning to English. Content learning was seen as mainly providing knowledge assessed on standardized exams which was required for graduation. This clearly illustrates competing goals of content learning and language acquisition. If it is assumed, as Mr. Williams does, that one is served at the expense of the other, then these goals will be in tension with each other. This emerged as a recurrent theme throughout the student and teacher interview data, providing the impetus for our study in Singapore, a second site to explore our research question.

8.5.2 Singapore

In Singapore, we were again concerned with the primary question of *how do ELLs use their L1 in a secondary science classroom?* Enabled by the survey and achievement data collected, we were interested in finding if relationships existed between language use and science achievement and if L1 usage influenced conceptual and linguistic development for students in Singapore.

8.5.2.1 Use of L1 in Secondary Science Classroom

The class we observed in Singapore also saw frequent use of the L1, but in different ways than what was observed in New York City. L1 use predominantly occurred during laboratory instruction rather than regular class time. We suggest that this occurred partly due to regular class time in Singapore tending to be predominantly teacher talk or teacher-guided discussion, leaving little room for group discourse. During laboratory activities, students were given greater freedom to talk through their L1. In the excerpt below, students were using microscopes to look for and identify the nuclei in plant cells:

Excerpt 5

Yiyuan:	Zhè dōngxī quándōu shì a.
	This thing, this is all it.
Zhuping:	A nà yuán de jiùshì nucleus.
	Oh, those round ones are nucleus?
Yiyuan:	Duì ā.
	Yup.
Zhuping:	Hǎo ě xīn.
	Very gross.
Yiyuan:	Nǎ gě dōng xī?
	What [is gross]?
Zhuping	Nucleus a? Mé yǒ nucleus a.
	Nucleus? There's no nucleus.

In the use of *nucleus*, the students were translanguaging, referencing English scientific vocabulary with their L1. However, here they were using the L1 in a more functional, social manner as part of the activity, rather than discussing *nucleus* as a concept. Conceptual information was mainly presented to students during the regular class period. The style of instruction observed in the regular classroom was predominantly lecture style with little peer-to-peer conceptual discussion. For this reason, we did not capture any instances of L1 use for conceptual discussion amongst peers. The students mainly used their L1 for translation of key vocabulary terms by asking peers or using Chinese-English dictionaries. This was identified during classroom observation and frequently mentioned throughout the interviews.

During interviews, students were asked whether or not using their L1 helps them in learning science. Students frequently referred to the use of dictionaries or the discussion of key terms with other students when there was a lack of understanding. As one student put it, "If you don't understand you can use the mother tongue to discuss with your friends." This was frequently mentioned during interviews and observed in-class. Although recorded examples of these interactions were hard to come by, interview and observational data provided instances of L1 usage in the class.

Similar to Freyan and Elias' responses in New York, the importance of using English is acknowledged, "We are learning our subject in English so when we discuss in English it's better." It is remarkable that such similar sentiments were independently expressed regarding the use of the L1 contrasting with the importance of using English to help acquisition. We do not think it is a stretch to suggest that tensions between competing goals of content and language learning exist at both sites, and this is shown by statements above.

8.5.2.2 Relationships Between Language Use and Science Achievement

We performed step-wise regression on students' science achievement scores ($n = 60$) using a wide range of predictor variables which were described earlier in the methodology section. We found two significant predictors of in-class science achievement: prior mathematics achievement ($p < .015$) and students' response to the question, "How comfortable are you with English?" ($p < .002$). The overall model was significant ($p < .002$) with an $R^2 = .203$, indicating 20% of the variation in science achievement can be explained by these two predictors. Comparison of standardized coefficients shows English comfort level (.368) a stronger predictor than prior mathematics achievement (.296). Although significant, the explained variance was not particularly large. However, these results do suggest the importance of students' comfort with English in the ELL science classroom. Importantly, this does not indicate a causal relationship, and correlated measures of English comfort, mathematics, and science achievement may reflect dimensions of underlying achievement. In-class science assessments were given in English, making language an inherent component. Nevertheless, there was good evidence

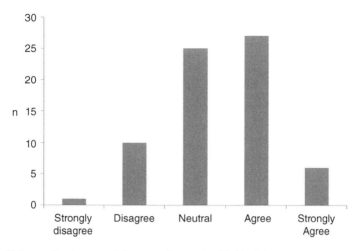

Fig. 8.1 Using another language helps me to learn scientific English

that use of English plays an important role in ELL science classrooms, leading us to investigate more specifically the relationship between L1 use and conceptual and linguistic development.

8.5.2.3 Influence of Native Language Usage on Conceptual and Linguistic Development

During survey administration, *scientific concepts* were defined as the ideas learned in science class. *Scientific English* was described as the language of science used in science class (Lemke, 1990). These terms were clarified by the researcher in this way during administration of the survey. Students in Singapore were asked to rate their agreement with the statement, "Using another language helps me to learn scientific English[1]" (see Fig. 8.1). About 50% of the students showed agreement (agree or strongly agree), 36% of students were neutral, and the remaining 14% disagreed (disagree or strongly disagree).

About half the class believed that using the L1 helped them with the acquisition of scientific English. At the same time, a nearly equally large portion of neutrality and disagreement reflected some split in opinion. Students were also asked to rate their agreement with the statement, "Using another language helps me to learn scientific concepts" (see Fig. 8.2). This resulted in a similar distribution, with about 50% of the students showing agreement, but with a greater proportion expressing neutrality.

---

[1]"Another language" refers to students' native languages or mother tongues. This was verbally clarified during survey administration.

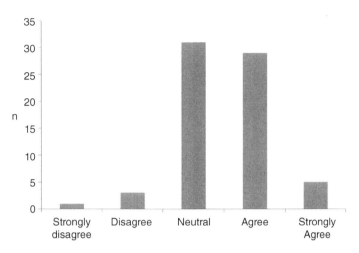

Fig. 8.2 Using another language helps me to learn scientific concepts

Overall, we saw sizeable agreement with the belief that using the L1 helps with learning scientific concepts and learning scientific English. However, in both measures there is also a general split in opinion. Some students support L1 use and some do not or are not sure. We suggest that this can be explained by the students' acknowledgement of L1 use contrasting with the acknowledged importance of using English during focus group interviews. It is reasonable to assume that these contrasting sentiments would appear in our survey data, providing another visible expression of the content-language tension.

It was important to determine whether agreement varied by levels of achievement, English proficiency, or majority/minority ethnic group status. Perhaps students with higher or lower language proficiency had different orientations toward language in class. Chinese-majority students had more opportunities to use their native language in class, which may influence beliefs about L1 use. Moreover, it is possible that students with lower science achievement or English proficiency would display stronger agreement as well.

No statistically significant differences in levels of agreement were found across groups of varying English proficiency, ethnic group status, or prior achievement. Conversely, students' beliefs regarding L1 use for learning scientific concepts or scientific English were not significant predictors of science achievement or language proficiency. Simply put, higher achievement was not associated with a particular orientation toward L1 use.

8.6 Discussion

We conducted studies of L1 use in ELL science classrooms in two distinct locations – New York City and Singapore. Both locations represent places with large immigrant communities fluent in a common language where learning English is a primary goal. In New York City, we collected data using classroom observation, audio recording of student discourse, and focus group interviews from one classroom. In Singapore, in addition to these methods, we collected classroom surveys, language proficiency data, and student achievement data from three classrooms.

Findings emerged to address the primary research question guiding both studies, "How is the L1 used in the secondary science classroom?" In New York, we see students using the L1 for conceptual discussion and the learning of content. At the same time, these students are divided on whether it is better to rely on the L1 in the classroom. In Singapore, we see similar contrasts between using the L1 with peers or via dictionaries versus using English to focus on acquisition. This is seen in both the student interview and survey data. The two settings in which the studies take place highlight the tension that exists between competing goals of content learning and language acquisition for ELLs. They provide evidence that the L1 is frequently and naturally used for content learning, in spite of the prevalent assumption that this interferes with English acquisition. Moreover, we see this phenomenon in two distinct sites, suggesting that such tensions may be found in a wide range of ELL science classrooms.

The students and classrooms in New York and Singapore do differ in many ways. Although we did not collect extensive background data, the students likely differ in socioeconomic status. Furthermore, the differences in context (e.g., language, instruction, culture) are not insignificant. However, like many other classrooms where linguistic minorities are learning science in a second language spoken by the dominant culture, tensions may arise as to the goals for learning. As eluded to by Mr. Williams, orientations toward English acquisition and assimilation likely add to the content-language tension that appears in such science classrooms.

For the students in both settings, the notion of becoming proficient in English as a matter of urgency is not to be downplayed. Passing standardized assessments in English is required for graduation at both sites. Their desire to learn English is so strong that they may prefer not to use the L1 to promote language learning; however, they do so out of necessity for passing standardized tests, which is secondary. The language burden extant in standardized assessments, as well as the practical needs of learning the majority language, underlies the goal of English acquisition. The L1 may be used as a tool to learn science content; however, pressures of language acquisition in scientific English have positioned it solely as a *transitional device* (i.e., to transition from the L1 to the L2) for the translation of key vocabulary.

8.7 Conclusion

A central theme emerged from this study of ELLs using the L1 in secondary science classrooms in two settings, New York City and Singapore. The L1 can help students learn scientific content, but may be perceived to slow their acquisition of English. In Singapore, students were evenly split regarding their opinions on whether L1 use supports learning scientific concepts and scientific English. This division was seen across all levels of student achievement and first and second language proficiency. Taken together, we propose that for these classrooms there exists a *content-language tension*, where competing goals of content learning and language acquisition are differentially supported by L1 use. Building scientific literacy for second language learners must take into account the differences between content and language learning. Practices with regards to the use of the L1 in science classrooms should consider individual differences as well as the content-language tension, and the role of the majority language must be considered carefully when addressing the goals of global scientific literacy. In other words, "it is important that teachers and students understand and use language in the classroom for acquiring the culture of power in its various forms" (Moore, 2007, p. 341), and that this should be considered when thinking about tensions in the classrooms due to language and learning.

References

Anstrom, K., DiCerbo, P., Butler, F., Katz, A., Millet, J., & Rivera, C. (2010). *A review of the literature on academic English: Implications for K-12 English language learners*. Arlington: The George Washington University Center for Equity and Excellence in Education.

Armour, M., Rivaux, S. L., & Bell, H. (2009). Using context to build rigor application to two hermeneutic phenomenological studies. *Qualitative Social Work, 8*(1), 101–122.

August, D., & Hakuta, K. (Eds.). (1997). *Improving schooling for language-minority children: A research agenda*. Washington: National Research Council, Institute of Medicine.

Boeije, H. (2010). *Analysis in qualitative research*. Thousand Oaks: Sage Publications.

Brown, B. A., & Ryoo, K. (2008). Teaching science as a language: A "content-first" approach to science teaching. Journal of Research in Science Teaching, 45(5), 529–553.

Bruna, K. R., Vann, R., & Escudero, M. P. (2007). What's language got to do with it?: A case study of academic language instruction in a high school 'English Learner Science' class. *Journal of English for Academic Purposes, 6*(1), 36–54.

Butler, F. A., Stevens, R., & Castellon, M. (2007). ELLs and standardized assessments: The interaction between language proficiency and performance on standardized tests. In A. L. Bailey (Eds.), *The language demands of school: Putting academic English to the test* (pp. 27–49). New Haven: Yale University Press.

Christian, D. (2001). Language policy issues in the education of immigrant students. In J. Alatis (Ed.), *Proceedings of the Georgetown University round table on languages and linguistics* (pp. 136–154). Washington: Georgetown University Press.

Collier, V. P. (1987). Age and rate of acquisition of second language for academic purposes. *TESOL Quarterly, 21*(4), 617–641.

Creese, A., & Blackledge, A. (2010). Translanguaging in the bilingual classroom: A pedagogy for learning and teaching? *The Modern Language Journal, 94*(1), 103–115.

Creswell, J.W. (2013). *Qualitative inquiry and research design: Choosing among five approaches*. 3rd ed. Thousand Oaks: Sage Publications.

Cummins, J. (1979). Cognitive/academic language proficiency, linguistic interdependence, the optimum age question and some other matters. *Working Papers on Bilingualism, 19*, 121–129.

Cummins, J. (1980). The construct of language proficiency in bilingual education. Current issues in bilingual education, 81–103.

Cummins, J. (1981). Age on arrival and immigrant second language learning in Canada: A reassessment. *Applied Linguistics, 2*, 132–149.

Driver, R., Asoko, H., Leach, J., Scott, P., & Mortimer, E. (1994). Constructing scientific knowledge in the classroom. *Educational Researcher, 23*(7), 5–12.

Echevarria, J., Short, D., & Powers, K. (2006). School reform and standards-based education: A model for English-language learners. The Journal of Educational, 95, 195–211.

García, O., & Wei, L. (2014). *Translanguaging and Education. In Translanguaging: Language, Bilingualism and Education* (pp. 63–77). London, UK: Palgrave Macmillan.

Genesee, F. (Eds.). (1999). *Program alternatives for linguistically and culturally diverse students* (Educational Practice Report No. 1). Santa Cruz: Center for Research on Education, Diversity & Excellence.

Hakuta, K., Butler, Y. G., & Witt, D. (2000). *How long does it take English learners to attain proficiency?* Santa Barbara: University of California Linguistic Minority Research Institute.

Hsieh, H. (2005). Three approaches to qualitative content analysis. *Qualitative Health Research, 15*(9), 1277–1288.

Janzen, J. (2008). Teaching English language learners in the content areas. *Review of Educational Research, 78*(4), 1010–1038.

Kearsey, J., & Turner, S. (1999). The value of bilingualism in pupils' understanding of scientific language. *International Journal of Science Education, 21*(10), 1037–1050.

Kieffer, M. J., Lesaux, N. K., Rivera, M., & Francis, D. J. (2009). Accommodations for English language learners taking large-scale assessments: A meta-analysis on effectiveness and validity. *Review of Educational Research, 79*(3), 1168–1201.

Krashen, S. D. (1981). *Second language acquisition and second language learning.* Oxford: Oxford University Press.

Lee, O. (2005). Science education with English language learners: Synthesis and research agenda. *Review of Educational Research, 75*(4), 491–530.

Lee, O., & Buxton, C. A. (2013). Integrating science and English proficiency for English language learners. Theory into Practice, 52(1), 3642.

Lee, O., Penfield, R., & Buxton, C. (2011). Relationship between "form" and 'content' in science writing among English language learners. *Teachers College Record, 113*, 1401–1434.

Lee, O., Quinn, H., & Valdes, G. (2013). Science and language for English language learners in relation to next generation science standards and with implications for common core state standards for English language arts and mathematics. *Educational Researcher, 42*(4), 223–233.

Lemke, J. L. (1990). *Talking science: Language, learning, and values.* Norwood: Ablex Publishing Corporation.

Lin, A. M. Y., & Wu, Y. (2014). "May I speak Cantonese?" – Co-constructing a scientific proof in an EFL junior secondary science classroom. *International Journal of Bilingual Education and Bilingualism, 18*(3), 289–305.

Maerten-Rivera, J., Myers, N., Lee, O., & Penfield, R. (2010). Student and school predictors of high-stakes assessment in science. *Science Education, 94*(6), 937–962.

Met, M. (1991). Learning language through content: Learning content through language. *Foreign Language Annals, 24*(4), 281–295.

Miles, M. B., Huberman, A. M., & Saldaña, J. (2014). *Qualitative data analysis: A methods sourcebook* (3rd ed.). Thousand Oaks: Sage Publications.

Moje, E. B., Collazo, T., Carrillo, R., & Marx, R. W. (2001). "Maestro, what is 'quality'?": Language, literacy, and discourse in project-based science. *Journal of Research in Science Teaching, 38*(4), 469–498.

Moore, F.M. (2007). Language in science education as a gatekeeper to learning, teaching, and professional development. *Journal of Science Teacher Education, 18*, 319–343.

Noble, T., Suarez, C., Rosebery, A., O'Connor, M. C., Warren, B., & Hudicourt-Barnes, J. (2012). "I never thought of it as freezing": How students answer questions on large-scale science tests and what they know about science. *Journal of Research in Science Teaching, 49*(6), 778–803.

Reinhard, B. (1996). How does the medium of instruction affect the learning of chemistry?. *School Science Review, 78*(283), 73–78.

Rollnick, M. (2000). Current issues and perspectives on second language learning of science. *Studies in Science Education, 35*, 93–121.

Rosebery, A. S., Warren, B., & Conant, F. R. (1992). Appropriating scientific discourse: Findings from language minority classrooms. *Journal of the Learning Sciences, 21*, 61–94.

Ruiz-de-Velasco, J., & Fix, M. (2000). *Overlooked and underserved: Immigrant students in US secondary schools*. Washington: The Urban Institute.

Scarcella, R. (2003). *Academic English: A conceptual framework* (Technical Report No. 2003-1, No. 1). Santa Barbara: The University of California Linguistic Minority Research Institute.

Snow, C. (2008). What is the vocabulary of science? In A. S. Rosebery, & B. Warren (Eds.), *Teaching science to English language learners: Building on students' strengths* (pp. 71–83). Arlington, VA: National Science Teachers Association.

Snow, C. E. (2010). Academic language and the challenge of reading for learning about science. *Science, 328*(5977), 450–452.

Tobin, K., & McRobbie, C. J. (1996). Significance of limited English proficiency and cultural capital to the performance in science of Chinese-Australians. *Journal of Research in Science Teaching, 33*(3), 265–282.

Torres, H., & Zeidler, D (2002). The effects of English language proficiency and scientific reasoning skills on the acquisition of science content knowledge by Hispanic English language learners and native English language speaking students. *Electronic Journal of Science Education, 6*(3).

Wojnar, D. M., & Swanson, K. M. (2007). Phenomenology: An Exploration. *Journal of Holistic Nursing, 25*(3), 172–180.

Wong Fillmore, L., & Snow, C. (2000). *What teachers need to know about language*. Washington: U.S. Department of Education, Office of Educational Research and Improvement.

Part 3
Science Classroom Literacy Practices

Chapter 9
A Case Study of Literacy Teaching in Six Middle- and High-School Science Classes in New Zealand

Aaron Wilson and Rebecca Jesson

Abstract This chapter reports a case study of the literacy practices and knowledge of six science teachers in Auckland, New Zealand (NZ). In NZ, the national curriculum requires that students develop sophisticated, subject-specialised literacy in science. However, little is known about actual patterns of literacy teaching and learning in NZ science classrooms. Participants were six teachers of science from schools serving low to middle socio-economic status communities. Two teachers taught Year 7 (students aged 11–12 years), two taught Year 9 (13–14 years) and two taught Year 11 (15–16 years). The data included observations of literacy teaching in science lessons, interviews with teachers and measures of teachers' subject literacy pedagogical content knowledge. Data from all three sources indicated that teachers considered vocabulary to be the key to literacy learning in science, and the literacy teaching observed was consistent with this. This vocabulary teaching tended to focus on definitions, supplied by the teacher and learned through repeated practice activities. Texts used in science lessons were most commonly short, teacher designed texts. Students had few opportunities to read science texts independently. We identify a need to expand the learning outcomes that are valued, from a primary focus on assessed science content to a broader focus that encompasses reading, writing, disciplinary and critical literacy outcomes. We see an opportunity to frame students, rather than teachers, as being responsible for the reading and writing of science text and to move from constrained to open-ended literacy learning tasks. Finally, we identify a need to move beyond short-term strategies towards a focus on generative teaching so that students are in a position to read and write the texts they need as citizens or as emerging science professionals.

Keywords New Zealand · disciplinary literacy · adolescent literacy · science · teacher observations

A. Wilson (✉) · R. Jesson
Woolf Fisher Research Centre, School of Curriculum and Pedagogy, Faculty of Education and Social Work, University of Auckland, Auckland, New Zealand
e-mail: aj.wilson@auckland.ac.nz

R. Jesson
e-mail: r.jesson@auckland.ac.nz

This chapter reports a case study of the literacy practices and knowledge of six science teachers in Auckland, New Zealand (NZ). The teachers in the case study taught in two high schools (Years 9–13) and two intermediate schools (Years 7–8) serving low- to mid- socio-economic status communities. The two intermediate school teachers taught Year 7 classes (comprising students of 11–12 years of age) and, of the four high school teachers, two taught Year 9 (13–14 years) and two taught Year 11 (15–16 years). Despite an increasing awareness of the importance of literacy in subject-areas such as science, little research had previously been conducted to investigate patterns of literacy instruction in NZ schools.

Our focus here is with students' developing knowledge of the language of science and how students learn to read, write, speak and listen to texts in science. Our focus is not on 'scientific literacy' which we take to refer to understanding of the natural world and key science concepts, principles and ways of thinking (Pearson, Moje, & Greenleaf, 2010). We do work from an assumption however that developing literacy in science contributes to the development of scientific literacy.

Students' ability to read, write, speak and listen to texts in science is considered to have a powerful effect on their overall science achievement. While a causal relationship between improved reading and improved scores on subject-based assessments has yet to be empirically established (Kamil et al., 2008), the correlations between PISA Reading scores with PISA scores in mathematics and science scores, for example, are 0.81 and 0.86, respectively (Kirsch et al., 2002). Moreover, like nations worldwide, meeting the challenges NZ faces as a society increasingly depends on its citizens' knowledge of science. A recent government report (Office of the Prime Minister's Science Advisory Committee, 2011), for example, argues that the objectives of science, for both 'pre-professional education' and 'citizen-focused education', require children to take 'an informed participatory role in the science-related decisions that society must make' and to 'distinguish reliable information from less reliable information' (p. 4). These objectives demand sophisticated forms of literacy specific to the teaching and learning of science.

There is consensus that the literacy and language demands of different subjects become more sophisticated and specialised as students move up the year levels (Lee & Spratley, 2010; Moje, 2008). The increased specialisation is closely related to the 'clearly demarcated subject orientation' (May & Wright, 2007, p. 374) of secondary schools, and their organisation into different, reasonably autonomous, subject-based departments (Hargreaves & Macmillan, 1995; Siskin, 1997). The texts that students encounter provide one instance of this subject specialisation. Mathematics word problems are almost exclusive to mathematics classrooms, historical documents to history classroom and scientific research reports to science classrooms (Lee & Spratley, 2010; Moje, Stockdill, Kim, & Kim, 2011).

Helping students develop knowledge to cope with increasing linguistic complexity and specialisation is the key feature of a body of recent research advocating for more attention to *disciplinary literacy* (Fang & Schleppegrell, 2010; Shanahan & Shanahan, 2008). From a disciplinary perspective generic reading skills such as decoding and general comprehension strategies are necessary but not sufficient for students to meet the sophisticated and specialised demands of reading in middle- and high-school science.

The statement of official policy that sets the direction for student learning in English-medium schools, New Zealand Curriculum (NZC), includes explicit messages about literacy learning in science (Ministry of Education, 2007). The vision for literacy in science expressed in NZC is an ambitious one. Students are expected not only to read texts for information but also to read texts that help them apply that knowledge to real world contexts. Students are expected to become critically literate users of popular and science texts: by Year 6, students are expected to begin to question the authors' purposes for constructing texts and, by Year 10, to use their understanding of science to evaluate both popular and scientific texts. The vision in NZC is consistent with a disciplinary literacy perspective insofar as students are expected to learn 'what science is and how scientists work' and 'how science ideas are communicated' (Ministry of Education, 2007, p. 28).

There is a range of instructional practices warranted in the literature as potentially effective in developing these valued literacy outcomes in science; our review of these practices shaped our inquiry. Firstly, there was a need to know about the amount and types of text use and the characteristics of those texts. Literacy learning in science is likely enhanced when students have plentiful opportunities to read appropriately challenging texts that have properties well aligned to curriculum expectations (Darling-Hammond & Bransford, 2005; Kuhn & Stahl, 2003). There is general agreement that the nature of the texts students encounter should change as they progress through secondary school. Texts are expected, across subjects, to become longer, and the words, sentences, structures and ideas within them more complex. Graphic representations increase in importance and there is more variety in texts across subject-areas (Carnegie Council on Advancing Adolescent Literacy, 2010). We see opportunities to read and write texts of the types valued in the discipline as a precondition of effective literacy instruction; no instructional effort to improve reading and writing can compensate for an absence of reading and writing.

Secondly, we were interested in the opportunities students had to discuss texts. One of the most powerful ways to raise students' subject-literacy is for them to engage in rich extended discussions about the texts that they read in different learning areas (Soter et al., 2008; Wilkinson & Son, 2009). Extended discussion contrasts with typical patterns of classroom discourse in that there is more time for open-ended discussion, greater use of authentic open teacher questions to explore rather than 'test' students' understanding, and attempts to increase 'uptake' whereby teachers prompt for elaboration and incorporate and build on students' ideas (Applebee, Langer, Nystrand, & Gamoran, 2003).

Thirdly, there was a need to investigate how science teachers developed students' knowledge of texts in order to navigate and make meaning from such texts. Given that every science topic presents students with a plethora of technical new terms to learn (Fang & Schleppegrell, 2010), and because students need to know a high proportion of words to comprehend a text (Lesaux, Kieffer, Faller, & Kelley, 2010), teaching of vocabulary is an important component of literacy instruction in science. Moreover, because science texts, such as science reports, differ from those of other subjects at the global text level (e.g. noncontinuous texts requiring students to move back and forward between running text, illustrations and diagrams) and the local text level (e.g. specialised types of graph and diagram), instruction to develop students' knowledge of structural or organisational features of texts is also warranted. Students would also benefit from knowing about specialised features of science texts at the level of sentences, particularly since many of the features identified as being common in science (Fang & Schleppegrell, 2010) are also identified as features known to inhibit comprehension (White, 2012). Such features include ellipsis, the use of lengthy noun phrases, complex sentence structures, passive voice and nominalisation.

Fourthly, it was important to investigate teaching to develop students' strategies for reading and writing. After all, the aim of disciplinary literacy instruction is to develop students' independent reading and writing in science. Such instruction might involve deliberate teaching of reading strategies (Conley, 2008; Pressley, 2004) or metacognitive reflection (Schoenbach, Greenleaf, & Murphy, 2012).

Finally, given the explicit statements about the importance of critical literacy in NZC, we needed to know about teaching practices to develop students' critical literacy. Such practices include teachers encouraging students to read from the perspective of someone with an opposing perspective, setting tasks where students produce counter texts, providing multiple perspectives on the same topic and conducting student-choice research projects (Behrman, 2006; Janks, 2013).

The research questions we addressed in this study were: 'What practices do teachers use in the teaching of science to support students' learning and achievement?' and 'What knowledge, beliefs and understandings do teachers have about the literacy and language of science?'

9.1 Methods

The settings of the study were six classrooms in three secondary and two intermediate schools in Auckland. Secondary schools in NZ serve students in Years 9–13 (approximately 13–18 years old) whereas intermediate schools are attended by students in Years 7 and 8 (approximately 11 and 12 years old). The schools served mid- to low- socio-economic status communities with ethnically diverse student populations. The six classrooms comprised two science classes from each of Years 7, 9 and 11.

The selection of schools and teachers was purposive; we wanted to investigate the literacy teaching practices and knowledge of science teachers identified as effective in literacy instruction and who worked in schools serving low- to mid-SES, ethnically diverse communities. One purpose of the study was to identify future directions for literacy and language pedagogy, and we reasoned that these would be most fruitfully built on a foundation of already effective teaching. We used publicly available data about schools' performance in national qualifications as the first step in identifying potential schools. The main school qualification in NZ is the National Certificate of Educational Achievement (NCEA) which is a standards-based assessment system. To gain an NCEA qualification students are assessed against a range of 'standards' in different subjects, each of which represents a particular skill, understanding or competency (New Zealand Qualifications Authority, n.d.). We focused specifically on standards identified by national subject experts as having a significant reading and writing component. First we identified all low- to mid-decile secondary schools in Auckland where more than 60% of the Year 11 roll were enrolled in these standards. We then ranked those 29 schools according to pass rates and invited the highest ranked schools to participate. The schools with the highest and fourth highest pass rates agreed to participate.

We then invited the principal and head of science to identify science teachers whom they judged to be particularly effective at developing students' literacy in science. We also asked the principal and head of science to nominate a local intermediate school which they judged to have been effective in preparing students for the literacy demands of high school science.

We used three sources of data: Teacher observations, a measure of teacher subject literacy pedagogical content knowledge (SLPCK), and teacher interviews. Members of the research team observed each teacher over three consecutive science lessons (typically 50–60 minutes) using an observation template. Researchers actively observed the teaching for 3 minutes then recorded their observations during the next 3 minutes, in a rotating cycle, enabling a 50% sample of each observed lesson. Instances of literacy or language instruction were coded according to their content focus and for what we called 'instructional depth'. The categories of content focus were vocabulary (words and sub-word parts), text structure (e.g. teaching students about organisational features of graphs such as titles and labels or about section-headings commonly used in science reports), language features (e.g. teaching students about passive voice) and spelling and punctuation. The categories of instructional depth were *item* (e.g. teaching an item of knowledge such as a definition), *activation of students' prior knowledge* (e.g. asking students to brainstorm synonyms for a word), *practice* (e.g. matching activities designed to reinforce understanding of meanings of previously taught words), *strategy*: developing students' metacognition/strategy use (e.g. discussing reading comprehension strategies; strategies for 'solving' unfamiliar words using sub-word parts such as prefixes and suffixes) and *critical literacy* (e.g. closely analysing a writer's word choices to identify bias). Detailed field notes were made about all texts and teaching and learning activities, particularly those that had a literacy or language focus.

Teacher observation data were analysed to determine the types of texts used, and the amount of time teachers spent engaged in different teaching activities. These quantitative data were supplemented with qualitative analyses of field notes using the same codes to give us a richer picture of how these forms of literacy teaching were enacted in classrooms.

To explore teachers' knowledge of features of a science text that might complicate comprehension, we employed a Subject Literacy Pedagogical Content Knowledge (SLPCK) tool (Wilson, Jesson, Rosedale, & Cockle, 2012) that was completed by the four secondary teachers. The SLPCK tool provided teachers with a science text to read (one used in a recent national external examination) and asked them to identify aspects of the language and literacy that might act as potential barriers for students' reading, and to suggest teaching moves they might make in response. Content analysis of the completed tools identified themes from the responses.

Teacher interviews occurred at the end of each observed lesson, when teachers were asked whether their learning goals for the lesson included a specific literacy goal and, if so, the methods used to assist the students in achieving that goal, and any measures they had used to understand whether students had achieved that goal. The researcher recorded a summary of the teacher responses, and these summaries were later qualitatively analysed to identify themes.

Qualitative data from the SLPCK measure and teacher interviews were then analysed in combination to identify the themes emerging from both data sources. Members of the research team met regularly to test thematic ideas. Percentages were then calculated to describe the relative frequency of each theme.

The study was approved by the University of Auckland Human Participants Ethics Committee. The participation of schools, teachers and students was completely voluntary and all participants were provided with a detailed Participant Information Sheet and signed a Consent Form.

9.2 Findings

We report firstly on the main activities that teachers and students were engaged in, regardless of whether or not these activities were focused on aspects of literacy or language. We then look more specifically at the frequency and properties of text use and literacy and language instruction observed in the lessons. Data from our analyses of the SLPCK tool and from the teacher interviews are used where appropriate to illustrate, nuance and explain observed patterns of teaching.

9.3 An Overview of Teaching and Learning Activities

Key finding: students' time in science lessons was divided fairly evenly between whole-class activities, such as question-and-answer discussions, and individual tasks such as completing worksheets. There were few opportunities for students to work in small groups or to engage in extended discussions.

For each 3-minute observation interval, observers made a judgement as to the main forms of teacher activity, student activity and student grouping.

Students in science lessons participated mainly in whole-class activities (50% of observed intervals) or in individual (albeit not individualised) tasks such as completing worksheets (43%). Although students commonly sat in groups, the tasks were rarely designed as group tasks (7%). The main teacher activities during whole-class sessions were question and answer sessions, which accounted for a third (32%) of all observed intervals, and lectures (17%). The overwhelming focus of these whole-class activities was on teachers explaining science concepts and terminology to students. The individual student activities consisted mainly of practice and reinforcement-type activities such as completing worksheets. While students worked on individual tasks, teachers divided their time between roving (20%), management (17%), conferencing (8%) and modelling (6%). Only one teacher (Teacher 1) incorporated practical science work such as experiments into their observed lessons. There was an absence of extended discussion (either about texts or about science more generally) in the lessons, with no intervals at all coded as having this as the main activity. Class and group discussions, when they took place, took the form of question-and-answer sequences characterised by teachers asking closed, checking-type questions about science ideas and students providing brief answers, often of just one or two words.

9.4 Opportunities for Reading

Key finding: the majority of texts observed in science lessons were short, teacher designed texts.

Working from an assumption that texts should be at the heart of literacy instruction, we analysed the frequency with which texts were used, and what the features of those texts were. We took a reasonably broad view of texts and included all texts with any written words, including diagrams with labels or headings and symbolic expressions, but did not count texts that had no written words whatsoever and were therefore solely oral or visual. In total, 82 texts were observed in science classrooms.

Texts were predominantly short, with the highest proportion (38%) of texts containing between 11 and 50 words and a quarter (26%) having 10 or fewer words. The majority of texts (72%) were created by the classroom teacher and presented as whiteboard or computer-projected notes or as worksheets. About a fifth (18.2%) of all texts were sourced from published print sources (books, magazines and newspapers) but only one-third of those (6.1%) were presented to students in their original published form; most of the texts sourced by teachers from published print sources were presented to students as photocopies or computer-projected copies. About half of the texts (49%) comprised written text only with the remainder of texts including at least one visual representation such as a diagram (20%), illustration (11%), photograph (6%) or table (5%). Forty-five per cent of the texts consisted mainly of running text in paragraphs, 27% consisted primarily of information presented in bullet points and 27% were predominantly visual representations of information.

There was little evidence of 'real world' texts or digital text. Newspaper or magazine articles were only observed in two intervals each. Also, all of the text use observed or discussed in the interviews was limited to single-texts; teachers made no mention of, and were not observed to use, multiple texts on a related topic or theme.

9.5 What Do Science Teachers Teach When Science Teachers Teach Literacy?

Key finding: teachers were very aware of the importance of specialised subject vocabulary and frequently taught new terminology. There were few instances of teaching to develop students' understanding of other aspects of literacy or language in science such as the structures or language features of texts.

Some form of literacy or language instruction was observed in 70% of the observed intervals. Vocabulary was by far the main focus of literacy and language instruction and this was observed in 62% of all observed intervals. Of the remaining 11 literacy-coded intervals, six included instruction about text structure, four about spelling and one about the audience and purpose of a text. The literacy focus of three of the teachers was solely on vocabulary whereas one teacher each also taught about structure or spelling, and one taught about vocabulary, structure and spelling. All of the intervals coded as relating to structure related to one form of representation within a text (such as a Punnett square) rather than to the structure of the text as a whole (for example about the challenges related to reading discontinuous text that incorporated running text, illustrations and specialised forms of visual representation). No intervals were coded as having any instruction about language features at the level of sentences. We observed little deliberate instruction directly related to reading or writing processes or strategies. Furthermore, no intervals were coded as mainly focussed on developing students' critical literacy, although there were instances of incidental teacher questioning to promote critical thinking (e.g. about possible explanations for unusual phenomena or unexpected results in experiments).

Unsurprisingly, given the high rates of vocabulary instruction we observed, teachers in the interviews viewed teaching vocabulary as an integral part of effective instruction in science. As one teacher put it, 'If they (students) don't have the vocabulary, they can't express ideas. Definition is the language of science' (Teacher 3). In the interviews, all six teachers articulated a learning goal related to vocabulary. In contrast, none of the six teachers identified a reading goal and although three stated a writing goal, these goals were expressed more as students *doing* writing than *learning about* writing: a typical writing goal was for students 'to be able to write a paragraph for the assessment' (Teacher 5). Consistent with the other data sources, teachers' responses to the SLPCK tool identified student knowledge of vocabulary as the most common barrier to text understanding and task completion.

Assessment success was a key rationale for vocabulary teaching at the upper levels. The two Year 11 teachers framed the importance of vocabulary instruction in terms of what students needed to know and do to succeed in high stakes qualifications. Teacher 5 said in the interview that it was important to focus on vocabulary 'as the markers in the assessments are tough on terminology' and Teacher 6 told her class that it was 'vital to use these words when it comes to assessments'.

The main vocabulary focus was on teaching specialised subject and general academic vocabulary. In the SLPCK measure for example, all four secondary school teachers cited academic topic words, such as *sterilise*, *process* and *microorganism*, as potentially problematic vocabulary for students in the provided text. Three of the four secondary science teachers also identified that students needed to recognise that general academic vocabulary items, specifically instructional verbs, such as *Name* and *Explain* as they provide clues as to the expected length and depth of student responses to the tasks. In contrast with the strong focus on academic vocabulary, no science teacher identified examples of more general, literate vocabulary related to the context of the text, such as *bottling* (meaning *preserving*), *air bubbles* or *refrigerated*, as potentially problematic for students. Consistent with this, there was no mention of non-scientific vocabulary in the interviews or observed lessons.

Teachers clearly valued and tried to promote the use of correct technical scientific vocabulary. A typical example of this occurred in one of Teacher 1's lessons with his Year 7 science class when a student spoke of a solid having 'hardness' and Teacher 1 responded that 'I'd rather we talked about density'. Another teacher said 'it's important students use the vocabulary correctly in context and to know why it's not appropriate to use the word *spin* instead of *rotate* say'.

Technical scientific vocabulary was seen as being particularly important in the first few lessons of a topic. One teacher described the key words introduced in the first lesson of a topic as providing 'a springboard to the concepts covered in the next lessons' (Teacher 1, Year 7) and another stated that vocabulary at the beginning of a topic because 'you want to sort it out first because you're going to use a lot more of these words throughout the next six weeks' (Teacher 5).

9.6 Teaching Approaches for Teaching Vocabulary

Key finding: the vocabulary instruction comprised explicit teaching of definitions of subject-specific items followed by multiple exercises of constrained vocabulary use. There were few observed instances of teaching to develop students' independent strategies for 'solving' and learning new vocabulary.

The typical sequence of vocabulary teaching consisted of explicit teaching of new words followed by constrained activities designed to reinforce students' knowledge of those words. Each of the 109 intervals that featured literacy instruction was coded according to its 'instructional depth'. The two most prevalent categories of instructional depth were *item teaching* (e.g. teaching word meanings) and *practice* (providing students with opportunities to reinforce content covered previously) which accounted for 31% and 48% of the literacy intervals respectively.

The bulk of literacy intervals coded for item teaching involved the explicit teaching of word meanings. Most commonly this was done through the teacher telling students the meaning of new words and/or writing new words and their definitions on the whiteboard for students to copy. In the interviews and SLPCK measure all teachers noted the importance of explicitly teaching key science topic words. There were no instances where the task specifically required students to infer the meaning of new words from texts, as they would likely have to do when reading texts independently. The absence of such opportunities is likely related to the very limited use of extended, contextualised and non-instructional text reported previously. There were teacher-designed activities however where students were given a term and definition and had to match it to a slightly reworded definition, for example, in Teacher 3's class when students were instructed to read a text that stated 'inside the nucleus are thread-like chromosomes' and then to write the name of 'the thread-like strands inside the nucleus of each cell'.

Teachers were very aware of the importance of repeated practice in developing and reinforcing students' vocabulary knowledge. One teacher in the interviews referred to such reinforcement as students needing to 'play' with new words. Teachers in the interviews and SLPCK measure identified a wide range of activities in which students 'played' with new vocabulary in written, visual and oral forms, including through cloze exercises, matching activities, poems, songs and crosswords. All of the teachers except Teacher 2 were observed using a variety of different activities to give students repeated exposure to new words. The most common approaches for reinforcing new vocabulary that we observed in the classroom lessons were matching activities (5/6 teachers) cloze activities (4/6 teachers), labelling diagrams and word finds (3/6 teachers each). In most cases, students matched words to definitions, although sometimes additional examples, including pictures, were required. In some cases the reinforcement teachers provided was embedded in activity but nevertheless very deliberate, for example when Teacher 1 was conferencing with a group during an experiment he said, 'How are you going to separate sand from the salt solution', emphasising *solution* through tone and volume.

While most of the item teaching we observed was limited to the teaching of word meanings, four of the teachers also were observed, or discussed the importance of, teaching students about morphology. For example, Teacher 6 (Year 11) drew students' attention to the prefix *bi-* when she introduced the term *binary fission* by saying: 'Bi meaning two, as in bicycle or bilingual'. Similarly, Teacher 1 explained to his class that the suffix *–ology* means 'The study of something'. Teacher 3 said in her interview that an important learning goal was on developing her students' 'word knowledge and breaking down word parts to find clues to scientific terminology'.

There was limited evidence of teaching of strategies students could use for 'solving' new words. The strategies observed consisted of morphemic strategies to break words down words to sub-parts to infer meaning and the use of mnemonics to memorise key terms. We did not observe instances where students were asked to articulate or reflect on vocabulary strategies they had used or could use independently, although one teacher (Teacher 6) tried to capitalise on students' knowledge of vocabulary from other subject-areas by asking her class what they knew about the prefix *–micro* from mathematics.

The instruction we observed was mainly focused on developing students' receptive understanding of new science words rather than on supporting students to use the words themselves in their speaking or writing. The absence of a strong focus on productive vocabulary in the observed lessons was at odds with the interviews in which five of the six teachers expressed both a receptive ('to understand') and productive ('to use') vocabulary learning goal in the interviews. A typical example of this type of goal was that 'they (students) have to develop definitions, they have to know what they are, and they should be able to write a paragraph using some of these words' (Teacher 6). Despite such goals, there were very few opportunities for independent and extended writing and the majority of tasks designed to develop students' proficiency in using new vocabulary were highly constrained. Most commonly, students at all levels were required to complete cloze (fill-the-gaps) activities or to use new words to label diagrams provided by the teacher with original labels blanked out. In a few cases this required students to write the missing words in gaps in a short paragraph but more often they only had to complete short sentences such as 'Particles in a – (solid) are close together'. In some cases, even at secondary school, the tasks were even more constrained: 'There are 365 ¼ d … in a y …, 8 p … in the s …. s …' (Teacher 4, Year 9).

In summary, data from all three sources indicated that teachers considered vocabulary to be the key to literacy learning in science, and the literacy teaching observed was consistent with this. This vocabulary teaching tended to focus on definitions, supplied by the teacher and learned through repeated practice activities. Texts used in science lessons were most commonly short, teacher designed texts. Students had few opportunities to read science texts independently.

9.7 Discussion

Literacy in science can be viewed as a valued learning outcome itself and as a vehicle for achieving other valued learning outcomes such as understanding science concepts. The combined results of this study indicate that understanding subject-content knowledge was the student outcome most valued by the science teachers; apart from vocabulary, there was little evidence of disciplinary literacy being viewed as a valued outcome itself. Neither was literacy seen as a primary means of developing students' subject-content knowledge; students had few opportunities to read, write, discuss or learn about the kinds of texts thought to be valued in the discipline, or indeed, the curriculum. The assumed shift from reading in primary to secondary school is commonly thought to be a shift from 'learning to read' to 'reading to learn' (Jetton & Alexander, 2004) but we saw limited evidence of either. Rather, science teachers typically provided content instruction, including vocabulary instruction, through telling, by adjusting or writing texts, and by providing constrained practice activities such as worksheets. In general this seemed to be a feature of teachers who compensated for perceived gaps in their students' literacy by identifying the science ideas and vocabulary students need to know, and summarising this in the form of teacher-made notes or modelling for students.

It is important to note that the general pattern of teaching we identified, while limiting from a disciplinary literacy perspective, was employed by teachers nominated as effective within schools that were demonstrably effective in promoting student success. This may be related to three potential issues with our sampling procedures. Firstly, it could be that the qualification standards which assess students' content knowledge *through* reading and writing are not a direct enough measure of students' reading and writing. Secondly, we did not have access to data at the level of classes and therefore the teachers we observed were not necessarily the teachers whose students achieved the highest results. Thirdly, it may be that the principals' and department heads' knowledge of literacy and effective literacy instruction in science may have been insufficient for them to accurately identify the science teachers who were the most effective teachers of reading and writing. It is possible too that although the rates of literacy instruction for the case study teachers were relatively low, they may have been higher than those of teachers more generally.

This apparent mismatch potentially challenges our assumption that high pass rates in science assessments would be associated with high rates of literacy instruction. It may be that students' relatively high pass rates in these Year 11 assessments happened because of, rather than despite, the teachers' efforts to ameliorate reading and writing demands; developing students' knowledge of science ideas in more direct ways may be more effective as well as more expedient than having them struggle to read challenging text. The evidence we have reported here does not refute that possibility.

Even if a focus on content at the expense of literacy might help students pass examinations in the earliest year of the formal qualification system, we are concerned for two reasons. Firstly, the teaching seems more geared to helping students answer specific test questions than towards the development of deep conceptual understanding of science in general or disciplinary literacy in science more specifically. Secondly, such practices may constrain students' future science learning, particularly when the time comes, as we hope it will, that students read science-related texts in situations where no teacher is available to mediate the texts, such as in the later years of schooling, at university, in the workplace, or in everyday life.

Fundamentally, literacy learning in science requires text access and use. There was little alignment between the types of texts students encountered in class and those that the curriculum implies would be important, and those that students will encounter in external examinations in later years, in the disciplines themselves and in 'real world' contexts. To become skilled users and producers of valued science texts, students need opportunities to read, write, think about and discuss those types of texts. We are not suggesting that written texts supplant other ways of teaching content or providing meaningful contexts to which students can apply their developing science knowledge, but we are suggesting that written texts should be used more often for these two purposes. There was ample scope for more time spent engaged in reading and writing texts. This is fundamental as until teachers expect students to regularly read and write valued texts in science, there will be little reason for them to teach students about the features of such texts or about strategies for reading and writing them.

Alongside text use, students need strategies to demand meaning from these texts. There were limited opportunities for students to develop knowledge about the specialised features of science texts, particularly those features identified in the literature as potential inhibitors of comprehension and meaning-making. The absence of teaching of such features is no doubt related to the absence of texts with these features in the observed classes. The teachers were however very aware of the importance of receptive vocabulary knowledge and invested considerable instructional time to this. The teachers taught students new words and their meanings and also designed repeated opportunities for students to play with new words and develop students' understanding of word parts. The pattern of high rates of vocabulary instruction but low rates of instruction about other aspects of texts does not support the conclusion that there was a generalised antipathy to the notion that science teachers address language and literacy, as has been found in much of the historical literature (O'Brien, Stewart, & Moje, 1995). It does suggest however limits to the role that language and literacy is understood to play in science learning; there was next-to-no instruction employing science texts or about purposes of science texts, how they are organised or about language features at the level of sentences. As well as employing texts with features that are valued in science, it is important that teachers have deep knowledge of how science texts work so they can anticipate or diagnose reading and writing problems, employ appropriate teaching strategies to address these problems and evaluate the effectiveness of these actions.

Finally, in line with both the citizenship and professional preparation roles of science advocated by NZ policy, students need to engage critically with the content and purposes of scientific (or purportedly scientific) texts. Critical literacy involves a shift away from 'getting the correct answer' to questioning the assumptions in texts, critiquing and challenging. In our observations of teachers we saw no evidence of any instruction that could be characterised as critical literacy. We would therefore argue, from a position of instructional depth, that students need opportunities to engage with issues, ideas and concepts, to challenge and critique them as part of deep learning in science.

In conclusion, we identify four opportunities for subject-specific literacy teaching to shift in ways that prepare students better to read and write science text. Firstly, we see a need to expand the learning outcomes that are valued, from a primary focus on assessed science content to a broader focus that encompasses using texts in science appropriate ways, with reading, writing, disciplinary literacy and critical literacy outcomes as central to these. Secondly, we see an opportunity to frame students, rather than teachers, as being responsible for the reading and writing of science text. Thirdly, there is a need to move from constrained to open-ended literacy learning tasks. Fourthly, there is a need to move beyond short-term strategies towards a focus on generative teaching so that students are in a position to read and write the texts they need as citizens or as emerging science professionals.

References

Applebee, A. N., Langer, J. A., Nystrand, M., & Gamoran, A. (2003). Discussion-based approaches to developing understanding: Classroom instruction and student performance in middle and high school English. *American Educational Research Journal, 40*(3), 685–730.

Behrman, E. H. (2006). *Teaching about language*, power, and text: A review of classroom practices that support critical literacy. *Journal of Adolescent & Adult Literacy, 49*(6), 490–498.

Carnegie Council on Advancing Adolescent Literacy (2010). Time to act: An agenda for advancing adolescent literacy for college and career success. New York: Carnegie Corporation of New York.

Conley, M. W. (2008). Cognitive strategy instruction for adolescents: What we know about the promise, what we don't know about the potential. *Harvard Educational Review, 78*(1), 84–106.

Darling-Hammond, L., & Bransford, J. (2005). *Preparing teachers for a changing world: What teachers should learn and be able to do*: Jossey-Bass Inc. Pub.

Fang, Z., & Schleppegrell, M. J. (2010). Disciplinary literacies across content areas: Supporting secondary reading through functional language analysis. *Journal of Adolescent & Adult Literacy, 53*(7), 587–597.

Hargreaves, A., & Macmillan, R. (1995). The balkanization of secondary school teaching. In L. S. Siskin & J. W. Little (Eds.), *The subjects in question: Departmental organization and the high school* (pp. 141–171). New York: Teachers College Press.

Janks, H. (2013). *Critical literacy*. Oxford: Blackwell Publishing.

Jetton, T. L., & Alexander, P. A. (2004). Domains, teaching, and literacy. In T. L. Jetton & J. A. Dole (Eds.), *Adolescent literacy research and practice* (pp. 15–39). New York: Guilford Press.

Kamil, M. L., Borman, G. D., Dole, J., Kral, C. C., Salinger, T., & Torgesen, J. (2008). Improving adolescent literacy: Effective classroom and intervention practices. IES Practice Guide. NCEE 2008–4027. National Center for Education Evaluation and Regional Assistance, 65.

Kirsch, I., de Jong, J., Lafontaine, D., McQueen, J., Mendelovits, J., & Monseur, C. (2002). *Reading for change: Performance and engagement across countries: Results from PISA2000*. Paris: OECD Online Bookshop.

Kuhn, M. R., & Stahl, S. A. (2003). Fluency: A review of developmental and remedial practices. *Journal of Educational Psychology, 95*(1), 3.

Lee, C. D., & Spratley, A. (2010). *Reading in the disciplines: The challenges of adolescent literacy*. New York: Carnegie Corporation of New York.

Lesaux, N. K., Kieffer, M. J., Faller, S. E., & Kelley, J. G. (2010). The effectiveness and ease of implementation of an academic vocabulary intervention for linguistically diverse students in urban middle schools. *Reading Research Quarterly, 45*(2), 196–228.

May, S., & Wright, N. (2007). Secondary literacy across the curriculum: Challenges and possibilities. *Language and Education, 21*(5), 370–376.

New Zealand Qualifications Authority. (n.d.). *NCEA*. Retrieved from http://www.nzqa.govt.nz/qualifications-standards/qualifications/ncea/

Ministry of Education. (2007). *The New Zealand curriculum*: Learning media Wellington.

Moje, E. (2008). *Foregrounding the disciplines* in secondary literacy teaching and learning: A call for change. *Journal of Adolescent & Adult Literacy, 52*(2), 96–107.

Moje, E., Stockdill, D., Kim, K., & Kim, H. (2011). The role of text in disciplinary learning. *Handbook of Reading Research, 4*, 453.

O'Brien, D. G., Stewart, R. A., & Moje, E. A. (1995). Why content literacy is difficult to infuse into the secondary school: Complexities of curriculum, pedagogy, and school culture. *Reading Research Quarterly, 30*(3), 442–463.

Office of the Prime Minister's Science Advisory Committee (2011). Looking ahead: Science education in the 21st century: A report from the Prime Minister's Chief Science Advisor. Retrieved from Office of the Prime Minister's Science Advisory Committee: http://www.pmcsa.org.nz/wp-content/uploads/Looking-ahead-Scienceeducationfor-the-twenty-first-century.pdf.

Pearson, P. D., Moje, E., & Greenleaf, C. (2010). Literacy and science: Each in the service of the other. *Science, 328*(5977), 459–463.

Pressley, M. (2004). The need for research on secondary literacy education. In T. L. Jetton & J. A. Dole (Eds.), *Adolescent Literacy Research and Practice* (pp. 415). New York: The Guilford Press.

Schoenbach, R., Greenleaf, C., & Murphy, L. (2012). *Reading for understanding: How reading apprenticeship improves disciplinary learning in secondary and college classrooms*. 2nd ed. San Francisco: Jossey-Bass.

Shanahan, T., & Shanahan, C. (2008). Teaching disciplinary literacy to adolescents: Rethinking content-area literacy. *Harvard Educational Review, 78*(1), 40–59.

Siskin, L. S. (1997). The challenge of leadership in comprehensive high schools: School vision and departmental divisions. *Educational Administration Quarterly, 33*, 604–623.

Soter, A. O., Wilkinson, I. A., Murphy, P. K., Rudge, L., Reninger, K., & Edwards, M. (2008). What the discourse tells us: Talk and indicators of high-level comprehension. *International Journal of Educational Research, 47*(6), 372–391.

White, S. (2012). Mining the text: 34 text features that can ease or obstruct text comprehension and use. *Literacy Research and Instruction, 51*(2), 143–164.

Wilkinson, I. A., & Son, E. H. (2009). A dialogic turn in research on learning and teaching to comprehend. *Handbook of reading research, 4*, 359.

Wilson, A., Jesson, R., Rosedale, N., & Cockle, V. (2012). *Literacy and language pedagogy within subject areas in years 7–11*. Wellington: Ministry of Education.

Chapter 10
Analyzing Discursive Interactions in Science Classrooms to Characterize Teaching Strategies Adopted by Teachers in Lessons on Environmental Themes

Ana Lucia Gomes Cavalcanti Neto, Edenia Maria Ribeiro do Amaral and Eduardo Fleury Mortimer

Abstract The chapter analyzes discursive interactions in science classrooms to characterize teaching strategies adopted by teachers when addressing environmental issues. We studied classes taught at three different public elementary and secondary schools in Escada, a town located in Pernambuco, Brazil. We analyzed six episodes extracted from 6 of 28 video-recorded lessons involving three science teachers and sixth and seventh grade students. Our analysis took into account discursive dynamics proposed by Mortimer and Scott. We also considered teaching strategy interventions whereby teachers exposed students to situations, phenomena, and scientific concepts to promote science learning and to engage students in decision-making processes. Our results show that the analysis of discursive interactions characterized various teaching strategies in classrooms and revealed different aspects of science teaching and learning that promote scientific literacy. For instance, interactive/dialogic communicative approaches seemed to encourage students to actively participate in classroom discussions and engage in meaning making in regards to scientific concepts and attitudes. Moreover, the content of classroom interactions involving different perspectives seemed to support learning beyond conceptual dimensions and motivate students to make decisions when faced with relevant socioscientific issues.

Keywords Teaching strategies · environmental issues · science education · classroom discourse

---

A.L.G. Cavalcanti Neto (✉)
Secretariat of Education, Escada, Pernambuco, Brazil
e-mail: analuneto@gmail.com

E.M.R. do Amaral
Federal Rural University of Pernambuco, Recife, Brazil
e-mail: edeniamramaral@gmail.com

E.F. Mortimer
Federal University of Minas Gerais, Belo Horizonte, Brazil
e-mail: efmortimer@gmail.com

10.1 Introduction

Environmental problems arising in the contemporary world have prompted science educators and researchers to consider that the educational process should focus more on environmental issues. A new rationale could lead to an emancipation of culture and humanization that may allow for the emergence of innovative forms of living around the world. Through this perspective, environmental education could affect individual ways of life and should be conducted to promote attitudes and skills such as awareness, knowledge, and capacity as defined by Medina (2003) for evaluation and critical action in different contexts.

Science and environmental education share several aims when teachers bring together environmental issues, scientific concepts, and models on the natural world. In science education, it is important to teach students to serve as active citizens, to fulfill certain roles, and to share responsibilities when faced with scientific and technological issues related to the environment and society (Cachapuz, Praia, & Jorge, 2002). According to Carvalho (2006), for environmental proposals, scientific knowledge of nature and of its technological applications constitutes an object of critical understanding as a form of cultural knowledge required to understand socio-environmental relationships. In this chapter, we bring together convergent perspectives on science and environmental education to identify teaching strategies and discursive dynamics in science classrooms that can facilitate scientific literacy.

10.2 Literature Review

Roberts (2007) considers that scientific literacy is related to curriculum goals, and it could characterize what school science should be all about and what school should emphasize about science. He lists the aims and purposes of science education, which generate conceptions of scientific literacy: (a) vision I: science education with an inward focus – products (laws and theories) and processes (hypothesizing and experimenting) and (b) vision II: science education involving situations wherein science plays a role, such as decision-making on socioscientific issues. For vision I, "goals for school science should be based on the knowledge and skill sets that enable students to approach and think about situations as a professional scientist would." For vision II, "goals for school science should be based on the knowledge and skill sets that enable students to approach and think about situations as a citizen well informed about science world" (Roberts, 2007, p. 9). From the latter perspective, science education must involve more than information and concepts, as science teaching is designed to address the formation of values and attitudes. The second perspective has informed most science curricula around the world, including the national curriculum used in Brazil. One goal of the Brazilian curriculum is to develop abilities that help students view nature as a complex system whereby individuals in society act as agents who live in relation to the environment and to other living beings, sharing responsibilities to make the world a better place (Brasil, 1998, 2006).

We argue that scientific literacy can complement perspectives on environmental and science education (e.g., related to the vision II, proposed by Roberts (2007)). In this case, the process of conceptualization involves fundamental relationships between individuals and society and social, cultural, economic, and political issues related to scientific knowledge. In this sense, environmental and science education can lead individuals to become more aware and to help transform their social conditions for the preservation and conservation of the environment. Norris and Phillips (2003) argued for distinctions to be made between fundamental and derived senses of literacy to show that conceptions of scientific literacy tend to neglect the fundamental sense of literacy associated with skills related to reading and writing scientific texts. Nevertheless, it is important to expand this concept toward a more holistic view of literacy that is related to knowledgeability, learning, and education. In this way, science education can promote scientific literacy when students engage in reading, writing, discussing, understanding, applying, and making decisions on scientific, environmental, and social issues.

With regards to teaching approaches in environmental issues, according to Cascino (2005), a naturalistic view of the environment often emerges in school contexts dedicated to environmental education. Pedagogical approaches tend to frame the environment as something to be understood based on laws of biology, chemistry, and physics while raising questions on the impact of human actions on nature (Carvalho, 2006). The naturalist view of the environment refers to the perception of nature as a biological phenomenon, where systemic interactions follow autonomously and independently of the social world, underpinning an understanding of a natural world in opposition to the social world (Carvalho, 2006). Carvalho (2006) states that a predominantly naturalistic view favors a limited understanding of the environment based strictly on physical and biological features despite interactions between the natural world and human culture. For us, it seems that the naturalistic view of the environment facilitates vision I approaches to science education as proposed by Roberts (2007).

In counterpoint to the naturalistic view, socio-environmental views are guided by a complex and interdisciplinary rationale that involves thinking of the environment not as untouched nature, but as a field of interactions among culture, society, and physical/biological dimensions of life processes, whereby all elements of such relationships mutually change dynamics (Carvalho, 2006). According to this perspective, humankind interacts with the environment as one participant of a social, natural, and cultural system of relations in which one component changes all others. In a similar way, we can view nature as a product of relationships of appropriation and transformation that humans form among themselves, which are mediated by work and development based on historical conditions (Tamaio, 2002). Socio-environmental views of the environment seem to favor vision II approaches to science education.

Grace and Ratcliffe (2002) argued that approaches to environmental issues require teachers to teach values that underlie science, environment, and society.

They claim that this challenges teachers to make a pedagogical shift, and many science teachers may find it difficult to do so. However, they cannot evade their responsibility to explain issues that fundamentally affect human health and the environment. Pedagogical strategies could lead students to learn scientific and environmental issues by articulating different dimensions for learning: conceptual, procedural, and attitudinal dimensions. Conceptual and procedural dimensions are related to the emphasis in scientific contents and procedures, respectively; and attitudinal dimension is related to the development of actions and values associated to the studied themes (Pozo & Crespo, 2009). Teachers must determine what students already know to design activities that challenge students, to create opportunities for discussion, to offer formative feedback, and to openly discuss their values and controversial issues (Dillon, 2012).

According to Haydt (1999), teaching strategies stand out as modes of intervention that contribute to teaching and that can expose students to scientific concepts, situations, or phenomena, thus enabling them to think about concepts, procedures, attitudes, and values depending on teachers' choices. Masetto (1997) highlights that teaching strategies function as tools that teachers use in the classroom to guide students toward learning outcomes, and then it gets success if they are embedding instructional value. The adoption of appropriate strategies favors pedagogical outcomes such as student participation and interest, group integration and cohesion, student motivation, attention to individual differences, and the expansion of learning experiences. In relation to critical environmental education, Jacobi (2005) states that teaching strategies can focus on changing habits, attitudes, and social practices; skills development; evaluative capacity; and student participation. Through such a process, discursive interactions established between teachers and students in the classroom play an important role in helping teaching strategies promote science learning and scientific literacy.

In analyzing teaching strategies, we consider an analytical framework on discursive interactions in science classrooms proposed by Mortimer and Scott (2002, 2003), allowing us to examine social interactions that occur between teachers and students in science classrooms and teachers' means of promoting such interactions. The analytical framework is based on five interrelated aspects that focus on the teacher's role, which are grouped into three dimensions: teaching focus (teachers' purpose, the content of classroom interactions), teaching approach (communicative approaches, patterns of interaction), and actions (teachers' interventions). We only discuss communicative approaches, patterns of interaction, and teachers' interventions in this chapter.

The communicative approach focuses on ways in which teachers work with students to address different ideas that emerge during lessons. Mortimer and Scott (2003) have identified four classes of communicative approaches, which are defined by categorizing the talk between teachers and students on two dimensions. The first dimension represents a continuum between *dialogic* and *authoritative* discourse, and the second dimension involves *interactive* and *noninteractive* talk. In a *dialogic* communicative approach, attention is placed on more than one point of view, more than one "voice" is heard, and an exploration or "interanimation"

(Bakhtin, 1934/1981) of ideas occurs. In an *authoritative* communicative approach, attention is placed on only one point of view, only one voice is heard and there is no exploration of different ideas. An important feature of the distinction between dialogic and authoritative approaches is that a sequence of talk can be dialogic or authoritative independent of whether it is uttered individually or between people. Thus, under the second dimension, *interactive* talk allows for the participation of more than one person, and *noninteractive* talk is performed by only one person. These two dimensions can be combined to create four classes of communicative approaches: (1) interactive/dialogic: teacher and students explore ideas; formulate authentic questions; and offer, consider, and work with different points of view; (2) noninteractive/dialogic: teacher reconsiders, in her speech, various points of view, highlighting similarities and differences; (3) interactive/authoritative: teacher generally guides students through a sequence of questions and answers, with the aim of reaching a specific point of view, typically one that supports school science; (4) noninteractive/authoritative: teacher presents a specific point of view, normally one that supports school science.

Patterns of interaction specify how a teacher and his or her students take turns in the classroom talk. It is helpful to evaluate whether interactions promote student engagement in classroom discourse. The most common patterns of interaction are I-R-E triads (Initiation by the teacher, Response by the student, and Evaluation by the teacher), but other patterns are also present in classrooms. In these patterns, a teacher offers a response to a student to prompt a further elaboration of their point of view and to thereby sustain interaction. In this way, the student is encouraged to elaborate on and explicitly outline their ideas. In some interactions, a teacher may prompt students to discuss through short interventions that often repeat part of what a student has just said or otherwise offer feedback for a student to explain his or her perspective further. These interactions generate chains of nontriadic turns (e.g., I-R-P-R-P … or I-R-F-R-F …) where P denotes a discursive action that prompts a student to talk and where F denotes feedback. Here, feedback is different from evaluation because it favors interactions between teacher and students to keep going. Evaluation, on the contrary, stops the chain of communication.

The final feature of the analytical framework presented by Mortimer and Scott (2003) focuses on ways in which a teacher intervenes to develop a scientific story and to make it available to all students in a class. In this chapter, we use these forms of intervention to characterize didactic strategies used by teachers in their classrooms. From Mortimer and Scott (2003), we characterize six forms of teacher interventions in terms of teacher focus and actions that correspond to: (1) shaping ideas, whereby a teacher's action can introduce a new term or paraphrase a student's response; (2) selecting ideas, whereby a teacher can focus attention on a particular student's response or overlook a student's response; (3) marking key ideas, whereby a teacher can repeat an idea; (4) sharing ideas, whereby a teacher can share individual ideas with a class; (5) checking students' understanding, whereby a teacher can solicit clarification on a student's idea; (6) reviewing, whereby a teacher can review activities of a previous lesson or the progress of the scientific story.

From the structure presented above, our analysis of discursive classroom interactions enabled us to characterize ways in which teachers interact with students in constructing meaning. This characterization seems to be essential to understanding how teaching strategies used by science teachers can promote scientific literacy.

10.3 Methodology

We employed a qualitative approach to our methodological design. The investigation involved three teachers of Biology (T1, T2, and T3), in three classes of sixth (T1), seventh (T2), and sixth (T3) grades, each one attended by about 25 students, in three different public schools across Escada, a town located in Pernambuco, Brazil. The teachers have the following professional profiles (Table 10.1).

We selected for analysis lessons in which teachers discuss environmental themes in the classroom. Data were collected from the video-recorded lessons. Various tables were constructed to illustrate the timing, activities, actions, themes/contents, and comments for each lesson. These tables provided an overview of the lessons, situating the analyzed episodes within full lessons. Our definition of episode is an adaptation of event definition in the tradition of interactional ethnography. Thus, an episode is defined as a coherent set of actions and meanings produced by the participants in interaction, which has a clear beginning and end and which can be easily discerned from the preceding and subsequent episodes. The episodes represent moments during the lessons whereby environmental themes emerged through discursive classroom interactions. From the selected lessons, six episodes were extracted and transcribed for analysis – two episodes for each teacher – and they were organized by numbering turns of speech. We refer to teachers (T1, T2, and T3) and students (S1, S2, S3, …) using initials and numbers. In Table 10.2, we present all of the episodes analyzed for this chapter.

In the study, we considered segments of episodes, represented by a set of turns, that depict different teaching strategies used by the teachers and discursive aspects that characterize interactions promoted during a specific moment of a lesson.

Table 10.1 Professional profiles of the teachers who participated in the study

Teacher	Formation: undergraduate/ specialization courses	Teaching experience	Number of analyzed lessons
T1	Science Teachers of Biology/ Environmental Science	Elementary and secondary school for 15 years	Two from 14 recorded lessons for grade 6
T2	Science Teachers of Biology/ Biological Sciences	Secondary school for 13 years	Two from six recorded lessons for grade 7
T3	Science Teachers of Biology/Science and Biological Teaching and Adult Education	Secondary school for 10 years	Two from six recorded lessons for grade 6

Table 10.2 Episodes analyzed

Teacher	ID episode	Discussion topic
T1	1.1	Soil degradation and agricultural practices
	1.2	Soil pollution and prevention measures
T2	2.1	Human effects on ecosystems
	2.2	Prevention of human effects on ecosystems
T3	3.1	School waste
	3.2	Environmental conservation actions

Due to space limitations, we only present the transcription for episode 1.1 in this chapter. However, all of the episodes are examined in our discussion of the results.

10.4 Results

We organize our results by presenting the episodes analyzed for each teacher and by then summarizing our overall analysis of the study data.

10.4.1 Teacher T1

In episode 1.1, T1 had the intention to introduce and develop scientific views by facilitating an understanding of soil degradation and pollution processes resulting from deforestation and burning. The teacher used, as teaching strategies, questioning, reading the textbook, and oral presentation (when teacher exposes contents to the students mainly by verbal language), as shown in the transcription for episode 1.1.

Episode 1.1 Discussion on soil degradation and agricultural practices[a]

Turns	Pattern of interactions
1. T1: How have human beings contributed to soil degradation? Can someone guess? Nobody knows ((The class is quiet; the teacher picks up a book)). Let's take a look at the textbook, let's go! (++++) ((Before the pause, the teacher asks the students to read the book excerpt)).	I – Initiation
2. READING FROM THE TEXTBOOK: Currently, ineffective agricultural practices degrade and pollute thousands of tons of soil worldwide.	
3. T1: Check this out briefly … right? Most soil degradation occurs when vegetation is removed. You see, there are agricultural practices.	I – Initiation

(continued)

(continued)

Turns	Pattern of interactions
Do you know what agricultural practices are? S1 says here ((points to student 1 and asks him to explain)).	
4. S1: Agriculture	R – Response
5. T1: Agriculture, isn't it? This involves the planting of foods that are essential to us, right? Soybeans, wheat, rice, etc. in many cases, right? This agricultural practice, right? It can harm and degrade the soil. What happened thousands of years ago? Were cities the same as they are today? No! Right? Long ago, going back in history, when Brazil was first discovered, I'm going to talk about Brazil, our country, when we arrived here in Brazil. When they (colonizers) arrived here, was Brazil the way it is today?	E – Positive evaluation I – Initiation
6. S: No	R – Response
7. T1: Was the population as large as it is today?	I – Initiation
8. S: No	R – Response
9. T1: No, and consequently our soils and natural environments weren't as they are today, right? The population grew, development occurred and people started to need more places to live – buildings and houses, right? Progress occurred – development and industry, right? Also, the need for agriculture and cattle ranching … what is cattle ranching? Can anybody tell me? Agriculture, (you) already know that involves plantation, but cattle ranching? (+++) Anyone remember? Have you never heard that word before? (++++) Okay, cattle ranching involves livestock on a farm. There can be cows, bulls, pigs, horses, etc., right? Often, vegetation is used as pastureland for these animals, and this can damage the soil. But back to what we were talking about before, check it out, what happened? There was a need … ((teacher points to someone near the door). There was a need for construction. Building became necessary to do … what? Someone must ….	E – Positive evaluation I – Initiation
10. S1: Plant trees	R – Response
11. T1: What was needed to plant trees?	Repeating the question, meaning E – Negative Evaluation
12. S2: Cutting down trees	R – Response
13. T1: Cutting down trees, is this clear? ((Teacher reinforces the student's response)). This raises the issue of deforestation, which is highlighted in your book and which also contributes to soil degradation. So is it deforestation? Is the meaning clear? What does this mean? What?	E – Positive evaluation I – Initiation
14. S3: Cutting down trees	R – Response
15. T1: Cutting down trees. And, many times….	E – Positive evaluation I – Initiation
16. S3: Destroying …	R – Response

(continued)

(continued)

Turns	Pattern of interactions
17. T1: Somehow, it is destroyed. So why does this happen? You see I'm not against progress. We cannot be against progress, as it is necessary. But unfortunately, it brings, in some ways, destruction to nature, and consequently to the soil. You see, as I was saying, this is the purpose of deforestation. For people to actually build on and populate a place, it must be cleared. This was a necessity, but we ended up destroying the soil. In addition to deforestation ((writing the word on the chalkboard)), what is the other item we have here? ((Referring to the textbook)) deforestation, what else? Another item? ((Asks students look at the textbook)).	E – Positive evaluation I – Initiation
18. S4: Erosion	R – Response
19. T1: Other point … erosion, this will come later (in the lesson). I am checking up there (the teacher points to a theme described in the textbook)). Deforestation we already see, but there is another point next to it.	F – Feedback
20. S: Forest burning	R – Response
21. T1: Exactly, forest burning. Actually, forest burning from deforestation, right? Why? Because people … ok. I am going to provide an example. Here in our town, what happens at a sugarcane plantation? What do people do (in soil) before cultivating sugarcane?	E – Positive evaluation I – Initiation
22. S: Burn	R – Response
23. T1: Burn. Do you think this is necessary? In some ways it is, but does it harm or benefit the soil?	E – Positive evaluation I – Initiation
24. S: Harm	R – Response
25. T1: It will cause harm because (burning) degrades more and more (soil), killing microorganisms. Soil supports many living things, right? Right? Many living things live in soil, and so burning forests kills these microorganisms and other animals, right?	E – Positive evaluation
26. S1: And (it) pollutes the air.	I – Initiation

[a]Conventions used in the transcription:
(+) – pauses;
( ) – insertions from authors;
(( )) – comments from authors;
…. – inconclusive speech or hesitation;
(…) – speech omission
CAPS LOCK – emphasis

In episode 1.1, the teacher focused the content of classroom interactions on a conceptual level by checking the students' comprehension of deforestation processes, agricultural practices, forest burning processes, etc. (turns 13–21), though not in the case of forest burning methods for cultivating sugarcane, for which a local case was highlighted (turn 21). This approach can be used to develop an understanding of the environment that is associated with the utilization of natural resources without considering an important dimension related to permanent interactions between the natural world and human culture.

The interactive–authoritative communicative approach was used in this episode. A focus on two enunciators – the teacher and textbook – contributed to an emphasis on the school science perspective in the discursive interactions. Student participation was restricted, as interactions permitted by the teacher only allowed students to speak briefly and to guess what she was thinking (see turns 9–12 and 17–21). In a general sense, students expressed ideas that reinforced what the teacher put forward in the discussion as supported by the textbook used. Patterns of interaction involved IRE triads (I-R-E (3-5); I-R-E (13-15), I-R-E (21-23) and I-R-E-I (23-26)) and one short chain (I-R-F-R-E (17-21)).

Several teacher's interventions were used to introduce agricultural practices as a cause of soil degradation. She asked the students to read an excerpt from the textbook featuring this idea (turn 2) and highlighted key points that support the view that soil degradation is a reflection of human actions. In the following turns (3–26), the teacher mainly focused on meanings related to scientific perspectives, highlighting key terms while repeating statements to the students (turns 5, 10, 13, 15, 21, 23, and 25), selecting meanings when responding to a student, and referring to another perspective expressed in the textbook before discussing the issue of "erosion" (turn 19). In this case, it seems clear that the textbook guided the teacher's discourse in the classroom.

It is important to highlight that the teacher drew attention to certain meanings (turns 11–17), leading students to think about negative effects of deforestation. She states that it is necessary to clear-cut areas so that humans can build homes, farms, offices, and so on. However, she does not facilitate dialogic interactions with the students when discussing opportunities for human beings to live in the natural world in a harmonious and respectful way. Despite the teacher's intention to pose questions on this particular theme, a naturalist view of the environment prevailed in the teacher's discourse.

10.4.2 Teacher T2

During the two lessons, the teacher described human effects on ecosystems, and students gave oral presentations on different roles or ways in which human beings can preserve the environment using posters that they had created.

For episode 2.1, we observed that the teacher mainly adopted oral presentation as a teaching strategy. During the episode, T2 presented a brief review of prior lessons and asked students to read and present textbook excerpts to facilitate class discussions. In doing so, it seemed that teacher's intention was to introduce scientific ideas on ecosystem degradation and to then explore the students' views through their presentations on these ideas. The content of classroom interactions addressed a conceptual dimension when the teacher cited excerpts from the textbook and a procedural dimension when students engaged planned actions related to environmental issues.

In the final moments of episode 2.1, the teacher seemed to reinforce a negative view of the relationship between human beings and the natural world. It is important to highlight that views that explain the origins of Earth, supported by faith in a Creator who conceived all the things in Universe (creationist view), seem to reinforce a negative view of human effects on the environment: "… we review how Earth was constructed … appeared. How God, the creator, gave this planet to man, right? We commented on all of these things … how we received Earth from God, and how science states how it was formed … It was beautiful … natural … without human effects. So what happened? God created man to master all things … and so now, Earth has adapted to this situation" (T2 in episode 2.1). This comment suggests that the teacher found it difficult to present a rational and critical view of environmental issues.

The communicative approach employed was predominantly noninteractive/authoritative, as the teacher adopted mainly oral presentation as her teaching strategy. When students presented ideas based on textbook excerpts, an interactive/dialogic approach was used in the classroom, as the teacher interacted with students during their presentations. At least two points of view emerged through the discursive interactions: the scientific view, which is represented by ideas presented in the textbook used, and the student's ideas. For example, S2 stated: "And so, with the way that mankind is damaging nature, even human beings could become extinct. But, if man does not make certain products like chairs, beds, and wardrobes, how could we survive without a seat or bed? That's one thing I want to know … ((asking the teacher))." The teacher addressed this question by describing ways in which humankind can intelligently use natural resources without damaging the environment. Despite the occurrence of student participation, interactions between the teacher and students followed triadic patterns of interaction, with more than one response provided by the students and with the teacher listening to them before closing the discussion through an evaluation. The students often limited themselves to expressing their ideas, but they did not comment on the teacher's evaluation. In this sense, scientific views prevailed in the face of student questioning or misunderstanding, and the textbook played a predominant role in the lesson.

In episode 2.1, T2's interventions involved reviewing the development of scientific ideas and sharing and selecting meanings oriented toward a view of the environment as separate from human issues. This view holds institutions responsible for addressing environmental issues without consideration of the roles played by individuals in this context.

In regards to episode 2.2, we highlight a moment when the students' presentations were concluding and the teacher tried to organize conclusions of the discussion raised in the previous lesson. Some concluding ideas appeared to emerge through a poster presented by student S4: "As we can see, here we have the first figure ((points to the figure)) of wheat crops. Wheat is very important to our lives. However, this is very different from the first frame ((points to the other figure)). Here, he (the farmer) is only clearing the forest. If he had already planted wheat with his wisdom, we can understand why he destroyed the forest and the trees of the forest to plant wheat. From his wisdom, he cleared trees and planted wheat (...)." It seems that the teacher argued that deforestation can be defensible if it involves growing crops, and she used this case as support for this claim. In the other hand, she highlighted the roles played by institutions and human beings in preserving the environment: "If we help human beings be conscious and aware of our negative effects on nature, we will have a better world." T2's position on the exploitation of natural resources by human beings is not clear. Opportunities and controversies involving human uses of natural resources were not addressed.

10.4.3 Teacher T3

In the two lessons, T3 explored issues of school waste and environmental conservation. In episode 3.1, the teacher returned to the theme of school waste and asked questions to have students reflect on causes of other environmental problems (e.g., forest burning, deforestation, poverty, disease, violence, waste, consumerism) and on ways to address and overcome such problems.

The teacher T3 adopted questioning as a teaching strategy throughout the episode. She guided and engaged students in a classroom discussion, thus encouraging them to think about environmental issues and social compromises. The content of classroom interactions predominantly focused on the attitudinal dimension, helping students make decisions and perform critical actions supported by concise arguments.

The communicative approach used in episode 3.1 was interactive/dialogic, as throughout the episode, the teacher and students expressed ideas, and different points of view were taken into account through a discussion. In this sense, much of the time, extended chains prevailed as patterns of interaction as shown in the excerpt from episode 3.1 (turns 4–10):

Excerpt from episode 3.1: Illustrating extended chains

Turns	Patterns of interaction
...	
4. S1: We have to collaborate.	R – Response
5. T3: We have to collaborate, but in what way?	P – Prompt
6. S2: By not littering?	R – Response
7. T3: But is just not littering collaborating? Could we do more?	P – Prompt
8. S3: By not polluting the rivers and air.	R – Response
9. T3: Yes – not polluting the air, not polluting the rivers, and not littering. We can do something to change this (situation), can't we?	P – Prompt
10. S: We can.	R – Response
...	

When the teacher offered her students feedback, she encouraged them to reply to questions on their responsibilities related to the environment. The teacher's interventions suggest that she framed meanings while the students expressed their ideas, identified key ideas (turn 5), shared meanings (turn 9), and measured the students' comprehension by asking them to elaborate on their ideas (turn 7).

In episode 3.2, the teacher posed questions about dengue fever and about areas where there is a higher incidence of this disease. In doing so, she prompted her students to think about causes and consequences of this social, health, and environmental problem. The teacher prompted interactions between the students by questioning and measuring their level of understanding. Her students then presented their ideas on environmental issues affecting daily life. In this case, they discussed dengue fever and school waste. Finally, the teacher asked her students to reflect on their ideas.

10.5 Discussion

Results point out particular characteristics for each teacher involved in this work. For teacher T1, our analysis of episode 1.1 shows that the didactic strategy adopted by this teacher mainly involved oral presentation to textbook content. In this case, the didactic strategy did not appear to promote effective discussion on themes introduced during the lesson. These discursive features characterize T1: the teacher's intention was to focus heavily on the presentation of scientific views on the themes, the content of classroom interactions was limited to the conceptual dimension, communicative approaches were mainly interactive–authoritative, and patterns of interaction were predominantly triadic (IRE) with only one short chain. In episode 1.2, T1 sought to enable students to reflect on negative effects of forest burning on soils. She discussed ways to prevent such environmental consequences, guiding students using scientific perspectives. In the discussion, there was an emphasis on pollution as a principal result of human actions related to garbage disposal. Again, the teacher adopted predominantly oral presentation strategies while maintaining a conception of society–nature relationships that was essentially naturalist, and even when she described behaviors that can promote environmental preservation. Patterns of interaction were, again, IRE triads and the content of classroom interactions was predominantly conceptual.

During the two lessons, we verified that teaching strategies adopted by T1 did not enable students to develop a greater appreciation for different ideas throughout the construction of meanings. The predominantly authoritative communicative approach emphasized the school science views on environmental issues, disallowing the emergence of different perspectives. This appears to hinder the development of educational services for citizens and goals of environmental education. Triadic patterns of interaction prevailed during these lessons. The teacher initiated all interactions, students were afforded few opportunities for participation and the

Table 10.3 Summary of the analysis of teacher T1: discursive aspects and teaching strategies

Teaching strategy	Oral presentation	Reading	Questioning
Main discursive features			
Teacher's intention	Introducing and developing a scientific perspective Promoting discussion Guiding students on scientific ideas	Introducing and developing a scientific perspective Promoting discussion	Introducing and developing a scientific perspective Promoting discussion Guiding students on scientific ideas
Content of classroom interactions	Conceptual	Conceptual	Conceptual
Communicative approach	Noninteractive/authoritative Interactive/authoritative	Noninteractive/authoritative	Interactive/authoritative
Teacher's interventions	Selecting meanings Marking key meanings Sharing meanings Shaping meanings		Sharing meanings Marking key meanings
Patterns of interaction	Episode 1.1 I-I-R-E (1-5); I-R-E (13-15); I-R-E (21-23) and I-R-E-I (23-26) – and one short chain – I-R-F-R-E (17-21) Episode 1.2 I-R-F-R-F-R-E (1-9); I-R-E (9-11); I-R-E (11-13); I-R-E (13-15); I-R-E (15-17); I-R-E (17-19); I-R-E (19-21); I-R-E (21-23); I-R-E (23-25); I-R-R-E (25-28); I-R-E (28-30); I-R-F-R-E (30-34); I-R-R-E (34-37); I-R-E (37-39); I-R-F-F-F-E (39-43).		

teacher's evaluation did not allow students to develop critical ideas. Table 10.3 shows a summary of the analysis on teacher T1.

In the two lessons, the teacher T2 used strategies that seemed to consider more than one point of view in discussions, as students were allowed to present their ideas. Effective student participation through oral presentations promoted an interactive/dialogic communicative approach. However, T2 did not promote a deep discussion on the themes, highlighting difficulties that can arise when developing values required to make critical decisions on environmental issues. This seemed to cause teacher T2 to focus the content of classroom interactions on conceptual dimensions and to limit interactions with students to triadic patterns. In this case, the interactive–dialogic communicative approach was limited, supporting weak interactions and superficial discussions. Table 10.4 presents a summary of the analysis on teacher T2.

In general, teacher T3 adopted teaching strategies that guided students through environmental education, questioning, study activity proposal, and supervision when she emphasized two dimensions in classroom interactions: conceptual and

Table 10.4 Summary of the analysis on teacher T2: discursive aspects and teaching strategies

Teaching strategies	Oral presentation	Oral presentation by students
Main discursive features		
Teacher's intention	Introducing and developing scientific ideas Supporting the student learning process	Exploring students' ideas
Content of classroom interactions	Conceptual	Conceptual Procedural
Communicative approach	Noninteractive/authoritative Noninteractive/dialogic	Interactive/dialogic
Teacher's interventions	Reviewing the progression of scientific ideas Sharing meanings Selecting meanings	Sharing meanings
Patterns of interaction	Episode 2.1 I-R-R-R-R-R-E (1-7); I-R-E (7-9); I-R-F-R-R-R-R-R-E (9-17); Episode 2.2 I-R-E (1-3)	

attitudinal. These teaching strategies promoted student participation and the discussion of different perspectives circulating through the classroom discussion. Such strategies also helped students make critical and sound decisions. Interactive/dialogic approaches were used in conjunction with interactive/authoritative communicative approaches during the analyzed episodes. Patterns of interaction predominantly included extended chains whereby feedback seemed to prompt students to think about the relevance of such themes and about their engagement in searching for solutions to environmental problems. Pedagogical positions related to socio-environmental views were adopted by teacher T3. Table 10.5 summarizes our analysis of episodes for teacher T3.

According to these results, the three teachers adopted different teaching strategies and discursive dynamics in their lessons. In addition, each teacher seemed to express a specific view on the environment. We summarize these results in Table 10.6.

Table 10.6 illustrates features related to teaching strategies and discursive dynamics in the analyzed lessons that facilitate or inhibit the development of scientific literacy in science and environmental education. In putting forward socioscientific and environmental issues in classroom discussions, the teachers did not necessarily help students develop skills and competencies associated with scientific literacy. In addition, it is not desirable for teachers' academic or scientific views prevail in discussions (see teacher T1). For scientific literacy in a fundamental sense (Norris & Phillips, 2003), it is not enough for students to read textbooks or make oral presentations, and it seems crucial to encourage critical debates touching on different points of view to achieve meaningful learning outcomes (see teacher T2). Finally, teach T3's socio-environmental views based

Table 10.5 Summary of the analysis on teacher T3: discursive aspects and teaching strategies

Teaching strategies	Questioning	Proposing and supervising study activities
Main discursive features		
Teacher's intention	Helping students engage in study activities Promoting discussions on environmental issues and social commitment Developing arguments to help students make decisions	Motivating students to plan actions
Content of classroom interactions	Conceptual Attitudinal	Conceptual Attitudinal
Communicative approach	Interactive/dialogic Interactive/authoritative	Interactive/dialogic Interactive/authoritative
Teacher's interventions	Shaping ideas Marking key meanings Sharing meanings Checking student understanding	Shaping ideas Sharing meanings Marking key meanings
Patterns of interaction	Episode 3.1 I-R-F-R-F-R-F-R-F-R-F-R-F-R-E (1-15); I-R-R-F-R´F-R-E (15-22); I-R-E (22-24); I-R-F-R-F-R-F-R-F-R-F-R-F-R-R-F-R-F-R-F-R-E (24-47). Episode 3.2 I-R-F-R-F-R-F-R-E (1-7); I-R-E (7-9); I-R-F-R-E (9-13); I-R-E (13-15); I-R-E (15-17); I-R-E (17-19); I-R-F-R-E (19-23); I-R-E (23-25); I-R-F-R-F-R-E (25-31); I-R-E (31-33); I-R-E (33-35); I-R-F-R-F-R-E (35-41); I-R-E (41-43).	

Table 10.6 Teaching strategies and discursive dynamics found for the three teachers

Teacher/environmental view	Didactic strategies	Content approach/patterns of interaction	Communicative approach
Teacher T1/naturalistic	Reading the textbook Oral discussion	Conceptual/ triadic	Interactive– authoritative
Teacher T2/Creationist view; the environment is separated from human beings	Oral discussion Oral presentations by students	Conceptual and Procedural/triadic	Noninteractive/ authoritative
Teacher T3/Socio-environmental	Questioning	Conceptual and Attitudinal/ extended chains	Interactive/ dialogic Interactive/ authoritative

on dialogic communicative approaches established from patterns of interaction in extended chains and based on questioning on conceptual and attitudinal dimensions of the content seemed to help students discuss environmental issues, fundamentally supporting scientific literacy.

Some teaching strategies (e.g., questioning) improved discursive interactions in science classrooms by creating opportunities for discussion and debate. Nevertheless, questioning does not guarantee that the students' points of view will be taken into account in classroom discourse. When questioning is based mainly on triadic patterns of interaction, as was the case for T1, communicative approaches employ a predominantly authoritative and interactive dimension and questioning serves mainly to measure and control meanings introduced in classroom discourse. By contrast, when questioning allows students to express their points of view, as was the case for T3, chains of interaction occur and communicative approaches are predominantly dialogic and interactive.

We highlight two relevant factors from the results of this investigation. First, the teachers' views on the environment – whether naturalistic, not well-defined, or socio-environmental – appear to guide teaching strategies in science classrooms, mainly regarding the content introduced and opportunities for students to express their ideas, as allowed by the teacher. There is not a necessary relationship between dialogic communicative approaches and socio-environmental views and between authoritative approaches and naturalistic views. As we have shown, T2 presented a not well-defined view but used a dialogic communicative approach. Nevertheless, the use of attitudinal content seems to improve opportunities for dialogic communication and brings about socio-environmental views in classroom discourse.

Second, interactive/dialogic communicative approach played a key role in engaging students in classroom discussions, and they appeared to favor the development of attitudinal dimensions for learning, although they did not guarantee such an outcome. In the same vein, we consider patterns of interaction in extended chains that promote dialogic interaction, which other works have examined (Aguiar, Mortimer, & Scott, 2009; Scott, Mortimer, & Aguiar, 2006).

10.6 Final Remarks

This work presents an analysis of didactical strategies and discursive dynamic adopted by teachers when they approach environmental issues in classroom, bringing together ways of integrating science and environmental education. In this sense, it seems necessary to engage students in dialogic and interactive discussions that offer them opportunities to learn, analyze, form positions, identify solutions, and make decisions in real life that are supported by scientific knowledge. According to this perspective, science curricula should not only be concerned with scientific content but also with values, cultural norms, ethics, policies, and social demands, guiding teachers and schools toward the development of scientific literacy in a fundamental sense.

References

Aguiar, O. G., Mortimer, E. F., & Scott, P. H. (2009). Learning from and responding to students' questions: The authoritative and dialogic tension. *Journal of Research in Science Teaching, 47*, 174–193.

Bakhtin, M.M. (1981). *The dialogic imagination* (ed. by Michael Holquist, trans. by Caryl Emerson and Michael Holquist). Austin: University of Texas Press.

Brasil – Ministério da Educação. (1998). *Parâmetros Curriculares Nacionais – Ciências Naturais*. Brasília.

Brasil – Ministério da Educação. (2006). *Orientações Curriculares para o Ensino Médio*. Brasília.

Cachapuz, A. F., Praia, J. F., & Jorge, M. P. (2002). *Ciências, educação em ciências e ensino de ciências*. Lisboa: Ministério de Educação.

Carvalho, I. C. M. (2006). *Educação ambiental: A formação do sujeito ecológico*. 2nd ed. São Paulo: Cortez.

Cascino, F. (2005). *Educação ambiental: Princípios, história, formação de professores*. 2nd ed. São Paulo: Editora SENAC.

Dillon, J. (2012). Science, environment and health education: Towards a reconceptualisation of their mutual interdependences. In A. Zeyer & R. Kyburz-Graber (Eds.), *Science/environment health: Towards a renewed pedagogy for science education* (pp. 87–101). Dordrecht: Springer.

Grace, M. M., & Ratcliffe, M. (2002). The science and values that young people draw upon to make decisions about biological conservation issues. *International Journal of Science Education, 24*, 1157–1169.

Haydt, R. C. C. (1999). *Curso de didática geral*. 6th ed. São Paulo: Ática.

Jacobi, P. R. (2005). Educação ambiental: O desafio da construção de um pensamento crítico, complexo e reflexivo. *Educação e Pesquisa, 31*, 233–250.

Masetto, M. T. (1997). *Didática: A aula como centro*. 4th ed. São Paulo: FTD.

Medina, N. M. (2003). *Educação ambiental: Uma metodologia participativa de formação*. 3rd ed. Petrópolis: Vozes.

Mortimer, E. F., & Scott, P. H. (2002). Atividade discursiva nas salas de aula de ciências: Uma ferramenta sociocultural para analisar e planejar o ensino. *Investigações em Ensino de Ciências, 7*(3), 283–306.

Mortimer, E. F., & Scott, P. H. (2003). *Meaning making in secondary science classroom.*. Maidenhead: Open University Press.

Norris, S. P., & Phillips, I. M. (2003). How literacy in its fundamental sense is central to scientific literacy. *Science Education, 87*(2), 224–240.

Pozo, J. I., & Crespo, M. A. G. (2009). *A aprendizagem e o ensino de ciências: Do conhecimento cotidiano ao conhecimento científico*. 5th ed. Porto Alegre: Artmed.

Roberts, D. A. (2007). Opening remarks. In C. Linder, L. Östman, & P. O. Wickman (Eds.), Promoting scientific literacy: Science education research in transaction. *Proceedings of the Linnaeus Tercentenary symposium* (9–17). Uppsala: Uppsala University.

Scott, P. H., Mortimer, E. F., & Aguiar, O. G. (2006). The tension between authoritative and dialogic discourse: A fundamental characteristic of meaning making interactions in high school science lessons. *Science Education, 90*, 605–631.

Tamaio, I. (2002). *O professor na construção do conceito de natureza: Uma experiência de educação ambiental*. São Paulo: Annablumme.

Chapter 11
Measuring Time. Multilingual Elementary School Students' Meaning-Making in Physics

Britt Jakobson, Kristina Danielsson, Monica Axelsson and Jenny Uddling

Abstract This chapter presents results from a study aiming at investigating multimodal classroom interaction and its contribution to multilingual students' meaning-making. The focus is on how science content is elaborated and negotiated through various semiotic resources. Data consist of video and audio recordings and digital photographs from a multilingual elementary school physics classroom during the unit "measuring time." Theoretically, the project takes its stance in social semiotics and pragmatist theory. Data are analyzed through systemic functional linguistics, multimodal analyses, and Dewey's principle of continuity. The results reveal that the teacher and the students were engaged in meaning-making activities involving a variety of semiotic resources with a potential to develop multilingual students' scientific literacy. However, some observations indicate classroom practices that might constitute a hindrance for meaning-making. The study has implications for ways of promoting scientific literacy, including learning science, competent action, and communicating through different modes.

Keywords Multilingual students · scientific literacy · multimodality · continuity · SFL

B. Jakobson
Department of Mathematics and Science Education, Stockholm University, Stockholm, Sweden
e-mail: britt.jakobson@mnd.su.se

K. Danielsson (✉)
Department of Swedish, Linnaeus University, Växjö, Sweden
e-mail: kristina.danielsson@lnu.se

M. Axelsson · J. Uddling
Department of Language Education, Stockholm University, Stockholm, Sweden
e-mail: monica.axelsson@isd.su.se

J. Uddling
e-mail: jenny.uddling@isd.su.se

The discourse of science comprises a specialized language often described in terms of linguistic density, a high degree of abstractness, and a need for students to handle multiple resources for meaning-making in parallel (Fang & Schleppegrell, 2008; Halliday & Martin, 1993; Lemke, 1998), all of which can be challenging for the learner. For multilingual students, science and its specialized language might be even more distant from their own lives if they learn science in their second language.

This chapter presents results from a project funded by the Swedish Research Council, aiming to study classroom interaction and its contribution to multilingual students' meaning-making in science. Our point of departure is the fact that meaning-making is always multimodal (Kress, 2010), not the least in science education (Danielsson, 2016; Kress, Jewitt, Ogborn, & Tsatsarelis, 2001; Lemke, 1998). Also, Lemke (1998) claims that a variety of semiotic resources need to be used in the science classroom, since each resource can contribute to the content in specific ways, and since a certain level of redundancy is needed in the learning situation. We consider a conscious use of various semiotic resources to be particularly important for enhancing multilingual students' opportunities to develop scientific literacy. Much of previous research has dealt with either multimodality in science classrooms or linguistic aspects of science learning, sometimes from a perspective of second language learning, while fewer studies combine these two perspectives (Zhang, 2016).

The specified research question for the present study addresses how science content is elaborated and negotiated multimodally in a multilingual classroom, with combinations of semiotic resources such as spoken or written language, models, and action. Through our multimodal perspective, we aim at giving a multifaceted characterization of classroom communication and the ways in which meaning-making takes place through the use of different resources. The results are discussed with regard to students' opportunities for developing scientific literacy, here including learning science, competent action in the science classroom, and communicating through different modes.

11.1 Theoretical Perspectives

Theoretically, the project takes its stance in social semiotics (Halliday & Matthiessen, 2004; Jewitt, 2017; Kress, 2010) and pragmatist theory (Dewey, 1938/1997; Wickman, 2006). From a social semiotic perspective, each choice of resource for meaning-making is seen as a result of social, cultural, and situational factors in the context in which the communication takes place, including participants and available semiotic modes and resources. The choice of resource concerns choice of semiotic mode (e.g., speech, writing, gesture), or combinations of modes, as well as choice of particular resource within each mode, such as choice of specific verbal formulation or gesture. A central concept in multimodal analyses is the notion of modal "affordance" (Gibson, 1977; Kress, 2010), here defined as the potential for meaning-making or potentials and limitations of modes (Kress, 2010, p. 84).

Dewey's (1938/1997) principle of continuity means that in any meaning-making situation, earlier experiences are reconstructed and transformed for a purpose, something which has consequences for the present and future situations. Hence, meaning-making is continuous, but might not always take the route intended by, for example, a teacher (Lave, 1996; Wickman, 2006). Continuity can be seen in how students proceed in action, using language and other resources, and how they relate this to the purpose of the learning activity. In accordance with Dewey's (1938/1997) principle of continuity, Johansson and Wickman (2011) have outlined the significance of purpose for learning science. They differentiate between ultimate (overall) and proximate (student centered) purposes. Thus, continuity can be analyzed as a function of the extent to which classroom interaction and the use of resources are coherent with the purposes of the activity.

11.2 Methodology and Analytical Framework

We present results from an 80-minute-long group session involving 9 students, 5 girls and 4 boys (out of 16), in a grade 5 Swedish elementary school classroom (students around 11 years old). The school is linguistically and culturally diverse, located in a suburban area. All students are multilingual with various linguistic backgrounds (e.g., Arabic, Somali, Turkish, Bulgarian) and varied proficiencies in Swedish. The language of instruction is Swedish (in the following, examples given have been translated into English). The teacher is an experienced elementary school teacher, educated in science teaching and Swedish as a second language. The analyzed lesson deals with the sundial, and it is the seventh in a total of 22 lessons within the unit *measuring time*, which was followed by the research team. The overall unit was structured in accordance with the Swedish version of Science and Technology for Children (NTA, 2005). The chosen lesson is representative of the unit as regards structure and ways of dealing with content and resources for learning. The teacher commonly recapitulated the prior lesson and connected to earlier discussions on the topic at hand. Thereafter, the teacher continued by using some artefacts to illustrate the actual phenomenon and introduced hands-on activities and/or reading and writing tasks.

The data consist of video and audio recordings, digital photographs, and students' written texts. The project adheres to the ethical principles outlined by the Swedish Research Council (2016) concerning information to the participants, the requirement of consent, confidentiality, and the use of data for research purposes only. For the sake of anonymity, all students have been assigned pseudonyms.

The data is analyzed through systemic functional linguistics (SFL), multimodal analysis, and Dewey's principle of continuity. On the basis of these analyses, we discuss the ways in which different resources are used in whole class communication and small group discussions, respectively, and to what extent teacher-led classroom discussions around the use of different semiotic resources take place.

In order to capture what resources are used for meaning-making, as well as how they are used, a framework developed for multimodal text analysis by Danielsson and Selander (2016) has been used. In this chapter, we focus on the aspects of the model concerning *general structure* and the *interaction between resources for meaning-making*. The model also includes how *values* are expressed through different resources. In this chapter, we comment on the *norms* about how to act that can be discerned through our analyses.

Our analyses start with the general structure of the lesson, including thematic orientation and sequencing (Danielsson & Selander, 2016). First, the ultimate (overall) purpose of the unit is described, followed by comments on the proximate purposes of the lesson (Johansson & Wickman, 2011). Then we describe the overall design of the lesson according to a number of activities that were noted. For each activity, we specify the semiotic resources used and the content made available through them.

In regard to the interaction between resources for meaning-making in the framework, we start off by presenting and analyzing *multimodal ensembles* (combinations of resources in different semiotic modes that form an entity) with special focus on a central learning sequence which included spoken and written language as well as hands-on activities. We specifically examine the process when students created multimodal texts. Then we examine, in turn, the use of *written* and *spoken* language, a *hands-on activity*, and *models and wordplay*. The use of gestures is commented on in relation to multimodal ensembles and a wordplay that the teacher introduced at the beginning of the lesson.

Regarding written and spoken language, analyses based on systemic-functional linguistics (Halliday & Matthiessen, 2004) were made to investigate how registers came into play in spoken and written texts. Register is defined as a "functional variety of language" related to a specific context (Halliday & Matthiessen, 2004, p. 27), characterized by certain lexico-grammar resources that are privileged and used for meaning-making in the context. Different subject domains, like science, have specialized registers. Learning can be seen as a development in register and a growing ability to "handle the meaning requirements of situations which are increasingly abstract and complex" (Macken-Horarik, 1996, p. 247).

Moreover, central to our analyses is to what extent different resources are continuous, or coherent with the purposes of the activity.

11.3 General Structure and Setting: The Lesson

The ultimate purpose of the lesson, which was expressed at the beginning of the unit *measuring time*, was for the students to learn about how people have measured time through history and to be able to account for the use of different tools. The teacher had intended to let the students investigate shadows caused by the sun by studying them outside, but since it was cloudy she let the students do indoor activities. During the lesson, several proximate purposes were expressed by the teacher. One was to learn how gnomons work and that they were tools for measuring time. However, as the weather was cloudy the measuring activity had to be performed in-doors. Accordingly, this proximate purpose was expressed as "we have to pretend a little that we've had it [the sun] inside here today". However, this proximate purpose was not continuous with the ultimate purpose "measuring time" as the ultimate purpose was not explicitly mentioned in this situation. Also, other proximate purposes concerned how to act, for example, writing scientific texts and doing science.

Table 11.1 gives an overview of the various phases of the lesson and the resources used, that is, a description of the lesson procedure. In regard to the

Table 11.1 Activities and resources used during the lesson (teacher T, students S)

Activity and content	Resources used for meaning-making
Recapitulating last lesson (photos of different sundials, gnomon, knowledge demands, central course content)	*Talking*: recapitulating last lesson (T) *Gestures*: enhancing spoken words (nonsubject-specific, e.g., "think" + "pointing towards head") (T) *Model*: gnomon (wooden stick on base) (T) *Writing*: content-specific (duplicates "gnomon") (T)
Introduction: Shows how gnomons work, talks about "the sun's movement across the sky"	*Talking*: explanations (e.g., the sun's movement across the sky, how to use stick and torch as sundial) (T) *Model*: sundial (gnomon, torch) (T)) *Gestures*: subject-specific (e.g., the sun's movement) (T & one S) and nonsubject-specific (e.g., think) (T) *Action*: how to make shadows with torch and stick (T & S)
Instruction: How to understand a prefabricated diagram and to transform it into table	*Talking*: explanations (T) *Diagram*: sun's shadows of gnomon *Writing* (multimodal): table with words and figures (T) *Gestures*: subject-specific (e.g., "north"+"moving arm upwards") nonsubject-specific (e.g., "adjust"+"jiggling one hand") (T)
Learning activity (individual): Transfer prefabricated diagram into table, analyze information from table	*Diagram*: sun's shadows of gnomon (S) *Other tools*: notebook, pencil, ruler (S) *Writing*: (multimodal) transfer content from diagram into table, generalize/analyze into written words (S) *Talking*: asking questions, answering teacher questions, comparing results (T, S)
Instruction: How to make a diagram similar to the prefabricated one	*Model*: sundial (gnomon, torch) (T) *Other artefacts*: flipchart to create diagram (T) *Talking*: explanation (T) *Action*: showing how to make shadows of gnomon (T)
Learning activity (pair work, hands-on): Make a diagram similar to the prefabricated one	*Model*: sundial (gnomon, torch) (S) *Action*: holding torch to create "correct" shadow (S) *Talking*: discussing and instructing (S), asking/answering questions (S, T)
Lesson ends: Students clear away	

thematic orientation of the lesson, we were able to note a movement from the students' previous school experiences and everyday knowledge, to the field of science. This was done by recapitulating what had happened during the last lesson, and then connecting to students' everyday experience of sunrise and sunset before introducing the learning activity around a prefabricated diagram representing the ways that the sun casts shadows on the ground at different times of the day. The learning activity involved both scientific content and scientific methods, in this case measurements (in a diagram) followed by analyses (in tables) and finally reaching a conclusion from the measurement (in written language). As a next step, a hands-on activity was performed.

11.4 Multimodal Ensembles

All human communication is to some extent multimodal (Kress, 2010). A typical example is when speech is combined with gestures, which was also the case in this classroom. What was striking in this regard was the extent to which the teacher used gestures in ways that could promote students' learning in a second language. Examples are when the teacher said "you have to remember," pointing toward her head, and when talking about "yesterday/…/or last week," gesturing backwards with her arm. Another example is when she combined gestures and speech to evoke a mnemonic strategy, which is commented on in relation to models and wordplay, below. In the following, we specifically comment on the learning activity where students created multimodal texts individually.

The whole process involved a number of "transductions," or processes where content was transferred, or "translated", from one semiotic mode into another, and where each resource was used in accordance to its modal affordance (Kress, 2010). The diagram (left, Fig. 11.1) was used to visualize a phenomenon in the outside world: how the shadows from a gnomon differ according to the sun's height in the sky at different times of the day. By transforming this visualization into a table, the level of abstraction and precision became higher. The next step

Fig. 11.1 Multimodal student text (Naihma)

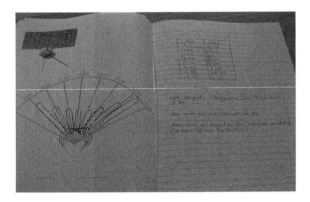

was to make written generalizations about the phenomenon from the table, with science having specific ways of structuring such texts. Given the notion of modal affordance, transductions can be challenging, since the "same" content cannot be expressed as accurately or as thoroughly through any resource. Thus, changes in form as well as in content will always be needed in transductions between modes.

When introducing this activity, the students were told by the teacher that since it was cloudy and consequently not possible to make a diagram of their own, they would use a prefabricated one. She then explicitly connected the diagram to sundials: "they have also drawn a scale with the time (points at time indications in the diagram) do you recognize it from sundials?". When instructing the students to make a table from the diagram, she modelled one on the flipchart: "out of that (starts drawing a table on a flipchart) /.../ you're going to make a table where it says time (writes 'TIME' in left column of table) and length (writes 'length' in right column)". Before the individual activity commenced, the teacher filled in the first figures under "TIME" with the help of the students, thus modelling the transduction process from diagram to table. She commented that if the timeline was not in the middle of the shadow, they would have to adjust the time: "shall we say nine fifteen? /.../ someone not satisfied with that?". She settled for 9.10 after some discussion. Later during the activity, the teacher had similar discussions with individual students. Thus, the first step of the activity was given considerable attention, with a focus on exactness.

The second step involved analyzing and transducting the content in the table into words. Here, the teacher gave short instructions to individual students or pairs (see Written language). During this process, she instructed the students to analyze and write claims from the table, and to use graphic features like bullets instead of "a mass of text".

Ammar and Ali finished the activity relatively quickly, and were asked to compare their analyses: "/.../ and then you discuss your analyses ... what you've arrived at ... okay?/.../ if you agree and if you can see some pattern ... if you can make some general statement". Instead of comparing the conclusions from the table, Ammar and Ali compared the time indications in their tables (see Spoken language), concluding that they had several similarities and a few differences. Given the teacher's previous focus on deciding exact times, the pupils' attention to similarities and differences regarding time indications is perhaps not surprising.

This learning activity continued for almost 40 minutes. There were a number of challenges for the students during the transduction process when creating a table from the diagram: (1) how to make a table, (2) how to decide the exact times, and (3) how to measure the length of the shadows. The teacher supported the students individually in these steps: (1) (to Mira): "count the number of shadows and then you know how many rows you will need ... and then you need /.../ another column for ... eh ... the heading", (2) (to Ali, who asked if he could write 12, 13) "you have to look here because here it might not be exactly thirteen ... here it might be a little bit more", (3) (to Suado): "and you have to remember that you're supposed to measure from the middle of the gnomon".

During the activity, it was obvious that the transduction processes were demanding, and the fact that the shadows in the diagram did not always coincide with exact hours created confusion. The teacher chose to stress exactness, which made the students spend too much time on details as hours and minutes. Accordingly, the proximate purpose of the activity was obscured; to act scientifically by transferring information from a diagram into a table and then to analyze data and draw conclusions about the surrounding world. Throughout the activity, the teacher emphasized norms of how to act in a science class, in this case focusing on exactness, meticulousness, and the forms related to writing.

When the students had finished creating their multimodal texts, a hands-on activity commenced, during which the students were supposed to create shadows similar to those in the prefabricated diagram, using a model of a sundial. This process is described below after the analyses of the use of written and spoken language.

11.4.1 Written Language

When the students had transformed the times and the lengths of the shadows from the prefabricated diagram into tables, the teacher instructed them individually to make a written analysis. Apart from giving hints about what to write in the analysis, the teacher told the students to use bullets to achieve a short and concise text: (to Muna) "have you begun your analysis? /…/ use bullets then", (to Ammar) "write down your thoughts … write them in bullets not in a mass of text … write it in bullets". Furthermore, students were instructed to write statements, draw conclusions, and generalize the result.

The students' writing varied in form and content. As to form, both bullets and running text were used, whereas the degree of detailed reporting, drawing of conclusions and generalizing varied, as shown in the following examples. However, all written analyses had a scientific structure in that they focused entirely on the scientific matter without mentioning any human actors. Likewise, many processes were relational (Halliday & Matthiessen, 2004), such as *was, became, got*.

One of the shortest and most generalized statements in bullets was written by Fatima. She refrained from writing statements on each shadow, and instead went directly to the final generalizations based on the extreme results:

Analysis of table

- When the sun is the highest in the sky the shadow gets the shortest.
- When the sun rises and when it goes down the shadows are the longest.

Ammar's report was written entirely in running text, and comprised a combination of detailed statements and generalizations:

Analysation

At 15.28 the shadow was at its longest, 8,7 cm was the result but at 12.00 the shadow became shortest, 5,0 cm. The results showed that when the sun is higher up in the sky at

12 o'clock the shadow is shortest. But the more the sun moves from the middle of the sky then the shadow becomes longer. The table shows that the sundial's result became the same 9.10 o'clock in the morning and at 15.28. The tests started at 8.00 until 16.00.

Before writing, Ammar apparently reconstructed prior experiences of how to write in science class and asked, "doesn't any analyzation have a precise order?" to which the teacher responded "not this one". After that, when discussing the order of the results with the teacher, he verbally created the nonidiomatic nominalization (Swe.) *analysering* (Eng. *analysation*), based on the verb *analyse* (Swe. *analysera*). This invented nominalization was used as headline in Ammar's report. A distinctive feature strengthening the scientific style in Ammar's report was the way he topicalized time and measure, *At 15.28 the shadow was ..., at 12.00 the shadow became ...* and *8,7 cm was the result ...*. Further adding to the specialized register in Ammar's report was the use of scientific participants instead of human actors, *The results showed ..., The table shows ..., The tests started ....* To express movement, Ammar used the more general and formal (Swe.) *förflyttar sig* (Eng. *move*) instead of the everyday *goes* in the sentence *But the more the sun moves from the middle of the sky*. Ammar used one extended nominal group in his report: *the sundial's result* and after the initial statements of the result, Ammar summarized them using a consequential explanation, *But the more the sun moves from the middle of the sky then the shadow becomes longer*.

A third example was Naihma's analysis (Fig. 11.1), reporting the hours for various shadow lengths without drawing general conclusions:

Analysis.

The longest shadows are 9.10 and 15.10.

The one which is shortest is 12.00.

Those which are about [the same] are (11.00–12.00. 13.00)

(10.00–14.00) (14.30–9.10)

When writing their analyses, the students were given an opportunity to develop their scientific register including an increasing abstractness. The nonspecific instructions by the teacher allowed for several solutions to the task. However, the activity was mainly continuous with the proximate purpose "how to write in science", in this case focusing on form rather than content. Thus, regarding purposes concerning scientific content, the activity was not continuous.

11.4.2 Spoken Language

During the lesson, the teacher used verbal strategies in ways that might have enhanced the students' possibilities to appropriate the specialized register. In doing so, the teacher moved between everyday and scientific registers (e.g., when referring to the gnomon: "stick" vs. "axis"), hereby increasing message abundance and channels for meaning-making (Brown & Spang, 2008; Gibbons,

2006). During these shifts, different strategies were used (Brown & Spang, 2008; Fang & Schleppegrell, 2008; Halliday, 1998/2004).

One strategy was *defining*, for example regarding the abstract term "general" when the students were supposed to analyze information in their tables: (to Emre and Muna) "general ... do you understand that? something that always applies to the shadows", (to Fatima) "general claims they always apply".

Another example was when Fatima asked about the term gnomon: "if I let my pencil stand here somewhere, like this ... then it's a gnomon?". The teacher recast the wordings into a more specialized register, specifying that "if you use it for measuring time by the help of the sun". The teacher's definition stressed the function of a gnomon; namely that a gnomon evaluates time with the help of the sun. Fatima, on the other hand, focused on the representation of a possible gnomon, here a standing pencil.

A second strategy was *exemplifying*. When the teacher had mentioned the points of the compass she exemplified: "you need to know where north is, where south is, where east is, where west is". When instructing Ammar on how to write the analysis from his table she gave examples of what analyses could consist of: "you should analyze the table and see what you can draw on or write statements out of the table". Ammar, who was familiar with the word "statements," used the same strategy when illustrating how to write his analysis: "at a certain time, hm, the shadow got longer". The teacher then gave further examples of statements: "then it was the longest and then it was the shortest it was just about as long at that and that time".

A third strategy was *unpacking* nominalizations and *packing* processes (Fang & Schleppegrell, 2008; Halliday, 1998/2004). In everyday language, processes are typically expressed in verbs (to *measure*), qualities in adjectives (*long* shadow), and things in nouns (a *shadow*). In more specialized language, processes and qualities are often expressed as nouns (*measurement, length*). This process is called nominalization. When "unpacking," *measurement* is expressed as *measure*, while "packing" is the reverse process. During her instruction in whole class, the teacher explained how to understand the depicted shadows in the prefabricated diagram:

> Teacher: and those who have been doing these measurements got this diagram [shows the prefabricated diagram] ... and they have also drawn a scale with hours labeled ... do you recognize from the sundials we have looked at?
>
> Ammar: from eight o'clock to four [looks at the diagram]
>
> Teacher: yes ... they've measured ... we're going to work with this, as we haven't got any of our own so we're going to work from this ... even though it may not be exactly as our measurements would have turned out right now ... but we'll come back to that later

During this dialogue, the teacher unpacked the nominalized process *measurement* into "they've measured." Then she packed the process, *measure*, into the abstract, nominalized process. Through the unpacking of the nominalization, the participants were made visible (*they have measured*). By using *measurement*, the students got a potential entity to compare, analyze, and reason about.

Later on, when Ammar and Ali were asked to discuss their results of the analyses, Ammar used the same process, *measure*, as the teacher and turned it into the nominalized process, *measurement*:

> Ammar: how did you get that result? did you measure it? and what was the result of your measurement?
>
> Ali: that it became [inaudible] like this
>
> Ammar: do you think that your result is reasonable?

Ammar's way of using the process *measure* as well as the nominalization, *measurement*, shows a scientific way of using the specialized language, presumably inspired by the teacher's verbal modeling, using *measurement* as something abstract to theorize about. Accordingly, in their communication, both teacher and students shifted between registers, as processes were unpacked or packed, and abstract terms were explained in everyday words and concrete examples. The shifts between the everyday and the scientific register can also be seen as examples of how continuity was created verbally.

11.4.3 Hands-On Activity: Creating Shadows According to the Diagram

After creating their texts, the students were asked to make shadows according to the prefabricated diagram, using a torch, a piece of white A3-paper and the stick on a holder (gnomon). The proximate purpose was expressed as: "you'll have an A3-paper and then you're going to draw a line around a gnomon ... and then you'll try to form the shadows and think about how you're going to hold the torch to get the shadows about the same way as in the picture". In addition, the teacher asked the students to mark the shadows "within the size of the paper" (Fig. 11.2). One example was Emre and Naihma who performed the task together.

When the teacher asked the students how to hold the torch in order to make the shadows as short as possible, Naihma immediately answered "up highest." Later on Naihma asked Emre to hold the torch "like this," in order to make the shadow as short as possible. Emre agreed, but obviously he was engaged in holding the

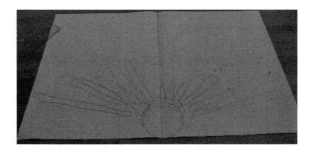

Fig. 11.2 Student diagram (Naihma)

torch in a way that resulted in Naihma hesitating, asking if they should "do like that", which Emre confirmed. Naihma then noticed that she could see "two shadows", which Emre agreed to by saying "we have two shadows". The torch was wider than the stick, and when holding it close to the stick, simultaneously keeping the whole shadow within the size of the paper, a true shadow and a half-shade occurred on the paper. This phenomenon resulted in the students concluding that there were two shadows. They did not question their observation, but proceeded with the activity. At the beginning of the next lesson, Naihma told the teacher that she had trouble holding the torch. Another student agreed, saying that it was "hard to find" and difficult knowing "how to make" the shadows. Accordingly, as the students' earlier experiences of shadows appeared to be limited, marking the shadows did not seem to clarify how shadows arise and change. Instead the students appeared to be occupied with the "doing" of the activity, in this case norms about how to act. Moreover, when recapitulating this lesson during the following lesson, the discussion dealt with the practice and not with what the students had observed.

11.4.4 Models and Wordplay

A model of the sundial was used during the hands-on activity and in the teacher's introduction to the lesson, dealing with "the sun's movement across the sky". During the recapitulation Fatima had asked: "what's the holder then?". As commented on above, the teacher explained that the holder was a base for the stick, resulting in Fatima's question: "if I let my pencil stand somewhere, like this… then it's a gnomon?" and the teacher's confirmation: "if you use it for measuring time by the help of the sun… then it's a gnomon". Accordingly, both of Fatima's questions were justified, clarifying the function of the different parts of the model. This appeared to make meaning for Fatima since she compared the stick to a pen as the vital part for measuring time. However, later on when instructing the students how to draw a contour around the holder on a piece of paper before creating shadows, the teacher explicitly said that "now you've got the basis for a sundial you may say". Fatima appeared to be puzzled by this comment, as the teacher previously had said that the gnomon was represented by the stick and not the holder. She commented: "it's only the holder… it's not the stick then?". Yet another puzzling circumstance occurred for Fatima as she could not relate the gnomon to measuring time as there were "no figures" on her piece of paper. Hence, using models in science class demands a thorough presentation concerning what models represent and how they can be used.

In her introduction to the learning activities, the teacher stressed the importance of knowing at what points of the compass the sun rises and sets. She began by stating that the sun rises at one point of the compass and sets in another. Then she used a mnemonic strategy, building on a wordplay, forming a multimodal ensemble with speech and gestures: "the sun sets (goes down) in the west as the tie goes down in the vest", continuously gesturing to make "vest", "tie" and "goes down" clear (in Swedish, the two words "west" and "vest/waistcoat" are homonyms: Swe. *väst*). The wordplay, "the tie goes down in the vest" appeared not to make meaning to the students. Fatima asked what it had to do with the sun and both Emre and Fatima exclaimed "south!" when the teacher asked the students where "the tie goes down" while simultaneously pointing downwards into an imagined vest. When the teacher repeated the question and the gesture, Fatima once again exclaimed "south".

For students reasoning about the directions of the compass, the downward gesture would be equivalent to "south" if you relate the gesture to south in a map. It is plausible that Fatima's and Emre's earlier experiences of maps were reconstructed and transformed in this situation, thus having consequences for their meaning-making. It was not until Muna exclaimed "vest vest... the sun goes down in the west!" that the discussion could proceed. Consequently, the wordplay was not continuous with learning where the sun rises and sets.

11.5 Discussion

During instruction, the teacher made use of several resources, often in multimodal ensembles, which increased the channels for meaning-making and message abundance. Also, the students were given opportunities to use different resources in a variety of semiotic modes in their meaning-making. Such teaching can promote students' development of scientific literacy. However, an implication of the present study is that teachers might need to develop their awareness of their own use of different resources as well as the ways in which they create opportunities for students to make meaning of the science content through a variety of semiotic modes. Students benefit from getting opportunities to reason about their observations in small groups or whole class, and from receiving instructions about both *how* and *what* to discuss. Furthermore, students would also benefit from discussions about modal affordances and how different resources are related in a given situation, for example, when involved in transductions from one semiotic mode to another. Such discussions can promote their disciplinary literacy, including learning science, competent action in the science classroom, and communicating through different modes.

In our study, the teacher and the students were engaged in meaning-making activities involving a variety of semiotic resources in ways that can enhance the development of multilingual students' scientific literacy. The communication in different registers, or everyday and specialized language, noted in this classroom function as additional resources to support such a development (Gibbons, 2006; Macken-Horarik, 1996). The analyses revealed that the students were given several learning opportunities during the lesson, through talking, writing, and performing hands-on activities. However, the focus was on norms of how to act in science class and what to include or exclude. In this classroom, "minds-on" in relation to the actual content, was shown in some students' written general statements about "the sun's movement across the sky". Continuity between activities and purposes (Jakobson & Axelsson, 2012) was shown in various ways. Examples are the students' creations of multimodal texts and the hands-on activity. Here, the students' undertakings were continuous with the proximate purpose about how to act in science class. When writing texts, this was shown in the expected meticulousness according to norms concerning form rather than content. During the hands-on activity, the norms appeared to concern "doing" rather than learning the science content.

A variety of semiotic modes and resources were used during the lesson, with different affordances, or potentials for meaning-making (Kress, 2010; Lemke, 1998). In short, models and diagrams were used to visualize and concretize abstract phenomena, whereas writing was used for generalizations. Speech was generally used for teacher instructions and explanations of the content, or for student discussions. In addition, the teacher used gestures as parts of multimodal ensembles to elucidate and reinforce her words.

In detail, the models of the sun and gnomon function as concretizations and opportunities to experience the phenomena in the classroom during the hands-on activity. The teacher was explicit that the torch and the stick on a holder were used as models. However, during the hands-on activity, it was obvious that the students encountered problems, and were not able to handle the torch to create shadows in accordance to the diagram showing different shadows at different times of the day. The wordplay vest–west was used as a mnemonic strategy for remembering where the sun sets and it was given as a multimodal ensemble with speech and gestures. Yet, in this case the modes contradicted each other, since the downward gesture indicated south rather than west. On other occasions with combinations of speech and gestures in multimodal ensembles, the modes boosted each other.

The students created a multimodal text involving a number of transductions, which is potentially challenging due to the different affordances of modes (Kress, 2010): from diagram into a table, and then into a written analysis which some of the students also discussed in pairs. Our analyses revealed that these transductions created problems for some of the students.

Some instances were noted where the participants brought to the fore some characteristics of what resources to use and how in science class. One example was when the teacher made explicit that the torch and stick were models. During other lessons in the unit, she also referred to linguistic specificities of the scientific register. A manifest initiation of a discussion on language use was made by a student, when Ammar, who was familiar with written genres, asked about the structure of the written analysis. Moreover, the teacher frequently used modelling during this lesson. One such example was when modelling how to make a table. Another example was when packing and unpacking nominalizations (e.g., *measure* > *measurement* > *measure*).

To conclude, using several resources increases the channels for message abundance and student meaning-making. However, teachers have to carefully consider the combination of multimodal ensembles as well as students' earlier experiences.

References

Brown, B. A., & Spang, E. (2008). Double talk: Synthesizing everyday and science language in the classroom. *Science Education, 92*(4), 708–732.

Danielsson, K. (2016). Modes and meaning in the classroom. *Linguistics and Education 35*, 88–99.

Danielsson, K., & Selander, S. (2016). Reading multimodal texts for learning. A model for cultivating multimodal literacy. *Designs for Learning, 8*(1), 25–36.

Dewey, J. (1938/1997). *Experience & education*. New York: Touchstone.

Fang, Z., & Schleppegrell, M. (2008). *Reading in secondary content areas. A language-based pedagogy*. Ann Arbor: The University of Michigan Press.

Gibbons, P. (2006). *Bridging discourses in the ESL classroom*. London: Continuum International Publishing Group.

Gibson, J. (1977). The theory of affordances. In R. Shaw & J. Bransford (Eds.), *Perceiving, acting and knowing: Toward an ecological psychology* (pp. 67–82). Hillsdale: Lawrence Erlbaum.

Halliday, M. A. K. (1998/2004). Things and relations: Regrammaticizing experience as technical knowledge. In J. Webster (Ed.), *The language of science* (Collected works of M.A.K. Halliday) (Vol. 5, pp. 49–101). London: Continuum.

Halliday, M. A. K., & Martin, J. R. (1993). *Writing science: Literacy and discursive power*. London: University of Pittsburgh Press.

Halliday, M. A. K., & Matthiessen, C. M. I. M. (2004). *An introduction to functional grammar*. London: Arnold.

Jakobson, B., & Axelsson, M. (2012). 'Beating about the bush' on the how and why in elementary school science. *Education Inquiry, 3*, 495–511.

Jewitt, C. (Ed.). (2017). *The Routledge handbook of multimodal analysis*. London: Routledge.

Johansson, A-M., & Wickman, P. O. (2011). A pragmatist understanding of learning progression. In B. Hudson & A. M. Meinert (Eds.), *Beyond fragmentation: Didactics, learning and teaching in Europe* (pp. 47–59). Leverkusen Opladen: Barbara Budrich.

Kress, G. (2010). *Multimodality. A social semiotic approach to contemporary communication*. London: Routledge.

Kress, G., Jewitt, C., Ogborn, J., & Tsatsarelis, C. (2001). *Multimodal teaching and learning: The rhetorics of the science classroom*. London: Continuum.

Lave, J. (1996). The practice of learning. In S. Chaiklin, & J. Lave (Eds.), *Understanding practice. Perspectives on activity and context* (pp. 3–32). Cambridge: Cambridge University Press.

Lemke, J. (1998). Multimedia literacy demands of the scientific curriculum. *Linguistics and Education, 10*(3), 247–271.

Macken-Horarik, M. (1996). Literacy and learning across the curriculum: Towards a model of register for secondary school teachers. In R. Hasan, & G. Williams (Eds.), *Literacy in society* (pp. 232–278). Harlow: Addison Wesley Longman.

NTA. (2005). *Mäta tid* [Measuring time]. Stockholm: Kungl. Vetenskapsakademien, Kungl. Ingenjörsvetenskapsakademien.

Swedish Research Council. (2016). *Humanistisk och samhällsvetenskaplig forskning*. Retrieved from http://www.codex.vr.se/forskninghumsam.shtml

Wickman, P. O. (2006). *Aesthetic experience in science education: Learning and meaning-making as situated talk and action*. Mahwah: Lawrence Erlbaum Associates.

Zhang, Y. (2016). Multimodal teacher input and science learning in a middle school sheltered classroom. *Journal of Research in Science Teaching, 53*(1), 7–30.

Chapter 12
Meaning-Making in a Secondary Science Classroom: A Systemic Functional Multimodal Discourse Analysis

Qiuping He and Gail Forey

Abstract The purpose of this chapter is to present a framework for examining meaning-making in the science classroom through a range of resources. Based on the notion of social semiotics from systemic functional linguistics, we propose a framework that examines the affordances of meaning in one mode, such as language, or gesture, or animation, and the multiplying of meaning across these modes. We argue that knowing what meaning can be afforded by a mode and the ways to communicate meaning across modes can enhance learning opportunities in the science classroom. We focus on three modes used in the science classroom, namely language, gesture, and animation, and propose a framework that helps unpack the meanings made. We draw on data collected from a 66-minute video recording of a Grade 9 class studying the process of digestion. We investigate the organization of meaning in the identified modes and the multiplying of meaning across modes in constructing explanations. While gestures and animation are found to make meaning through the logics of time and space, language plays a significant role in mediating the technicality of scientific knowledge. The findings also identified two ways of multiplying meaning across modes, namely, creating multimodal links and reiterating organizing structures. The complex mediation of meaning within each mode and across modes highlights the need for explicit instruction by the teacher to support and highlight how meaning is made in science and other teaching and learning contexts. We suggest that the findings are relevant for apprenticing learners into the world of science, and also apprenticing scientists into the world of teaching.

Keywords Systemic functional linguistics · multimodal discourse analysis · semiotic affordances · constructing explanations

---

Q. He (✉) · G. Forey
Department of English, The Hong Kong Polytechnic University, Hung Hom, Hong Kong
e-mail: ares.he@connect.polyu.hk

G. Forey
e-mail: gail.forey@polyu.edu.hk

© Springer International Publishing AG 2018
K.-S. Tang, K. Danielsson (eds.), *Global Developments in Literacy Research for Science Education*, https://doi.org/10.1007/978-3-319-69197-8_12

12.1 Introduction

In apprenticing learners into the world of science, teachers often use a range of resources[1] to make meaning in the classroom. In a science classroom, the resources such as a gaze, a verbal comment, an action, or a touch can carry certain meanings. From a social semiotic perspective, each choice of meaning-making is socio-culturally shaped and is dependent on the conventions of a community (Halliday & Matthiessen, 2014). The notion of modes or modalities refers to the resources to represent meaning, as opposed to the resources to disseminate meaning (Kress, 2010). Multimodal research studies the meaning-making beyond language and incorporates the meaning represented by other modes, such as diagrams, gestures, postures, and positions. Different modes afford different kinds of meaning, that is, they have different semiotic affordances. The notion of semiotic affordances examines the possibilities and constraints of the potential meaning that one mode can make (Kress, Jewitt, Ogborn, & Tsatsarelis, 2001). The study of semiotic affordances of different modes helps reveal the efficiencies and deficiencies in a specific mode. For instance, the mode of speech[2] unfolds over time and thus has rich potential for construing temporal relations. This contrasts with the mode of visual image, the elements of which are displayed simultaneously through spatial relations and thus efficiently portray spatial relations (Kress, 2010). Meaning can be multiplied through an integration of modes, where the meaning made through such integration is greater than the sum of the meaning in each mode, which is coined by Lemke (1998) as multiplicity or multiplying of meaning. Multiplying of meaning may occur when a science teacher uses a gesture to visualize the movement of muscles accompanying the verbal explanation that "the muscle squeezes." In this case, both gesture and speech contribute to the understanding of muscle contraction by relating the abstract process to the observable hand gesture and unpacking the technical term into commonsense wording. While research on science education has pointed to the need of analyzing the meaning-making from a multimodal perspective (Lemke, 1998; Norris & Phillips, 2003), little research illustrates the relation between the meaning afforded in each mode and the multiplying of meaning across modes. The purpose of this chapter is to present a framework that examines both the semiotic affordances of individual mode and the multiplying of meaning across modes.

The focus of this chapter is on the teacher, and the meanings made by the teacher. As pointed out by Hattie (2003), the teacher is a significant factor leading to student achievement. During the study, the science teachers reported that their understanding of disciplinary literacy had improved dramatically, and that this

---

[1]The notion of resource in this chapter is general, which covers mode, the resource to represent meaning, and medium, the resource to disseminate meaning (see Kress, 2010, for a detailed discussion on modes and medium).

[2]Speech and writing are considered as two different modes, although both speech and writing belong to linguistic resources.

new found knowledge had helped them to improve their teaching, and had an impact on their students. However, due to time and space, the impact on the learner is beyond the scope of this chapter (see Humphrey, 2017; Polias, 2016; Rose & Martin, 2012 for a discussion of the impact on the learner). We begin with a brief review of research on scientific literacy before presenting the framework for multimodal classroom interactions, which is illustrated through a case study. Finally, the chapter concludes with the implications for future research on multimodal classroom interactions.

12.2 Multimodal Research in Science Education

Although the term "scientific literacy" is frequently mentioned in research and education domains, the precise meaning of this term remains contested among scholars (Osborne, 2002). In this chapter, we adopt Norris and Phillips' (2003) definition of *scientific literacy* which comprises the fundamental sense of scientific language and the derived sense of science content knowledge. As advocated by Halliday and Martin (1993, p. 212), "the evolution of science was, we would maintain, the evolution of scientific grammar." Lemke's (1990) analysis of science classroom dialogue demonstrates that science learning is the learning of specialized scientific language through which scientists perform their social activities. Halliday and Martin (1993) further investigated the nature of scientific language and shed light on its unique features, such as technical taxonomies, abstraction, and nominalization.[3] In addition to language, science discourse has been noted for construing meaning via multiple modes such as images, diagrams, charts, and symbols (Kress et al., 2001; Lemke, 1998, 2002).

Kress et al. (2001) described the complex meaning-making in science classrooms through the teachers' adoption of multiple representative resources. Lemke, in his study of science journal articles (Lemke, 1998) and multimedia resources in science education (Lemke, 2002), concludes that science communication is "close and constant integration and cross-textualization among semiotic modalities" (Lemke, 1998, p. 27). Findings from recent studies on the modes of written and spoken language (Tang, Delgado, & Moje, 2014), images (Waldrip, Prain, & Carolan, 2010), gestures (Jaipal, 2010), apparatus, mathematics, and activities (Airey & Linder, 2009) have urged for an awareness of the relationship between different modes that support successful learning.

While research on science education has pointed to the importance of analyzing the meaning-making of different modes (Ainsworth, 2006), limited research incorporates the examination of meaning potential in one mode (i.e., semiotic affordances) and the tracking of meaning integration across modes (i.e., multiplicity). In this chapter, we present a framework that examines semiotic affordances and

---

[3]Nominalization is the grammatical choice of taking a verb (v.) and packaging it as a noun (n.), for example digest (v.) and digestion (n.).

multiplying of meaning in classroom interactions. This is achieved by examining how a specific concept is constructed in different modes and tracking how the meanings are integrated across modes. The theoretical background and the framework are presented in the following section.

12.3 Developing a Framework for Multimodal Classroom Interactions

The analysis of choices used to make meaning in the science classroom in this chapter is informed by Halliday's (1978) social semiotic perspective and Systemic Functional Linguistics (thereafter SFL). In systemic-functional theory, semiotics refers to the study of sign systems embedded in social situations and contexts (Halliday, 1978). A sign system, such as language, is social in that it operates in social cultural settings, and is semiotic in that it makes meaning within these social contexts. An SFL approach to language is based on three organizing principles: stratification, instantiation, and metafunction (Halliday, 1985; Halliday & Matthiessen, 2014). From the perspective of stratification, language system consists of the strata of semantics, lexicogrammar, phonology, and phonetics. From the perspective of instantiation, a text can be regarded as a particular instance of a semiotic system. From the perspective of metafunction, meaning can be understood if we deconstruct it based around three social functions, three metafunctions:

1. the ideas represented — ideational;
2. the relationship developed — interpersonal; and
3. the organization of the message — textual (Halliday & Matthiessen, 2014).

The ideational metafunction concerns how language construes experience of the external world and the internal mental world as well as the logical relations. The interpersonal metafunction enacts social roles and relationships. The textual metafunction organizes the ideational and interpersonal meanings into a meaningful message in the context (see Halliday & Matthiessen, 2014, for a detailed discussion of metafunctions).

Although the primary focus of SFL is on language, these organizing principles have been extended and adapted to other modes. The approaches to multimodal resources informed by SFL can be summarized into two general trends (Jewitt, 2014). One pioneered by Kress and van Leeuwen (2006) is the social semiotic approach, which emphasizes the context of communication and the ideology found within signs. The other is systemic functional multimodal discourse analysis (SF-MDA), approaching one particular "discourse" at a microlevel, with an emphasis on metafunctional systems and systematic choices (Baldry & Thibault, 2006; O'Halloran, 1998; O'Tool, 2010). Despite some subtle differences, such as different levels of delicacy in terms of rank and stratification, these approaches are compatible (see Jewitt, 2014, for detailed discussions). A common agreement in all multimodal analysis is that as different semiotic systems are employed to make meaning, then we need a different metalanguage to describe, discuss, analyze, and investigate these systems.

Lemke (1998) extends the principle of metafunction from language to multimodal texts and proposes a typology to explain the interaction between images and written texts. He argues that the three metafunctions are construed simultaneously during the meaning-making process where the modalities of visual images and written texts "are deployed, singly or jointly" (Lemke, 1998, p. 91). The correlation between Lemke's terminology and Halliday's terminology is illustrated in Table 12.1.

In this chapter, we extend Lemke's framework to incorporate a range of resources available to the science teacher in the classroom. Figure 12.1 presents the framework for this study using the resources of language, gesture, and animation as an example. The presentational meaning is defined as the conceptual aspects of teaching and learning science, for example, the concept of digestion and the elements, processes involved as outlined in Fig. 12.1; the orientational aspect considers the authority building of science and the teacher–students relations, for example, how the teacher realizes the meaning and develops the relationship with the learners, through, for example, relating the meaning to the learners'

Table 12.1 Halliday's metafunction and Lemke's typology

Halliday's theory of language	Ideational metafunction	Interpersonal metafunction	Textual metafunction
Lemke's typology of printed documents	Presentational function	Orientational function	Organizational function

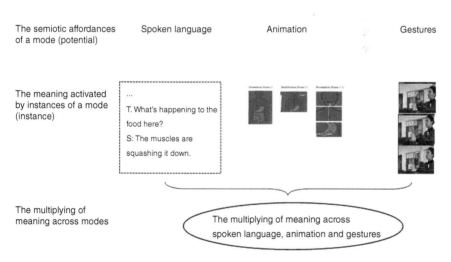

Fig. 12.1 The framework for multimodal classroom interactions

backgrounds and experiences; and the organizational function refers to the structure and the sequence of different resources in teaching a topic or a concept, for example, the sequence of activities and information that unfold throughout the lesson. As shown in Fig. 12.1, the teacher asks the students a question after showing an animation and the ideational concept represented in the animation, the contraction of the esophagus is complemented by the teacher's hand gesture of squeezing. The semiotic affordances of one mode consist of the potential meaning that can be made in presentational, orientational, and organizational aspects. The choices made by the teacher in her use of spoken language, activate certain presentational, orientational, and organization meanings from the potential pool of language system. Similar activation happens to the choices found in animation and gestures, where particular meanings are activated by instances of a mode. These activated meanings from different modes can be multiplied to construct a meaning that is greater than the sum of the meaning in one mode. The multiplying of meaning across modes incorporates the multiplicity of presentational, orientational, and organizational functions of these modes. Figure 12.1 illustrates how three different modes, what the teacher says, the animation being played, and the gestures made by the teacher co-occur at one point in a lesson to reinforce and multiply the meaning.

This framework enables the mapping of the potential meaning, that is, the semiotic affordances of an individual mode that co-occur at the same time as other affordances from different modes and together they reinforce and multiply meaning across modes. The meaning afforded by one mode (i.e., semiotic affordances) is instantiated through particular cases (i.e., instances) to achieve a particular communicative purpose. Therefore, identifying the instances of one mode (i.e., instances) provides access to the overall meaning that one mode can afford (i.e., semiotic affordances). The instance, the instantiation of verbal language *the muscles are squashing it down*, is simultaneously mirrored by the instantiation of animation, the instantiation of the hand gesture. The meanings made through the collection of these modes are inter-dependent, and contribute to the understanding of digestion. When multiple modes are deployed for a particular communicative goal, the organizational, orientational, and presentational meanings in each mode are integrated and multiplied, and the integrated meaning is greater than the sum of the meaning of one individual mode on its own (i.e., multiplying of meaning).

The framework is valuable in (a) examining both semiotic affordances and multiplicity of meaning in science communication, and (b) extending discourse analysis to incorporate a multimodal perspective. In the next section, this framework is demonstrated through a case study. Due to the limited space, we report findings from the organizational function with reference to presentational function when appropriate. The range of affordances available in any classroom varies and a study of all affordances available in the science classroom is beyond the scope of this chapter. In this chapter, we limit our analysis to focus on the modes of spoken language, gestures, and animation in terms of the semiotic affordances of organizational meaning in each mode and the multiplying of organizational meaning across three modes.

12.4 Methods

12.4.1 Data Collection and Selection

The data used in this study is from an in-service professional development (PD) project conducted in Melbourne, Australia in 2012. One of the aims of this PD project was to support the teachers' disciplinary literacy development in ways of representing and communicating science knowledge using language and other modes (Forey & Polias, 2017). The project included a three-day workshop led by a Professional Development Consultant (PDC), and one day of lesson co-planning with the PDC, after which teachers receive further support through email communication which included feedback on lesson plans and responses to any questions raised and a classroom observation. In this chapter, we focus on the data from one classroom observation, which includes the video recordings of the lesson (66 minutes), the immediate feedback by PDC members (20 minutes), and the reflections of the teacher being observed (16 minutes). This extract from the classroom is selected for the richness found in both the meaning-making through multimodal resources and the teacher's reflection. The lesson was taught in English by Ms. Grace (a pseudonym), an experienced science teacher, who previously worked in the pharmaceutical industry in Egypt before immigrating to Australia. There were 23, Year 9 students (14–15 years old) in her classroom, among whom 17 used English as a Second/Additional Language (ESL/EAL). The aim of this lesson was to understand the function of digestive system by constructing explanations for the processes in each organ (i.e., mouth, oesophagus, stomach, small intestine, and large intestine).

Informed consents to video-tape the lesson and to use the data for research purposes were obtained from the teacher, the students, and their parents in advance. To the extent possible, personal information of the participants was anonymized and only those directly involved in the research could access to it. The lesson was then video recorded mainly focusing on the teacher, and the semiotic affordances mobilized by the teacher. The video recording of the lesson formed the primary data source. Additional data sources were observation notes, informal discussions with other teachers from the school and a discussion with the teacher after the lesson. The researcher (one of the authors of this chapter) was the principal investigator for the PD research project and a participant observer throughout the data collection.

12.4.2 Data Analysis

The video data was reviewed repeatedly to search for representative modes for analysis. Although multiple modes were used in the classroom, language (both verbal and written), hand gesture, and animation were used most frequently by the teacher in communicating scientific knowledge and were selected for further analysis. The data were transcribed and analyzed in three steps. First, the video data was segmented into five consecutive teaching units to reveal the structure of the lesson: digestion in the

Table 12.2 Teaching units based on concepts and types of modes

Concepts	Modes
Digestion in mouth	(Fragments of the video) 1. Writing
Digestion in the oesophagus	1. Animation/visual images 2. Speech 3. Hand gestures 4. Writing
Digestion in the stomach	1. Animation/visual images 2. Speech 3. Writing
Digestion in the small intestine	1. Animation/visual images 2. Speech 3. Writing
Digestion in the large intestine	1. Animation/visual images 2. Speech 3. Writing

organs of mouth, oesophagus, stomach, small intestine, and large intestine, respectively. The modes contributing to the scientific knowledge building were identified and counted accordingly to identify segments for detailed analysis (see Table 12.2). The unit of digestion in the oesophagus was selected because this unit contained examples of all modes of interest (i.e., language, hand gesture, and animation).

The second step of analysis focused on the structuring of meaning in one mode to highlight the semiotic affordances of organizational meaning in each mode. The talk between the teacher and students was transcribed based on the unit of an exchange. The visual mode of animation was transcribed as a sequence of frames, each frame being captured in one second. The visual mode of gestures was transcribed at the unit of a stroke (McNeill, 2005). The unfolding of organizational meaning in each mode is related to generic structures: the unfolding stages and phases in "a staged, goal-oriented social process" (Martin & Rose, 2008, p. 5). The stages are orientation (to orient the taxonomy), identification (to identify an element in the taxonomy), processes presentation (to demonstrate processes), and explanations construction (to form explanations). The stages of processes presentation and explanations construction potentially have several composing sequences (see Section 12.5 for details). The organizational function of animation is further examined by conducting multimodal analysis of salience (Kress & van Leeuwen, 2006). Salience refers to the prominence of the elements in the frame. The salience of each element can be decided depending on whether it is foregrounded or backgrounded, its relative size, its color, and shape, etc. The hand gestures were categorized based on McNeill's (2005) classification of hand gestures into iconic, metaphoric, deictic, and beat gestures. Both iconic gestures and metaphoric gestures are gestures with pictorial content: iconic gestures represent concrete objects or events whereas metaphoric gestures represent abstract ideas or categories. Deictic gestures are pointing actions and beat gestures are rhythmic actions. This chapter focuses on examining the organizational meaning in the teacher's metaphoric gestures that are used to represent the digestive process in the oesophagus.

The third step of analysis examined the multimodal structuring of the discourse to identify the multiplying of meaning across modes. The teaching sequence was mapped with the sequencing of modes to highlight the cross-modal links in the lesson. The following section presents findings and discussions.

12.5 Findings and Discussions

12.5.1 Generic Stages and Phases in Animation, Gesture, and Language

12.5.1.1 Animation

The animation sequence of digestion is from a short video produced for educational purposes. The sequence about the digestion in the oesophagus consists of five frames (see Fig. 12.2).

Frame 1 informs the constituents of digestive system, orienting the taxonomy in digestive system. Frame 2 identifies a particular organ (i.e., oesophagus) to be presented by highlighting the organ in orange. The following frames zoom into the organs of mouth, oesophagus, and stomach sequentially. These frames present the digestive process in the oesophagus: while Frame 3 and Frame 5 show the initial state (mouth in Frame 3) and the final state (stomach in Frame 5), Frame 4 present the digestion in the

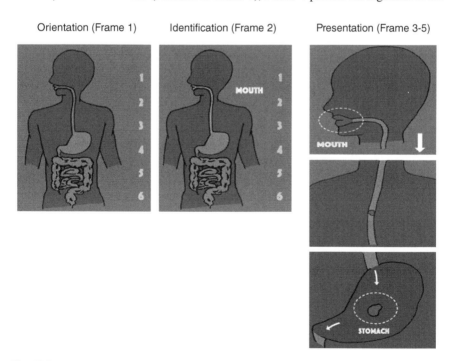

Fig. 12.2 The frames in the animation sequence (recreated based on the original animation)

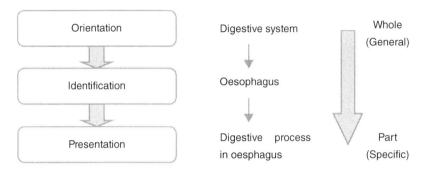

Fig. 12.3 The generic stages of the animation sequence

oesophagus. The interpretation of Frame 3, Frame 4, and Frame 5 presenting the process rather than presenting the organs is based on the salient elements found in these frames. The salient elements in these frames are organs comprising digestive system and the food being digested. Prominence is achieved through contrastive colors and the contrast of a moving object versus a static one. While the use of contrastive colors can be identified in all frames, the contrast of the moving and static object is only found in Frame 4. In Frame 4, the food is represented as a small grey mass that moves downwards in the oesophagus, which is shown as a static tube in light blue. Both food and oesophagus are highlighted by using colors that are different from the background of human body. The food is further accentuated through the movement, which makes the static oesophagus the background. Kress et al. (2001) point out that there is a constant foreground and background of modes in science teaching. This is reiterated in our findings, where even within one mode (in this case, animation), there are shifts of foreground and background of elements in the information flow.

As shown in Fig. 12.3, three generic stages are identified from the succession of these five frames: orientation (showing taxonomy of digestive system), identification (identifying the underlying organ of oesophagus), and presentation (presenting the underlying process of digestion in the oesophagus). The animation sequence unfolds from the digestive system, the general system, to one component of the system and then to the specific digestive process in this organ, suggesting a focalization on presenting the process.

12.5.1.2 Gesture

While gesture can refer to any willful bodily movement, this chapter focuses on the teacher's metaphoric gestures that are used to represent the digestive process in the oesophagus. Metaphoric gestures are gestures with pictorial content used to visualize and construe abstract entities (concepts) or processes (actions) (McNeill, 2005). The metaphoric gestures identified in the data are shown in Fig. 12.4.

The metaphoric gestures are used to visualize two processes: the contraction of muscle and the process of food moving downwards through oesophagus.

Fig. 12.4 The metaphoric gestures representing the digestion in the oesophagus

Ms. Grace opened her hands in a natural position and gently squeezed both hands to imitate the contraction of muscles in the oesophagus. Then she clenched her right hand and moved her left hand downwards to represent the movement of food in the oesophagus. Finally, she combined the first and the second gesture, squeezing her right hand, clenching her left hand and moving downwards to repeat the processes. In this sequence of gestures, the logic of space and time are combined. While the spatial position of both hands shows the spatial relation of stomach and oesophagus, the sequential movements of hands and arms represent the sequence of two processes (i.e., muscle contraction and food movement). The generic stages of the presentation of gesture can be summarized as Presentation of composing processes ^ presentation of overall process.

The first gesture (imitation of the contraction of muscles) and the second gesture (visualization of food movement in the oesophagus) are presented sequentially to represent two processes which constitute another process visualized by the third gesture (visualization of food movement with the contraction of muscles). Such organization highlights the relation between the digestive organ and the digestive process. The link between processes (i.e., muscle contraction and food movement) is created through the sequential display of gestures. While the animation sequence include the orientation of digestive system and the identification of oesophagus before presenting the digestive process, the metaphoric gestures represent the digestive process. This set of metaphoric gestures represents the movement of food (Gesture 1) due to the contraction of muscles (Gesture 2) while only the food movement is presented in the animation (Frame 4). The findings suggest that the modes of gesture and animation combine the logic of time and space (Kress, 2010).

12.5.1.3 Language

The language used in the classroom mainly serves two purposes: either to regulate the classroom or to engage students in the disciplinary knowledge (see Christie & Martin, 2005). This chapter focuses on the scientific knowledge building, specifically on the construction of an explanation of the digestion in the oesophagus. The excerpts from transcription contributing to the explanations of the digestion in the oesophagus are shown below. The teaching sequence is mapped to the interaction between the teacher (shown as T) and one single student (shown as S) or the whole class (shown as Ss).

Seq	Spk.	Extracts from transcription
1	T	We need to look at that part, the tube-like thing in the picture, which we call oesophagus.
2	T	So that's the second part of the digestive system. (T regulates the class) Now, the second part or the second organ of the digestive system we call … is the oesophagus, so we can see if s like a pipe, it's like a tube, OK?
3	T	And, what is this tube doing? What's happening to the food here?
	S	The muscles are squashing it down.
4	T	Good, so the muscles are squashing it down. So, it's basically in the … oesophagus, the food is ….What's happening to the food if we start with the same statement like similar to what we did under for the mouth? (T regulate the class) Now we're going to frame a statement similar to what we did for mouth. So, what's happening to the food in the oesophagus? Food is …?
	S1	… going down …
	S2	… being processed …
5	T	[Writes "Food is…"] So it's being pushed down or …
	S	… processed …
6	T	And how is it pushed down
	S	By the muscles.
7	T	Good, so as the muscles are squashing or squeezing, can we have another word for that?
	S	(…)
8	T	It's just like the lollies that you eat — you push them and … those pipe things that you people eat sometimes. Remember?
	S	Yes
	T	So you push them and you get the lolly. So the same thing is happening in the oesophagus as well.
	S	Oh, sugar straws!
	T	Yeah. So the same … the same thing is happening over here as well. (T regulates the class). So what's happening in the oesophagus, these muscles are …?
	S	Pushing
9	T	Squeeze … or what could be the other word? What happens when we are squeezing something? Can anyone think of another word?
	S	The space gets smaller.
	T	Con…?
	S	Contracts.
10	T	Contracts, good. So the muscles, as the muscles contract, the food is pushed down through the oesophagus. So in the oesophagus, what's happening? The food is pushed down. Where is it pushed down?
	S	in the stomach
	T	in the stomach. And how as the muscles contract? So if we are looking at … if we write down "as the muscles contract." Can we look at some process … what would the process be? Just like digest, it's digestion; contract … what would the process be?

(continued)

(continued)

Seq	Spk.	Extracts from transcription
	Ss	Contraction.
	T	Contraction. Contraction of what?
	Ss	(...)
11	T	Contraction of...?
	S	The muscles
12	T	Contraction of muscles. Good. Food is pushed down in the stomach by the contraction of muscles. [Writes "Food is pushed down in the stomach by the contraction of muscles."]

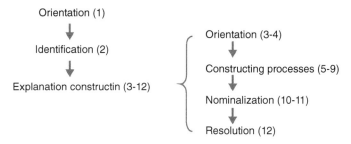

Fig. 12.5 The generic stages of verbal language and teaching sequence

Following Martin and Rose's (2008) stages and phases in reports and explanations, we identified the generic stages in teacher–students verbal interactions: orientation ^ identification ^ explanation construction, as shown in Fig. 12.5. The teaching sequence is shown in parenthesis.

The first stage of orientation introduces the name of oesophagus and relates oesophagus to an everyday item of tube (*we need to look at that part, the tube-like thing in the picture, which we call oesophagus*). In the second stage, oesophagus is identified in digestive system through verbal statement (*the second part or the second organ of the digestive system*). The shift to the third stage is provided by an interrogative question (*what is this tube doing?*). This signals a shift from the taxonomy of digestive system to the explanation of the digestive process in the oesophagus. The use of metaphor here (*this tube*), instead of the technical term (oesophagus), enables the control of new information in manageable chunks and reinforces the shape of oesophagus. Compared with the first two stages, the third stage is more complex in structuring, comprising four phases: orientation, constructing processes, nominalization, and resolution. The first phase of orientation starts with a WH-interrogative question (*what's happening to the food if we start with the same statement like similar to what we did under for the mouth?*), followed by a declarative clause (*now we're going to frame a statement similar to what we did for mouth*). The processes to be constructed in the following phase are food movement and muscle contraction. The statement for food movement is provided by one student (*Food is going down*) to answer the question (*What's happening to the food here?*). While several trials are made by the teacher and students, only the fourth one is successful. These trials use and recycle synonymies (*pushed down, going down, is being processed*), synonyms (*squashing, squeezing*) and verbal metaphors (*lollies, pipe things*) to relate digestion to everyday examples (*you push them and you get lollies, so the same thing is happening in the oesophagus*) as well as using an incomplete utterance to prompt (*con …*). The answers given by students in each trial are (i) *Oh, sugar straws!*, (ii) *pushing*, (iii) *the space gets smaller*, and (iv) *contract*. This suggests that the learners are gradually engaged in the science domain. The difficulty in constructing the process of contracting can be attributed to the abstract nature of this process and the dearth of presentation in the animation sequence, where only the process of food movement is presented. The construction of this process involves several shunts along a register continuum from one end representing everyday concrete knowledge to the other end of abstract scientific knowledge. By moving along the continuum of everyday knowledge and scientific knowledge, students are apprenticed into the world of science (Forey & Polias, 2017; Kress et al., 2001; Polias, 2016).

In the next phase, the process *contract* is changed into a noun *contraction*. The conversion of a verb, a doing process, to a noun, a thing, is referred as nominalization. This is achieved by students deducing the analogy "contract-contraction" from the example given by the teacher "digest-digestion." The nominalized form *contraction* was then extended to *contraction of muscles* by adding the actor of this process (*muscles*). This is achieved by a WH-question (*Contraction of what?*) by the teacher to suggest the linguistic form. In the last phase of resolution, the scientific explanation is constructed: *food is pushed down in the stomach by the contraction of muscles*. In the last two phases, the features of scientific language are highlighted, such as abstraction and nominalization (Halliday & Martin, 1993). The generic structures of language, animation, and gesture are summarized in Fig. 12.6.

The stage of explanation construction in language involves both process presentation and process construction, which are often intertwined with each other. Similar generic stages are identified in both animation and language, namely, taxonomy orientation, oesophagus identification, and processes presentation, where processes presentation is included in the stage of processes construction in language. For both animation and language, an overview of the digestive system is presented before identifying the oesophagus as the element to be studied, which is followed by presenting the processes through visualization (in animation) or linguistic resources. The processes presentation involves two related processes: food movement and muscle contraction. While only food movement is presented in the animation, both food movement and muscle contraction are presented sequentially with hand gestures. To some extent, the gestures provide more information about the movement of food compared with the animation. While the animation only presents the movement itself, the gesture reveals that such movement is sequentially linked to the contraction of muscles. With the help of language, the relation

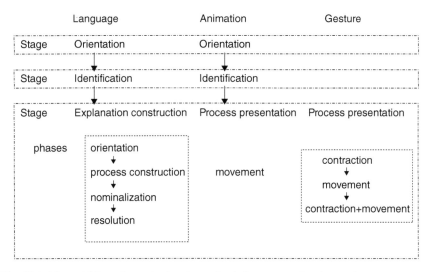

Fig. 12.6 The unfolding of organizational meaning in language, animation, and gesture

between food movement and muscle contraction is able to move between temporal (as what is shown in animation, gestures and the third statement) and causal (as the contraction of muscles becomes the trigger for the movement of food to happen in the fourth statement). The findings highlight different semiotic affordances of language, gestures, and animation. Gestures and animation are found to make meaning by combining the logics of time and space (Kress, 2010). The findings also highlight the affordances of language to unpack and repack abstract technical knowledge, which involves constant shunts between everyday knowledge and science knowledge (Halliday & Martin, 1993; Lemke, 1998).

12.6 Multiplying Meaning: Mapping Between Teaching Sequence and Multiple Modes

In this section, the multiplying of meaning across modes is examined to identify typical patterns in the discourse of science education. The teacher oriented the topic by projecting an animation sequence of the digestion in the oesophagus onto the screen. This was followed by the identification of oesophagus in digestive system, using verbal language and a pointing gesture. The teacher and students then co-constructed explanations for the digestion in the oesophagus, where verbal language and metaphoric gestures were co-deployed to unpack and repack the scientific knowledge.

Three themes are introduced sequentially in the animation: (i) the taxonomy of digestive system, (ii) the identification of oesophagus in this system, and (iii) the process of digestion in the oesophagus. These themes are recycled during the lesson through verbal language and hand gestures. For instance, the taxonomy of

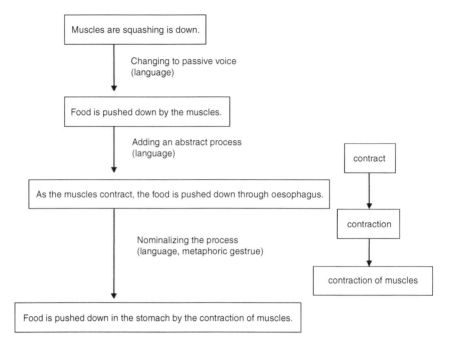

Fig. 12.7 The development of technicality in explanations

digestive system is reinforced by the teacher's gesture pointing to the system shown in Frame 1 in animation (see Fig. 12.2). Similarly, the identification of oesophagus is reinforced with a deictic gesture, pointing to the oesophagus shown in Frame 2 of the animation (see Fig. 12.2) and the verbal statement of it as *the second organ of the digestive system*. It can be argued that the recycling of these themes prepares the students to construct this process in an explanation.

The aim of the lesson is to construct an explanation of digestion. Four explanations were constructed to account for the digestion in the oesophagus, with collective effort from the teacher and students (see Fig. 12.7).

As shown in Fig. 12.7, the technicality accumulates in these explanations and achieves what appears to be more scientific language in the statement *food is pushed down in the stomach by the contraction of muscles*. While language is used to aid the accumulation by changing the voice and adding an abstract process, both language and metaphoric gestures are used to nominalize the process. Nominalization enables more abstract and complex relations (e.g., causal, classification) to be constructed between processes. The accumulation of technicality through nominalization is achieved in two steps. In the first step, *contract* is nominalized as *contraction* through the use of analogy and metaphoric gestures (see related discussions in the previous section). In the second step, *contraction* is extended to a nominal group *contraction of muscles*, by using metaphoric gestures and interrogative question (*contraction of what?*). The meaning-making of language and gestures are mutually supportive and create multimodal links to the scientific knowledge. As several studies (Airey & Linder, 2009; Jaipal, 2010) have shown, the co-deployment of language, gesture, and other visual presentations can elucidate how scientific knowledge is gradually and progressively recontextualized in a sequence of episodes. This suggests the importance of using multimodal representations to create multimodal links in the classroom (Tang et al., 2014; Waldrip et al., 2010).

The multiplying of meaning across modes can also be identified through the relationship between animation and language. Two types of participants are introduced in the animation: (1) digestive organs — mouth, stomach, and oesophagus and (2) the food being digested. Digestive organs can be considered as either the participant involved in the process of digestion or the circumstance where this process happens. In the animation, digestive organs are shown as static objects, while the food is shown as a moving object. Such contrast suggests that digestive organs are more conceptual and backgrounded as circumstances while food is more material and foregrounded as a participant of the process. This correlates with the configuration of meaning in the fourth explanation, where food is the participant and stomach is the circumstance.

Another way of multiplying the meaning across modes is the reiteration of organizing structures. Similar generic stages are found in animation and speech; namely, taxonomy orientation, oesophagus identification, and process presentation (see Fig. 12.6). In each stage, different modes are used to reinforce the taxonomy of the digestive system and the process of digestion in the oesophagus. It is noted that clear signals are provided when shifting to the next stage via verbal orientation and deictic gestures. It can be argued that using similar organizing structures in different modes provides more opportunities for students to enhance their understanding of scientific knowledge.

How meaning is configured and multiplied across modes is summarized in Fig. 12.8, where the mode explicitly adopted by the teacher is labeled "scaffold" and the mode that is not directly adopted yet provides relevant information is labeled "support."

As shown in Fig. 12.8, the stage of explanation formation is divided into two substages of processes presentation (to foster students' understanding of the process) and processes construction (to form the explanation about digestion). Such distinction is made in order to underscore the important role of language in unpacking and repacking the abstract technical knowledge. In the orientation stage, animation and verbal language are used to establish the taxonomy of digestive system. In the identification stage, the focus moves from digestive system to oesophagus. This is achieved through the teacher's deictic gesture (pointing to the oesophagus) accompanied by verbal orientation to oesophagus. The animation also supports the identification of oesophagus in that it provides prior knowledge about the taxonomy of digestive system. In the stage of processes presentation, language and metaphoric gestures are co-deployed to scaffold decomposing the digestion in the oesophagus into two processes (i.e., muscle contraction and food movement). This stage is also supported by the animation (see the previous sections for details). The last stage of processes construction is scaffolded through language, whose role is significant in constructing the explanations.

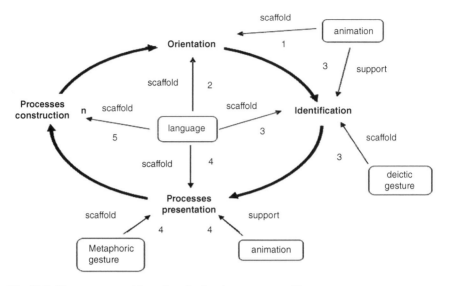

Fig. 12.8 The sequence and interplay of animation, gesture, and language

It is important to note that the boundary between processes presentation and processes construction in language is vague as they are interwoven with each other. This is an example of the complexity of classroom interaction, where the patterns of instantiation need to be seen as potentiality rather than certainty (Halliday & Matthiessen, 2014). The complex mediation of meaning occurs within one mode and across modes, and highlights the need for explicit instruction to orient the flow of meaning. While research on explicit instructions in the classroom focuses either on linguistic modes (Rose & Martin, 2012) or on visual modes (Ainsworth, 2006), findings from this chapter call for explicitness and a greater understanding of the links across linguistic and visual modes. For instance, following similar organizing patterns across modes is helpful for recycling the information, which could reconsolidate and even enhance the students' understanding of scientific knowledge. The similarity of generic stages across modes also demonstrates how meaning tends to be organized in science discourse. When reflecting on her lesson, Ms. Grace stated that she was not conscious of the similar generic stages in the animation and language. This suggests that even an experienced teacher may not be fully aware of her choices in the classroom (Ainsworth, 2006; Waldrip et al., 2010). There are instances where meaning-making could have happened but it did not. For example, Ms. Grace did not provide explicit instructions to scaffold the interpretation of the animation in terms of (i) the salience of each frame (ii) the choice of color and (iii) the sequential natural of digestive process. The assumption that students are able to notice, interpret, and link the meaning across modes independently is fraught with problems. It is possible that without the teacher's help, some students would not be able to notice the multimodal links and stumble in their literacy and content knowledge development (Tang et al., 2014; Waldrip et al., 2010).

12.7 Conclusion

This chapter demonstrates how the multimodal analytical framework is useful in examining the meaning-making from two perspectives: the semiotic affordances of an individual mode and the multiplicity of meaning across modes. The implications of this study encompass theoretical and pedagogical issues. In terms of theory, this chapter contributes to the literature on science discourse in three particular ways. First, it substantiates the literature related to scientific language (Halliday & Martin, 1993; Lemke, 1998) by highlighting the significant role of language in constructing abstract technical knowledge in science classrooms. It also contributes to the literature on the multimodal features in science classrooms (Airey & Linder, 2009; Jaipal, 2010) by examining both the semiotic affordances of one mode and the multiplying of meaning across modes. Finally, whereas research on explicit instruction in the classroom focuses on linguistic modes (Rose & Martin, 2012) and downplays other modes, this chapter highlights the need of providing explicit instructions on multimodal resources rather than language alone. We believe that developing the teacher and, to a limited extent, the learner's understanding of how multimodal resources are used to construct meaning in the classroom can enhance teaching and learning. If the teacher has access to understand the affordances of each mode and the ways to multiplying meaning across modes, this will provide greater opportunity for learning to occur in the classroom.

In terms of pedagogy, this chapter targets educational researchers and teaching practitioners on the importance of understanding and investigating the meaning-making in the classroom. The similar generic stages (orientation ^ identification ^ processes presentation) identified in the animation and language can be applied to other contexts and disciplines for validation. Findings from this study also show that even an experienced teacher may not be fully aware of the meaning potential of different modes. Therefore, more attention should be given to the choices of modes and the ways of multiplying meaning when choosing and organizing the modes in the classroom to support science teaching and learning.

References

Ainsworth, S. (2006). DeFT: A conceptual framework for considering learning with multiple representations. *Learning and Instruction, 16*(3), 183–198.

Airey, J., & Linder, C. (2009). A disciplinary discourse perspective on university science learning: Achieving fluency in a critical constellation of modes. *Journal of Research in Science Teaching, 46*(1), 27–49.

Baldry, A., & Thibault, P. J. (2006). *Multimodal transcription and text analysis: A multimedia toolkit and coursebook*. London: Equinox.

Christie, F., & Martin, J. R. (2005). *Genre and institutions: Social processes in the workplace and school*. London: Continuum.

Forey, G., & Polias, J. (2017). Multi-semiotic resources providing maximal input in teaching science through English. In A. Llinares & T. Moton (Eds.), *Applied linguistics perspectives on CLIL*. Amsterdam: John Benjamins.

Halliday, M. A. K. (1978). *Language as social semiotic: The social interpretation of language and meaning*. London: Arnold.
Halliday, M. A. K. (1985). *An introduction to functional grammar*. London: Arnold.
Halliday, M. A. K., & Martin, J. (1993). *Writing science: Literacy and discursive power (Critical perspectives on literacy and education)*. London: Falmer Press.
Halliday, M. A. K., & Matthiessen, C. M. I. M. (2014). *Halliday's introduction to functional grammar* (4th ed.). Abingdon: Routledge.
Hattie, J. (2003, October). Teachers make a difference: What is the research evidence? Paper presented at the Australian Council for Educational Research Annual Conference on Building Teacher Quality, Melbourne.
Humphrey, S. (2017). *Academic literacies in the middle years: A framework for enhancing teacher knowledge and student achievement*. New York: Routledge.
Jaipal, K. (2010). Meaning making through multiple modalities in a biology classroom: A multimodal semiotics discourse analysis. *Science Education, 94*(1), 48–72.
Jewitt, C. (Ed.) (2014). *The Routledge handbook of multimodal analysis* (2nd ed., Vol. 22). London: Routledge.
Kress, G. (2010). *Multimodality: A social semiotic approach to contemporary communication*. London: Routledge.
Kress, G., Jewitt, C., Ogborn, J., & Tsatsarelis, C. (2001). *Multimodal teaching and learning: The rhetorics of the science classroom*. London: Continuum.
Kress, G., & van Leeuwen, T. (2006). *Reading images: The grammar of visual design* (2nd ed.). New York: Routledge.
Lemke, J. (1990). *Talking science: Language, learning, and values*. New Jersey: Ablex Publishing Corporation.
Lemke, J. (1998). Multiplying meaning: Visual and verbal semiotics in scientific text. In J. R. Martin & R. Veel (Eds.), *Reading science: Critical and functional perspectives on discourses of science* (pp. 87–113). London: Routledge.
Lemke, J. (2002). Multimedia semiotics: Genres for science education and scientific literacy. In M. J. Schleppegrell, & M. C. Colombi (Eds.), *Developing advanced literacy in first and second languages: Meaning with power* (pp. 21–44). Mahwah: Erlbaum.
Martin, J. R., & Rose, D. (2008). *Genre relations: Mapping culture*. London: Equinox.
McNeill, D. (2005). *Gesture and thought*. London: University of Chicago Press.
Norris, S., & Phillips, L. (2003). How literacy in its fundamental sense is central to scientific literacy. *Science Education, 87*(2), 224–240.
O'Halloran, K. L. (1998). Classroom discourse in mathematics: A multisemiotic analysis. *Linguistics and Education, 10*(3), 359–388.
O'Tool, M. (2010). *The language of displayed art* (2nd ed.). London: Routledge.
Osborne, J. (2002). Science without literacy: A ship without a sail? *Cambridge Journal of Education, 32*(2), 203–218.
Polias, J. (2016). *Apprenticing students into science: Doing, talking & writing scientifically*. Melbourne: Lexis Education.
Rose, D., & Martin, J. R. (2012). *Learning to write, reading to learn: Genre, knowledge and pedagogy in the Sydney school*. Bristol: Equinox.
Tang, K.-S., Delgado, C., & Moje, E. B. (2014). An integrative framework for the analysis of multiple and multimodal representations for meaning-making in science education. *Science Education, 98*(2), 305–326.
Waldrip, B., Prain, V., & Carolan, J. (2010). Using multi-modal representations to improve learning in junior secondary science. *Research in Science Education, 40*(1), 65–80.

Part 4
Science Disciplinary Literacy Challenges

Chapter 13
Literacy Challenges in Chemistry: A Multimodal Analysis of Symbolic Formulas

Yu Liu

Abstract This chapter aims to advance the understanding of literacy challenges in chemistry by analyzing symbolic formulas from a multimodal perspective. Through the analysis of the different formulaic representations of chemical compounds (e.g., copper peroxide), the present study demonstrates that symbolic formulas have evolved an array of grammatical resources not found in natural language. These grammatical resources, on the one hand, effectively transmit essential information to experts including the quantitative composition and reaction behaviors of substances. On the other hand, the maximally condensed structure of symbolic formulas causes comprehension barriers for young learners. It is pointed out that a functionally oriented multimodal analysis carries implications for chemistry education, as this approach is able to make explicit the literacy challenges posed by chemical symbolism, offer useful clues to address the challenges, and provide concrete methods to probe into students' alternative conceptions.

Keywords Literacy challenges · multimodal · symbolic formulas · grammar · meaning

13.1 Introduction

The past three decades have witnessed an increasing emphasis on the medium for teaching and learning science (Prain & Hand, 2016). While pioneering studies focused on language as the central meaning-making resource to shape young learners' specialized forms of consciousness (Halliday & Martin, 1993; Lemke, 1990), recent research asserts that language is always nested with other forms of representation in knowledge production and communication (Danielsson, 2016; Liu, 2011; Tang, Delgado, & Moje, 2014). It therefore has served as an aspect of scientific literacy achievement that students are able to engage with the

---

Y. Liu (✉)
College of International Education, Sichuan International Studies University, Chongqing, China
e-mail: liuyunus@gmail.com

multimodal representations employed in science communities and to relate the different forms of signs to scientifically accepted ideas (Tang & Moje, 2010).

Whereas multimodal representations provide experts with the most effective means of communication (Yore & Hand, 2010), the semiotic complexity of scientific discourse poses literacy challenges for novices. For example, it has been frequently reported that both school and university students had difficulties in understanding chemical formulas despite the extensive application of formulaic expressions in science classrooms (Marais & Jordaan, 2000; Nemeth, 2006; Taskin & Bernholt, 2014). On the one hand, symbolic representation made a high demand on learners' calculation abilities (Scott, 2012). On the other hand, even those who succeeded in solving numeric problems (e.g., balancing a chemical equation) might not understand what happened at the molecular or atomic levels (Smith & Mertz, 1996; Yarroch, 1985).

The present study aims to explore the literacy challenges of chemical formulas by analyzing how scientific knowledge is symbolically represented. To that aim, I start with an introduction of the multimodal theory through which symbolic expressions are conceptualized as semiotic resources to create particular domains of experience. Then the analytic tools of transitivity and rank are proposed to demonstrate how chemical formulas exploit a number of unique grammatical resources to represent the quantitative and the qualitative aspects of chemical substances, and why students may have a limited conceptual understanding despite their mastery of mathematical operations. I also discuss the implications for teaching and learning chemistry.

13.2　A Multimodal View of Literacy Challenges in Chemistry

This chapter uses the term "multimodal" for two main purposes. First, it serves to highlight the representational nature of chemistry education as a semiotic ensemble, where curricula contents are represented and communicated in diverse modes in addition to natural language alone. Second, this notion points to a specific theory underpinning the present study, that is, the social semiotic interpretation of all communicative signs as meaning-making resources to constitute human culture (Halliday & Hasan, 1985).

From a social semiotic perspective (Kress, Jewitt, Ogborn, & Tsatsarelis, 2001; Lemke, 1998), all forms of representation simultaneously fulfill three generalized functions: to represent what is going on or happening in the world (ideational meaning); to take an attitude toward the presentation and make an evaluation (interpersonal meaning); and to organize related elements into a coherent message (textual meaning).

With a focus on the ideational meaning, Lemke (1998) endeavored to elucidate the nature of scientific discourse as representational hybrids by analyzing the semantic patterns produced within individual modes and across different modes. Each mode has its own functional specialization, which cannot be exactly duplicated by other forms of representation, so semiotic resources need to complement each other to cover the whole set of meanings of a scientific concept. Lemke (1998) emphasized that semiotic resources in the same multimodal representation do not simply have their individual meanings added together, but can modulate each other's meanings to capture the complexity of scientific ideas.

The meaning-based approach to scientific discourse offers an explicit account of students' learning difficulties. Whereas young learners' limited understanding of scientific concepts tends to be attributed to the lack of meta-visualization capabilities (Gilbert, 2005, p. 15), multimodal research does not regard novices' alternative conceptions about chemistry as internal cognitive deficits, but as difficulties in interpreting, relating, or translating external representations in the same way as scientists do (Cheng & Gilbert, 2009, p. 59). Following Lemke's (1998) semantic model, these representational difficulties can be further re-conceptualized as literacy challenges in grasping the scientific meanings made within the same semiotic resource and across different forms of representation. Accordingly, a multimodal analysis of symbolic representation is a promising means to illuminate the semantic complexity of chemical formulas and the corresponding literacy challenges.

13.3 Data and Analytic Methods

The data for analysis are chemical formulas collected from introductory science textbooks for secondary school students, which seem to become a major cause of learning challenges (Taskin & Bernholt, 2014). While formulaic expressions may exploit visual-spatial resources (e.g., in structural formulas) to communicate significant information, the present study focuses on linear symbolic signs and only analyzes empirical formulas and molecular formulas. The empirical formula differs from the molecular formula in that the former represents the simplest ratio in which atoms combine, while the latter indicates the actual number of atoms that interact to form a molecule (Gallagher & Ingram, 1997, pp. 68–70). For example, the empirical formula and the molecular formula for Ethane are CH_3 and C_2H_6, respectively.

Two main kinds of symbols can be found in a modern chemical formula. In the expression CuO_2, for example, Cu and O are symbols standing for chemical elements whereas the subscript 2 is a mathematical symbol. Admittedly, the symbolic representation is assumed to include both chemical and arithmetic signs (Talanquer, 2011, p. 184), and substantive chemistry education studies (Liu & Taber, 2016; Taber, 2009) have demonstrated their close and effective cooperation in the construction of scientific knowledge. However, this research makes a distinction between element symbols and mathematical expressions in a formula with the aim of revealing how the two kinds of signs mutually modulate each other's meaning. The semantic modulation across different modes has been identified as a major source of learning difficulties for novice students in science education

Table 13.1 Example of the transitivity analysis

Iron(III) oxide		Is reduced	To iron
Function	Participant	Process: material	Circumstance
Class	Nominal group	Verbal group	Prepositional phrase

(Lemke, 1998), but remains underexplored in the field of multimodal research (Jones, 2006, p. 55).

To explore the functions of chemical formulas, the present study adopts a social semiotic method of grammatical analysis (Halliday & Matthiessen, 1999, 2004). Within this perspective, grammar is not confined to a system of formal linguistic rules of correctness, but serves as an essential resource to construe ideational, interpersonal, and textual meanings (Kress et al., 2001). This study mainly examines the symbolic construction of scientific knowledge and thus the analysis of chemical formulas is restricted to the ideational function fulfilled.

One grammatical resource to construe ideational meaning is "transitivity"[1] (Halliday & Matthiessen, 2004, p. 170), which provides a means to demonstrate how different domains of experience (i.e., "doing," "being," "saying," "thinking," "behaving," and "existing") are represented in the grammar of a clause. For instance, a transitivity account of "Iron(III) oxide is reduced to iron" does not merely analyze this clause as a sequence of formal categories like "nominal group ^ verbal group ^ prepositional phrase."[2] Rather, the transitivity system models the experience of "doing" or "happening" construed in this clause as a material process, and links the grammatical units in the clause to their respective functions in the process.

As Table 13.1 indicates, a process type consists of two central functional components: a Process, which is typically realized by a verbal group (e.g., "is reduced") and one or more Participants involved in the process, which is typically realized by a nominal group (e.g., "iron(III) oxide"). Further to this, a process type may include an element of Circumstance in the grammatical form of a propositional phrase (e.g., "to iron"). The three components in Table 13.1 are combined to form an instance of the material process, representing the experience that something undergoes a change and thus plays a different role.

---

[1]According to Halliday and Matthiessen (2004), the transitivity system discriminates six types of process to sort out a wide range of experience. The material process is the resource to represent actions and events (e.g., "Molten ionic compounds conduct electricity"). The relational process is the resource to classify and define (e.g., "Covalent compounds are normally gases"). The mental process represents perception, cognition, and affection (e.g., "Students need to understand basic chemistry"). The verbal process construes the experience of "saying" (e.g., "The teacher explained the effect of concentration"). The behavioral process construes people's physiological and psychological behavior (e.g., "The teacher frowned at the student"). The existential process represents occurrence and typically employs the structure "there is/was ..." (e.g., "There are several definitions of oxidation").

[2]The ^ sign indicates a sequence, and it means "followed by."

Table 13.2 Example of the rank analysis

Rank scale	Examples						
Clause	Iron (III) oxide is reduced to iron						
Word group/phrase	Iron (III) oxide			is reduced		to iron	
Word	Iron	(III)	oxide	is	Reduced	to	iron

Another analytic means adopted in this study is the grammatical constituency of rank. Following Halliday and Matthiessen (2004), a clause can be analyzed into word groups or phrases at a lower rank, which can be further analyzed into words at an even lower rank (see Table 13.2). It is important to note that from a social semiotic perspective (Halliday & Matthiessen, 2004), the grammatical labeling of language is functionally oriented and thus a letter like "i" in "iron" does not count as a constituency unit, for it has no semantic significance in this word.

Also noteworthy is that the function of a grammatical unit (e.g., a clause) can be performed at a lower rank (e.g., a word group or a phrase) through the mechanism of rankshift or downranking, which, according to Halliday (1998), has the meaning-making power to facilitate scientific evolution. For example, when the experience of the clause "iron(III) oxide is reduced to iron" is compacted in a word group "the reduction of iron(III) oxide to iron," a previous Process in the form of a verbal group (i.e., "is reduced") becomes a new entity in the form of a noun (i.e., "reduction"). As nouns are the primary resource to construe phenomena into classes (Halliday, 1998), the mechanism of rankshift therefore enabled scientists to categorize a particular type of chemical change and set up a technical taxonomy (e.g., the contrast relationship between "reduction" and "oxidization" in modern chemistry theories, see Liu & Owyong, 2011).

While the social semiotic methods of grammatical analysis were mainly used to explore the functions of language in scientific discourse (Halliday & Martin, 1993; Lemke, 1990), recent research has successfully extended the analytic means like transitivity and rank to demonstrate the meaning-making patterns of mathematical symbolism (O'Halloran, 2000, 2005) and chemical signs (Liu & Owyong, 2011; Liu & Taber, 2016). Furthermore, a growing amount of chemistry education research (Jacobs, 2001; Sanger, 2005) tends to identify chemical symbolism as a specialized language. A semiotic model is therefore needed to compare symbolic expressions with natural language, for it provides a common platform for conceptualizing the two modes.

13.4 Grammatical Construction of Chemical Formulas

As Taasoobshirazi and Glynn (2009) observed, chemical formulas have developed into a remarkably short and reduced mode of representation, which facilitates efficient communication among experts, but sets up comprehension barriers against novices. From a functional standpoint (Liu & Owyong, 2011; Liu & Taber, 2016), reduction of formulaic expressions may stem from the operation of several grammatical resources employed in chemical symbolism to encode particular domains of experience. The following multimodal analysis focuses on the evolving representations of copper peroxide, one of the first compounds to be symbolized in modern chemistry. It aims at demonstrating how and why chemical formulas historically developed a range of grammatical resources in the construction of scientific knowledge.

13.4.1 The Operative Process in Chemical Formulas

Modern symbolic formulas were first designed by the Swedish scientist Jacob Berzelius to represent compounds in the 1810s when chemists were intensely interested in the quantitative composition of chemical substances (Brock, 1993). According to Klein (2003), Berzelian formulas provided an effective means to record the outcome of experimental analysis and demonstrate the numeric relations between elemental constituents in a compound. For example, Berzelius (1814, as cited in Klein, 2003, p. 10) symbolized copper peroxide as $Cu + 2O$ and explained that it was composed of one portion of metal and two portions of oxygen.

From a social semiotic perspective, Berzelian formulas originally incorporated a specialized transitivity configuration from the symbolic discourse of mathematics: the operative process (O'Halloran, 2000). To illustrate, the symbolic expression for copper peroxide comprised two operative processes. The first one represented an arithmetic process of adding, where both Cu and $2O$ functioned as the Participants with the addition operator as the Process. The second one was an arithmetic process of multiplying. This operative process consisted of two Participants: the mathematical symbol 2 and the element symbol O, while the Process in the form of the multiplication sign was omitted.

Although the experience of "calculating" can be regarded as one kind of "doing," the operative process in symbolic expressions differs from the material process in natural language. For instance, in the material process "The boss counted his money," humans (e.g., "the boss") and everyday objects (e.g., "his money") can enter into the transitivity configuration as Participants. By contrast, the operative process $Cu + 2O$ restricts its scope of Participants to algebraic expressions (e.g., 2) and chemical elements (e.g., Cu, O), thereby creating a reality that goes beyond the realm of mundane experience. Further to this, based on the improved experimental measurement of the masses of reacting substances, Berzelius assigned "a determinate quantity" to the element symbol for each substance in 1813 (Klein, 2003, p. 16). Then the operative process in symbolic formulas made it possible to calculate the weight of chemicals needed in a reaction, which contributed to both chemistry research and chemical manufacture.

13.4.2 The Reactive Process in Chemical Formulas

It is important to note that in Berzelius' original symbolic representation for compounds, the plus sign not only represented the additivity of numbers but also signified the chemical combination between elemental constituents (Klein, 2003, p. 19). According to Brock (1993), Berzelius used the plus sign for a second purpose of applying electrochemical theories to explain reactivity. For example, Cu + 2O indicated that one portion of copper was electropositive whereas two portions of oxygen were electronegative, and copper peroxide was formed by the mutual neutralization of opposite charges (Brock, 1993, p. 154).

The chemical combination meaning of the plus sign might have given birth to another specialized process type in Berzelian formulas: the reactive process (Liu, 2009). While the reactive process resembles the material process in construing the experience of "doing," their transitivity configurations have two major differences. First, the meaning scope of the reactive process is considerably reduced to the electrochemical interaction between specialized entities like elements, atoms, and electrons. Second, the material process may take a one-Participant transitivity configuration to describe chemical reactions, but the reactive process must consist of at least two Participants (Liu & Taber, 2016). As shown in Table 13.1, "iron(III) oxide" is the sole Participant to actualize the material process "Iron(III) oxide is reduced." For novice learners who do not understand that the technical term "reduce" implies the presence of an agent to donate electrons, the one-Participant material process could cause misconceptions. However, the equivalent reactive process (e.g., Fe_2O_3 + CO) seems to provide a more transparent account of the reaction, for it has two equally important Participants (i.e., Fe_2O_3 and CO).

Berzelius' symbolization of compounds brought about a semiotic transition from a word group (e.g., copper peroxide) to a symbolic representation (e.g., Cu + 2O). This transition was not merely a process of substituting names, but involved significant semantic shifts (Liu & Owyong, 2011). Whereas the word group "copper peroxide" was likely to stand for a stable entity containing two constituents, the reactive process Cu + 2O enabled experts to conceive the compound from a dynamic perspective as an ongoing interaction between two elements. This dynamic interpretation opened a new avenue for further development of modern chemistry theories such as valency and bonding.

13.4.3 Parentheses in Chemical Formulas

The dual functions of the plus sign made it possible to have the operative process combined with the reactive process in the same chemical formula so as to represent a compound's quantitative and qualitative composition in an economical way. However, scientists in the early 19th century found it necessary to distinguish mathematical additivity from chemical reactivity (Klein, 2003). Parentheses were therefore employed as a grammatical device to highlight the chemical behavior of elemental constituents in a compound.

For example, European chemists designed different formulaic representations for alcohol in the 1830s. Berzelius symbolized alcohol as (2C + 6H) + O, whereas Dumas proposed the representation (2C + 4H) + (2H + O). In terms of the arithmetic operation of additivity, there existed little difference between the two expressions. Both of them showed that alcohol consisted of two portions of carbon, six portions of hydrogen, and one portion of oxygen (Klein, 2003). On the other hand, the two models implied significantly different chemical meanings. From a social semiotic perspective, parentheses in chemical formula serve as a crucial resource to perform two functions: to group chemical entities into different reactive processes, and to arrange the reactive processes in a specific order similar to the use of brackets in mathematics (O'Halloran, 2000, p. 366).

To illustrate, the parenthesis in Berzelius' representation (2C + 6H) + O demonstrates that this formula possessed two reactive processes at different levels. At the first level, the reactive process had two Participants: 2C and 6H. At the second level, the Participant in the form of 2C + 6H and the Participant O entered the same reactive process. In contrast, Dumas' formula (2C + 4H) + (2H + O) incorporated three reactive processes at two levels. At the first level, there existed two reactive processes: the Participants 2C and 4H entered one reactive process while the Participants 2H and O entered the other. At the second level, the reactive process consisted of two Participants: one in the form of 2C + 4H and the other in the form of 2H + O.

The grammatical analysis suggests that Dumas was likely to interpret alcohol as a structurally more complex substance than Berzelius, for Dumas' model demonstrates that the formation of this compound involved three (rather than two) different chemical combinations. Further to this, it appears that the two scientists did not agree with each other on the elemental constituents' behaviors and properties in the compound. For instance, Berzelius' expression shows that two portions of carbon were chemically combined with four (rather than six) portions of hydrogen.

13.4.4 Structural Condensation of Chemical Formulas

When Berzelian formulas won increasing popularity among scientists in the early 19th century, their Process/Participant configurations (i.e., the operative process and the reactive process) were gradually condensed into a short and reduced form, possibly in order to represent structurally more complex compounds. For example, Berzelius symbolized copper(II) persulfate as $CuO^2 + 2SO^3$ and explained that this compound was composed of one portion of copper peroxide and two portions of sulfuric acid (Berzelius, 1813, as cited in Klein, 2003, p. 10). Subsequently, the original expression for copper peroxide (i.e., Cu + 2O) changed to CuO^2 through the use of superscripts and the ellipsis of the plus sign.

Admittedly, the two grammatical strategies introduced by Berzelius for structural condensation of chemical formulas were similar to those found in mathematical symbolism (O'Halloran, 2000, p. 366). However, it should be kept in mind that Berzelius' use of the two devices was not consistent with the conventions of algebraic notations (Klein, 2003), which may cause comprehension difficulties for novices. First, Berzelius' ellipsis of the plus sign was the same as the ellipsis of the multiplication sign in mathematics (Whewell, 1831, as cited in Klein, 2001, p. 28). In addition, Berzelius' use of superscripts in chemical formulas was likely to be confused with the use of exponents in mathematical operations. To address this issue, Liebig recommended subscripts instead in 1834, and thus CuO^2 was remodeled as CuO_2 in modern usage (Brock, 1993).

13.4.5 The Historical Rankshift of Chemical Formulas

As might be clear from the preceding account, Berzelius and other scientists designed and redesigned the symbolic representation for chemical substances during the evolution of chemistry. From a social semiotic perspective (Liu & Owyong, 2011; O'Halloran, 2000), the symbolic expressions for compounds like copper peroxide performed their functions at three constituency levels of chemical symbolism analogous to the ranks of word, clause and phrase respectively in natural language (see Fig. 13.1).

In the first stage, two major kinds of symbols were employed to represent a compound. The first kind included the symbols for elements (e.g., Cu, O) while the second kind comprised mathematical signs (i.e., 2, +, ×), though the plus sign also stood for chemical combinations. From a functional standpoint, both the two kinds of symbols appear to operate at the constituency level analogous to the rank of word (Liu & Taber, 2016). For example, the element symbols are similar to nouns in that they stand for chemical substances. Furthermore, the plus sign and the multiplication sign resemble verbs for representing chemical reactions and arithmetic operations.

A number of chemistry educators and philosophers of science (Markow, 1988; Restrepo & Villaveces, 2011; Weininger, 1998) claim that element symbols are equivalent to letters in language. Despite their similarity in terms of graphology

Fig. 13.1 Historical evolution of the symbolic representation for copper peroxide

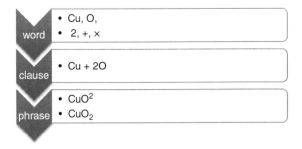

(e.g., the element symbol for oxygen O and the letter "o"), such a comparison may have problems. First, not all element symbols consist of only one letter. The symbol for copper Cu, for instance, comprises two letters. Second, a letter itself (e.g., "o") has little meaning in a word (e.g., "oxygen"). By contrast, a symbol like O is a meaningful unit in the discourse of chemistry, for it has the potential to denote a chemical element and one atom of this element. Accordingly, element symbols are functionally similar to nouns in language, especially abbreviations (e.g., "Dr." for "Doctor"), as both of them have meanings compacted in a reduced form.

In the second stage, scientists combined the symbols together following their empirical research to represent how a compound was formed out of chemical and arithmetic operations. These operations were symbolized as the reactive process and the operative process at the rank of clause. For Berzelius, the clausal configuration Cu + 2O not only demonstrated the numeric relation between the constituents of copper and oxygen (e.g., their combination ratio is 1:2), but it also indicated that the two elemental constituents were electrically combined in a compound (Klein, 2003).

In the third stage, the linear structure of Berzelian formulas was maximally condensed to represent the qualitative and the quantitative composition of a compound in an economical way. The short and reduced form made a formulaic expression in modern usage (e.g., CuO_2) look like a word in language (e.g., "cut"). Thus, recent chemistry education studies (Goodney, 2006; Jacob, 2001; Nemeth, 2006) drew an analogy between a chemical formula and a word in natural language. This analogy, however, does not fit from a functional standpoint. For instance, in the word "cut," none of the individual letters (i.e., "c," "u," "t") plays any semantic role. By contrast, the element symbols (i.e., Cu, O) and the subscript (i.e., "2") in the formula CuO_2 carry essential qualitative and quantitative information about the compound, as they all performed the function of Participants in the previous reactive and operative processes of Cu + 2O. Therefore, a chemical formula has multiple semantic cores, and it is functionally analogous to a phrase in language following Halliday and Matthiessen's (2004) linguistic rank scale framework.

Compared with the original symbolic components in the first stage and the symbolic representation for copper peroxide in the second stage, the Berzelian formulas in the third stage had the multiplication operator and the plus sign elided, and employed the spatial resource of superscripts and later subscripts instead. From a functional perspective (Halliday & Matthiessen, 2004; Liu & Owyong, 2011), the semiotic transition from Cu + 2O to CuO_2 served as an instance of rankshift in symbolic expressions, because a previous clause was restructured to function at the lower rank of phrase. Through the mechanism of rankshift, experts gained an efficient means to represent the formation of chemical compounds. On the other side, chemical formulas have become a source of learning difficulties for novices (Taskin & Bernholt, 2014), especially when young learners are not explicitly taught why and how formulaic expressions in modern usage developed a reduced form for the accumulated empirical information.

13.5 Implications for Chemistry Education

The multimodal analysis of chemical formulas carries pedagogic implications. First of all, the unique meaning-making patterns found in the preceding grammatical account are useful to illuminate the literacy challenges posed by chemical formulas. Yarroch (1985) reported that many high school students who successfully balanced a symbolic equation were not able to explain the reaction on the submicroscopic scale. This learning difficulty may be attributed to the semantic complexity of chemical formulas, which are capable of incorporating two different types of transitivity configurations (i.e., the operative process and the reactive process). Accordingly, novices' mastery of the operative process does not guarantee their mastery of the reactive process. For instance, to successfully calculate the relative formula mass of CuO_2, students only need to perform the basic arithmetic operations of multiplication and addition: $64 + (16 \times 2)$, even if they do not know how this compound is formed through ionic bonding. To resolve this learning difficulty, teachers should put an emphasis on both the numeric and the conceptual sides of symbolic formulas when students are assigned to balance chemical equations. Likewise, young learners will gain a clearer understanding of the functional difference between an empirical formula (e.g., CH_3) and a molecular formula (e.g., C_2H_6) if the teacher can clearly explain that the former expression only comprises the operative process whereas the latter combines both the operative process and the reactive process.

A multimodal analysis of chemical formulas also yields useful clues to address the literacy challenges. One formidable challenge is the use of subscripts in chemical formulas. Arasasingham, Taagepera, Potter, and Lonjers (2004, p. 1521) found that only half of the university students in their study understood that subscripts could represent the reaction ratios of elements in a compound. The analysis of the historical signs for copper peroxide in this study provides evidence that a chemical formula (e.g., CuO_2) actually evolved from an operative/reactive process (e.g., $Cu + 2O$). Thus, the subscript in a modern chemical formula was previously the stoichiometric coefficient in the symbolic representation for reactions. If students are taught the historical context for how chemical formulas have come out, they may have less difficulty in grasping the meaning of notations like subscripts.

Third, social semiotic multimodal research provides the analytic methods to probe into students' (mis)understanding of chemical symbolism. A multitude of empirical research (Arasasingham et al., 2004; Keig & Rubba, 1993; Yarroch, 1985) has demonstrated that novices often interpret symbolic expressions by relating a group of symbols to particular meaning. This way of interpretation is therefore similar to a functionally oriented analysis of chemical signs, although young learners' mapping of connections between grammar and semantics may differ radically from experts'. For example, Smith and Mertz (1996, p. 235) noticed that students were prone to regard $NiCl_2$ as a substance containing diatomic chlorine Cl_2. It is therefore likely that the learners identified $NiCl_2$ as the reduced form of the symbolic representation $Ni + Cl_2$. In terms of its transitivity configuration, Ni and Cl_2

function as two equal Participants to actualize the reactive process. This analysis reveals that the novice students had a very limited understanding of bonding. They seemed to believe that a nickel atom could be bonded to a diatomic molecule of chlorine to form nickel(II) chloride and there existed covalent bonds between the chlorine atoms in the compound. In contrast, experts tend to analyze $NiCl_2$ as the condensed expression of the reactive process $Ni^{2+} + 2Cl^-$, which clearly shows that the compound of nickel(II) chloride is formed through the ionic bonds between the constituent elements. Thus, it is recommended that science teachers encourage students to expand structurally condensed formulas like $NiCl_2$ to explain their own understanding of a scientific concept. It is also worthwhile for science teachers to learn basic multimodal analytic frameworks such as transitivity to compare students' self-generated representations with the authorized ones in textbooks.

13.6 Conclusion

This chapter is meant to make contributions to both multimodal studies and literacy research for chemistry education. The multimodal analysis of chemical formulas has shown how different modes such as the element symbols and the mathematical symbols modulate each other's meaning through a number of grammatical resources. For instance, while the incorporation of the operative process and the reactive process into chemical formulas efficiently represents both the quantitative and the qualitative composition of a compound, its use in science education increases the semantic complexity of symbolic expressions and thus may cause comprehension problems for novice learners, as this grammatical device is likely to blur the distinction between mathematical additivity and chemical reactivity.

The present study also advances the understanding of literacy challenges posed by chemical formulas. As briefly exemplified in the preceding discussion, the functional account of formulaic expressions has a great potential to make sense of students' learning difficulties and to develop effective pedagogy. Yet further research is needed to explore how the nexus between the research on multimodality and scientific literacy serves to improve the design of curriculum and instruction for chemistry education.

References

Arasasingham, R.D., Taagepera, M., Potter, F., & Lonjers, S. (2004). Using knowledge space theory to assess student understanding of stoichiometry. *Journal of Chemical Education, 81*, 1517–1523.

Brock, W. (1993). *The Norton history of chemistry*. New York: Norton.

Cheng, M., & Gilbert, J. K. (2009). Towards a better utilization of diagrams in research into the use of representative levels in chemical education. In J. K. Gilbert & D. F. Treagust (Eds.), *Multiple representations in chemical education* (pp. 55–73). Dordecht: Springer.

Danielsson, K. (2016). Modes and meaning in the classroom: The role of different semiotic resources to convey meaning in science classrooms. *Linguistics and Education, 35*, 88–99.

Gallagher, R., & Ingram, P. (1997). *GCSE chemistry*. Oxford: Oxford University Press.

Gilbert, J. K. (2005). Visualization: A meta-cognitive skill in science and science education. In J.K. Gilbert (Ed.), *Visualization in science education* (pp. 9–27). Dordrecht: Springer.

Goodney, D. E. (2006). Acid–base chemistry according to Robert Boyle: Chemical reactions in words as well as symbols. *Journal of Chemical Education, 83*, 1001–1002.

Halliday, M. A. K. (1998). Things and relations: Regrammaticizing experience as technical knowledge. In J. R. Martin & R. Veel (Eds.), *Reading science: Critical and functional perspectives on discourses of science* (pp. 185–235). London: Routledge.

Halliday, M. A. K., & Hasan, R. (1985). *Language, context and text: Aspects of language in a social-semiotic perspective*. Geelong: Deakin University Press.

Halliday, M. A. K., & Martin, J. R. (Eds.). (1993). *Writing science: Literacy and discursive power*. London: Falmer.

Halliday, M. A. K., & Matthiessen, C. M. I. M. (1999). *Construing experience through meaning: A language-based approach to cognition*. London: Cassell.

Halliday, M. A. K., & Matthiessen, C. M. I. M. (2004). *An introduction to functional grammar*. 3rd ed. London: Arnold.

Jacob, C. (2001). Analysis and synthesis: Interdependent operations in chemical language and practice. *International Journal for Philosophy of Chemistry, 7*, 31–50.

Jones, J. (2006). *Multiliteracies for academic purposes: A metafunctional exploration of intersemiosis and multimodality in university textbook and computer-based learning resources in science*. Unpublished Ed.D. thesis. University of Sydney, Sydney.

Keig, P.F., & Rubba, P.A. (1993). Translation of representations of the structure of matter and its relationship to reasoning, gender, spatial reasoning, and specific prior knowledge. *Journal of Research in Science Teaching, 30*, 883–903.

Klein U. (2001). Berzelian formulas as paper tools in early nineteenth-century chemistry. *Foundations of Chemistry, 3*, 7–32.

Klein, U. (2003). *Experiments, models, paper tools: Cultures of organic chemistry in the nineteenth century*. Stanford: Stanford University Press.

Kress, G., Jewitt, C., Ogborn, J., & Tsatsarelis, C. (2001). *Multimodal teaching and learning: The rhetorics of the science classroom*. London: Continuum.

Lemke, J. L. (1990). *Talking science: Language, learning, and values*. Norwood: Ablex Publishing Corporation.

Lemke, J. L. (1998). Multiplying meaning: Visual and verbal semiotics in scientific text. In J. R. Martin, & R. Veel (Eds.), Reading science: Critical and functional perspectives on discourses of science (pp. 87–113). London: Routledge.

Liu, Y. (2009). Teaching multiliteracies in scientific discourse: Implications from symbolic construction of chemistry. *K@ta, 11*, 128–141.

Liu, Y. (2011). *Scientific literacy in secondary school chemistry: A multimodal perspective*. Unpublished PhD thesis. National University of Singapore, Singapore.

Liu, Y., & Owyong, M.Y.S. (2011). Metaphor, multiplicative meaning and the semiotic construction of scientific knowledge. *Language Sciences, 33*, 822–834.

Liu, Y., & Taber, K. S. (2016). Analyzing symbolic expressions in secondary school chemistry: Their functions and implications for pedagogy. *Chemistry Education Research and Practice, 17*, 439–451.

Marais, P., & Jordaan, F. (2000). Are we taking symbolic language for granted? *Journal of Chemical Education, 77*, 1355–1357.

Markow, P. G. (1988). Teaching chemistry like the foreign language it is. *Journal of Chemical Education, 65*, 346–347.

Nemeth, J. M. (2006). Translating a linguistic understanding of chemistry to outcome achievement and interdisciplinary relevance in the introductory classroom. *Journal of Chemical Education, 83*, 592–594.

O'Halloran, K. L. (2000). Classroom discourse in mathematics: A multi-semiotic analysis. *Linguistics and Education, 10*, 359–388.

O'Halloran, K. L. (2005). *Mathematical discourse: Language, symbolism and visual Image.* London: Continuum.

Prain, V., & Hand, B. (2016). Learning science through learning to use its language. In B. Hand, M. McDermott, & V. Prain, (Eds.), *Using multimodal representations to support learning in the science classroom* (pp. 1–10). Dordrecht: Springer.

Restrepo, G., & Villaveces, J.L. (2011). Chemistry, a lingua philosophica. *Foundations of Chemistry, 13*, 233–249.

Sanger, M. J. (2005). Evaluating students' conceptual understanding of balanced equations and stoichiometric ratios using a particulate drawing. *Journal of Chemical Education, 82*, 131–134.

Scott, F. J. (2012). Is mathematics to blame? An investigation into high school students' difficulty in performing calculations in chemistry. *Chemistry Education Research and. Practice, 13*, 330–336.

Smith, K. J., & Metz, P. A. (1996). Evaluating student understanding of solution chemistry through microscopic representations. *Journal of Chemical Education, 73*, 233–235.

Taasoobshirazi, G., & Glynn, S.M. (2009). College students solving chemistry problems: A theoretical model of expertise. *Journal of Research in Science Teaching, 46*, 1070–1089.

Taber, K.S. (2009). Learning at the symbolic level. In J. K. Gilbert & D. F. Treagust, (Eds.), *Multiple representations in chemical education* (pp. 75–108). Dordrecht: Springer.

Talanquer, V. (2011). Macro, submicro, and symbolic: The many faces of the chemistry "triplet". *International Journal of Science Education, 33*, 179–195.

Tang, K. S., Delgado, C., & Moje, E. (2014). An integrative framework for the analysis of multiple and multimodal representations for meaning-making in science education. *Science Education, 98*, 305–326.

Tang, K. S., & Moje, E. (2010). Relating multimodal representations to the literacies of science. *Research in Science Education, 40*, 81–85.

Taskin, V., & Bernholt, S. (2014). Students' understanding of chemical formulas: A review of empirical research. *International Journal of Science Education, 36*, 157–185.

Weininger, S. (1998). Contemplating the finger: Visuality and the semiotics of chemistry, *International Journal for Philosophy of Chemistry, 4*, 3–27.

Yarroch,W. L. (1985). Student understanding of chemical equation balancing. *Journal of Research in Science Teaching, 22*, 449–459.

Yore, L., & Hand, B. (2010) Epilogue: Plotting a research agenda for multiple representation, multiple modality and multiple representational competency. *Research in Science Education, 40*, 93–101.

Chapter 14
Gains and Losses: Metaphors in Chemistry Classrooms

Kristina Danielsson, Ragnhild Löfgren and Alma Jahic Pettersson

Abstract This chapter reports on findings from classroom communication in secondary chemistry teaching and learning. The data was analyzed qualitatively regarding the use of metaphors and analogies in relation to atoms and ion formation, with an intention to shed light on students' scientific understanding as well as on their enculturation into the disciplinary discourse. Theoretically we draw on social semiotics, which allows analyses of language use in its widest sense, comprised of verbal language, images, action, gestures, and more. In our data, we identified common disciplinary metaphors in science, as well as metaphors connected to everyday life. Through the analyses based on systemic functional linguistics (SFL), we also identified anthropomorphic metaphors, with particles, atoms, and ions being humanized with intentions and feelings. Linguistic choices signaling metaphoric language were mainly noted in relation to quite obvious metaphors whereas no such signals or explanations were noted in connection to anthropomorphic metaphors. The study has implications for the design of classroom practices, including the use of discussions to enhance a more reflective use and understanding of the gains and losses around metaphors.

Keywords Chemistry teaching · classroom practices · metaphors · analogies · systemic functional linguistics

---

K. Danielsson (✉)
Department of Swedish, Linnaeus University, Växjö, Sweden
e-mail: kristina.danielsson@lnu.se

R. Löfgren · A.J. Pettersson
Department of Social and Welfare Studies, Linköping University, Linköping, Sweden
e-mail: ragnhild.lofgren@liu.se

A.J. Pettersson
e-mail: alma.pettersson@liu.se

14.1 Introduction and Background

The complex nature of scientific language and its representations in different semiotic modes (such as verbal language, diagrams, 3D models, e.g., Jewitt, 2017; Kress, Jewitt, Ogborn, & Tsatsarelis, 2001) has proven to be challenging for students (Fang, 2005; Halliday & Martin, 1993; Lemke, 1990, 1998; Norris & Phillips, 2003; Yore & Treagust, 2006). The present study focuses on one aspect of the discourse of science and representations in areas of science, namely the use of metaphors. Metaphors are integrated in the discourse of science as a way of creating new knowledge (Ogborn, 1996; Sutton, 1992) and in educational contexts, analogies and metaphors are suggested as tools to make abstract or complex content more accessible (Aubusson, Harrison, & Ritchie, 2006; Sutton, 1992). Also, they can be expressed through different modes of representation. However, metaphors can be challenging for students. Therefore, teachers need to find ways of creating classroom practices that develop both students' understanding of content and their disciplinary literacy, that is, the ability to interact and communicate through the specialized language in a wide sense, comprising verbal language and various forms of representations, which are sometimes based on metaphors.

All disciplinary areas constitute social, historical, and cultural practices. From a sociocultural perspective, a consequence of this is that learning science also involves understanding and appropriating specific norms, meanings, concepts, and ways of communicating and acting, all of which combine to create meaningful activities in the science classroom (Lave & Wenger, 1991). Aikenhead (1996) claims that a "border-crossing" from everyday subcultures to the subculture of science needs to be made explicit to the students, not the least for students with competing or conflicting world views, for whom the border-crossing can be an extra challenge (Brown & Kelly, 2007; Tan, Calabrese Barton, Kang, & O'Neill, 2013).

In order to participate in a meaningful way in a science practice for students with various worldviews, Rogoff (1995) stresses three aspects: apprenticeship, guided participation, and participatory appropriation. Apprenticeship corresponds to the community processes which refers to the specific nature of activity as well as relations to institutional structures in which it occurs. Guided participation refers to an interpersonal plane and attempts to describe what happens "between people as they communicate and coordinate efforts while participating in a culturally valued activity" (Rogoff, 1995, p. 142). Guidance refers to a direction of the activity offered by the cultural and social values. Wenger (1998) points out that the participation dimension must be balanced by the reification dimension, that is, the processes where a common understanding is turned into meaningful artifacts. Accordingly, artifacts such as tools, models, procedures, stories, and language will reify some aspects of its practice. Wenger claims that participation and reification form a duality, with the two aspects being analytically inseparable from each other (*ibid.*). Similar ideas justify teaching practices where the teacher assists students in their transition from the everyday story (an everyday point of view) to the scientific story (a scientific point of view) (Mortimer & Scott, 2003). This, in its turn, is in accordance with the notion of the importance of disciplinary literacy for a person to be able to participate in the knowledge construction and social practices of the discipline (Moje, 2007). Aikenhead's assertion that the border-crossing needs to be made explicit is parallel to claims that teachers must help students translating between the everyday and the scientific language to keep the varieties "straight for the students" (Lemke, 1990, p. 173), thus making the register of science (how to communicate within the field) explicit.

As mentioned, the focus of this chapter is the use of metaphors in science classrooms. There are several ways of defining "metaphor," although a central aspect is that metaphors create the "possibility of activating two distinct domains" (Cameron, 2002, p. 674). Similar to Cameron, in this study we include analogies in the concept. Furthermore, we here include metaphors expressed both linguistically and through other semiotic modes, such as images, or in combinations of modes (Danielsson & Selander, 2016). When new nature phenomena are discovered, this creates a need for new concepts to communicate about and further develop the field. For example, the discovery that electrons move at approximate distances to the nucleus has generated the two metaphors "electron orbit" and "electron cloud". These metaphors meet a need to take different perspectives depending on what to emphasize: either a well-organized structure, which can explain chemical bonding, or the proximal distances of quickly moving electrons. Thus, it could be argued that science cannot be understood without metaphors. As regards metaphors employed for pedagogical reasons, some have come to be used quite frequently and across cultures, such as the analogy between electronic movement and "planetary orbits", while others are more provisional, and in some cases dependent on the language in use (e.g., the apple as an analogy to the atom, which was found in our data). As opposed to more obvious scientific terminology (like "protons" or "neutrons"), which does not build on everyday concepts, metaphoric expressions – including metaphors that have been integrated in scientific discourse, such as "electron cloud" – can be incorrectly understood in their everyday sense, thus creating unexpected obstacles for a novice in the field (Askeland & Aamotsbakken, 2010; Golden, 2010). Furthermore, metaphors might not only be challenging from the linguistic perspective, but also in regard to the relation between the source and the target, or in relation to the reach of the metaphor, since the parallelism between the concepts is always partial (Haglund, 2013). A specific type of metaphor used in science is anthropomorphisms, where scientific concepts are described as having intentions, feelings, and desires (Tibell & Rundgren, 2010, p. 28). Studies in science education have reported that students sometimes actually hold anthropomorphic views as true (Tibell & Rundgren, 2010). Hence, there are both potential "gains" and "losses" (i.e., possibilities and hindrances)[1] in relation to metaphor use in education. One way of dealing with these possible

---

[1] The discussion around "gains and losses" is informed by the concept of semiotic *affordance*, introduced by Gibson (1977) and later used in social semiotic perspectives of multimodality (e.g., Kress, 2009; Danielsson, 2016). In short, *affordance* concerns the meaning making potential of a resource in a specific communicative situation.

gains and losses with metaphors in science education could be to work in explicit ways, for instance, to discuss the reach of a metaphor, or what aspects of the nature phenomenon the metaphor emphasizes, something which is also suggested as an effective way of helping students with the border-crossing into the scientific field (Tobin, 2006).

The use of metaphors is thus central to the scientific disciplines as such, as well as for science education. At the same time, metaphoric language is a phenomenon which has gained attention in linguistic research (Lakoff & Johnson, 1980; Svanlund, 2007). Hence, metaphoric language can be described from a linguistic point of view, and possible obstacles in relation to learning can be discerned from such a perspective. However, to fully understand the gains and losses with metaphoric language in science, and to develop teaching practices within the area, there is a need for cross-disciplinary research and cross-domain collaboration among science teachers, science educators, and linguists (Tibell & Rundgren, 2010). This chapter is an attempt to contribute to the field meeting these requirements, as one of the authors is a linguist focusing on disciplinary literacy within the science field, one is a science educator (chemistry), and one is a PhD student in science education, with a background as a science teacher (chemistry).

In this chapter, we present results from analyses of metaphor use in different modes in teachers' and students' meaning making in secondary school science classrooms. The study draws on multimodal perspectives of social semiotic theory (Jewitt, 2017). From this perspective, each choice in meaning making is viewed as a result of social, cultural, and situational factors in the context of the communicative situation. A foundation of the social semiotic perspective is systemic-functional linguistics (SFL) (Halliday, 1978; Halliday & Matthiessen, 2004). According to SFL each communicative event contains three perspectives, corresponding to three basic meta-functions of language use: the content that is conveyed (ideational metafunction), how the content is organized (textual metafunction), and how relations are created through interaction (interpersonal metafunction). Analyses based on these meta-functions yield different perspectives on the ways in which the communicative resources are used. The framework was developed for linguistic analyses, but has later been used for other modes of communication (Kress & van Leeuwen, 2006; Martinec, 2004; O'Halloran, 1999) or combinations of semiotic resources in different modes (Danielsson, 2016; O'Halloran, 1999). In the following, we will present results of analyses related to the ideational meta-function.

14.2 Aims

The aim of this study is to describe and analyze how teachers and students use metaphors when working with chemistry content in secondary science classrooms, in this case to explain and discuss the atomic structure and ion formation. We will also describe if, and how, metaphors are made explicit for the students through classroom discussions. We relate the results to "gains and losses" (see above) in the learning process, for example, in relation to the reach of the metaphors.

An important aspect of the study is whether we can discern possibilities or hindrances for students' understanding of content and for their border-crossing and participation in a science practice, including an enculturation into the disciplinary discourse.

Our research questions are:

1. What characterizes metaphorical language (in a wide sense) used by teachers and students in chemistry classrooms focusing on the atomic structure and ion formation?
2. What possibilities and hindrances for students' understanding of the content and for their border-crossing and participation in a science practice can be discerned in relation to the use of metaphorical language?

14.3 Data

The present study builds on data collected within the multidisciplinary project *Chemistry texts as tools for scientific learning*, financed by the Swedish Research Council (Eriksson, 2011). That project followed a series of lessons in a number of Swedish (in Sweden) and Finland-Swedish (in Finland)[2] secondary school chemistry classrooms (grade 8–9, students aged 14–15 years) dealing with the atomic model, the periodic table of elements, and chemical bonds. To delimit, yet at the same time ground the present study on a relatively rich set of data from the overall project, video recordings of communication in three classrooms were used: one Swedish and two Finland-Swedish classrooms, with between 20 and 25 students in each class. In the present study, we have analyzed teaching episodes where teachers explain the structure of the atom, the relative stability of different substances in the periodic table, and ionic formation. Each of these episodes lasted for around 10 minutes, and occurred during various parts of the unit followed by the overall project. In the following, we present results from qualitative analyses of metaphors used in relation to the atomic model or ion formation. The research team in the overall project did not express an interest in metaphors, so any such occurrences in the data are metaphors that the participants themselves chose to use in the communicative situation.

The study aligns to the ethical principles outlined by the Swedish Research Council (2016) concerning informed consent and confidentiality. Student names are avoided and teachers have been assigned pseudonyms. The Swedish teachers

---

[2]Finland-Swedish is a variety of Swedish spoken in Finland. Apart from Finnish, Swedish is an official language in Finland and it is the first language of around 5% of the population. In areas with a high proportion of Swedish speakers, some schools use Swedish as the language of instruction.

were given names beginning with /s/ while Finland-Swedish teachers were given names beginning with /f/. The data presented in the following derive from Sture's (Swedish), Fred's and Fredrika's (both Finland-Swedish) classrooms.

14.4 Analytical Methods

The video recordings of the teaching episodes were transcribed. The transcriptions included spoken communication alongside with comments on gestures, blackboard notes, or the use of artifacts, such as concrete models or the periodic table of elements. After that, the team conducted a thematic analysis in order to get an overview of the data (Bryman, 2012). Instances of metaphoric use were grouped into the different areas that the metaphors could be connected to, in this case the scientific or the everyday life domains, or anthropomorphisms.

To answer our first research question, detailed analyses of the occurrences of metaphoric language were made using the systemic functional linguistics (SFL) framework (Halliday & Matthiessen, 2004). In the present study, we have analyzed our data in relation to the ideational meta-function and specifically processes, participants, and circumstances, which are basic parts of the transitivity system (Halliday & Matthiessen, 2004, p. 302, for a model). Through transitivity analyses the type of process (that something happens, is said, is, or is perceived), what participants are involved in the process (who or what does, is, says, owns what, etc.), and the circumstances around the process (where, how, when, etc. the participant did, thought or said something) are identified. In line with a number of Scandinavian scholars (Andersen, Petersen, & Smedegaard, 2001; Holmberg & Karlsson, 2006), we have considered four basic process types, namely relational, material, verbal, and mental processes. In the framework of Halliday and Matthiessen (2004), there are two more process types: experiential and behavioral. These two types were merged with relational and material processes, respectively, due to their close relationship to these process types (Holmberg & Karlsson, 2006).

The process types can be connected to three different "worlds of meaning": a world of abstract relations (relational processes), a mental world (mental and verbal processes), and a physical world (material and verbal processes) (Halliday & Matthiessen, 2004, p. 172). Relational processes typically deal with the ways in which something is related to something else, such as "the atom *consists* of different particles". Material processes typically demand some kind of energy from the first participant, which can be seen as an actor, such as "the electrons *move* quickly around the nucleus". Verbal processes imply that something is said, for example, "we *say* that the nucleus is positively charged". Mental processes are inner processes, and they imply some kind of mental activity of the first participant, who functions as an experiencer: "before, scientists *thought* that the atom was the smallest particle". In the participant analysis, you also differ between first and second participants: "Mendelejev (first participant) developed the periodic table of elements (second participant)". By performing transitivity analyses, patterns in the use of metaphorical language can be distinguished, for example, to determine the areas that the metaphors could be connected to. Also, the parts of the transitivity system that are used to express the metaphor can be discerned. This in turn can be used as a basis to discuss what connections the students need to make in order to understand the metaphor.

In order to analyze participation and meaning making in scientific practices (our second research question), we have used Rogoff's (1995) three aspects of analysis: apprenticeship, guided participation, and participatory appropriation. In this study, we focus on the guided participation referring to the interpersonal plane in order to describe and analyze how students' participate and reify aspects of the scientific practice in relation to the sub-micro level of the atom and ion formation. The "participation" focus on how students get involved in different activities, with whom and with what materials. The "guidance" focus on the direction of the activity in which the students are engaged (*ibid.*). In this study, the direction of the activities is connected to the different areas, or source domains, that the metaphors could be connected to: the science domain, the everyday life domain, or anthropomorphic metaphors. By investigating guided participation in the communication around metaphors, both obstacles and instances of "signs of learning" (Kress, 2010) can be discerned.

14.5 Results

The metaphors used in the teachers' expositions were noted both in connection to the atomic structure (including electronic movement and electronic configuration), and ionic bonds. As mentioned in the description of our analytical methods, the source domains could become salient through any part of the transitivity analysis (process, participants, circumstances), or in combinations of them. In the following, the results are presented with regard to the following source domains: the science or everyday life domains, and anthropomorphic metaphors.

14.5.1 The Science Domain

Not surprisingly, throughout our data, we found quite a few examples of metaphors that are more or less integrated into the disciplinary discourse. Examples of this are *nucleus* (Sw. *kärna*),[3] *electron clouds* or *shells*, and *attraction* between ions. Such metaphors can be challenging both regarding their reach and in relation

---

[3]The fact that *nucleus* and *seed* is the same word (*kärna*) in Swedish makes the Swedish term more obviously metaphoric than the English *nucleus*, which is probably perceived as nonmetaphoric for most English speakers.

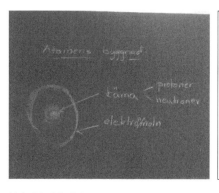

Fig. 14.1 Model of the atom, Fred's classroom

to the fact that the expressions have an everyday sense that students might already be familiar with, leading to risks of misinterpretations. Some metaphors were connected to other parts of the scientific field, like *planetary orbit* (Sw. *planetbana*, where *bana* also corresponds to Eng. *track*), or *circulating satellites*. Here, we can also note the use of concretizing models in the classrooms (Fig. 14.1).

The SFL analysis revealed that many of these metaphors functioned as participants, like "the *electron cloud /…/* swirls around". At times, they were also expressed through material processes, like "ions with opposite charges *attract* each other". In other cases, the metaphor was expressed in circumstances, like "electrons *on a shell*". They could also be introduced in circumstances, like "as a kind of first explanation we say that they [the electrons] are *in the electron cloud*". By stating "we say that", some kind of metaphoric expression is indicated. Explicitly expressed analogies were also found, like "Electrons circle around the nucleus in orbits (Sw. *banor*) *similar to satellites circulating around the earth*".

When first introducing the atom, many teachers chalk-talked (Artemeva & Fox, 2011), drawing a model on the blackboard (or equivalent) at the same time as they talked about the different particles.

Figure 14.1 shows Fred's drawing during one of the first lessons in the content area. Here, he used two disciplinary metaphors, *nucleus* (Sw. *kärna*) and *electronic cloud* (Sw. *elektronmoln*). When making the drawing, he tried to explain what is meant by "electron cloud":

> there are two central parts of the atom /…/ something in the middle and something around it /…/ the electronic cloud which is around this atom … swirls around at a high speed … not in well-defined orbits but at a certain distance … they swirl around there freely … around and around" [makes a circular gesture with the chalk around the electronic cloud in the drawing] (Fred, lesson 2)

Sture introduced the atomic model during one of the first lessons. During the third lesson, he returned to the atomic model. He chalk-talked, using Lithium as an example, trying to involve the students in deciding where to draw the electrons. After having drawn two circles around the nucleus, he indicated the numbers of electrons on each shell, writing 2⁻ on the first circle and 1⁻ on the second circle, commenting:

> two electrons on one *ring*. or you usually say a *shell* ... I will write that here [draws a line from the outer circle and writes "electron shell" at the end of the line] ... *electron shell* (Sture, lesson 3)

A few minutes earlier, a student had come up with the word *ring* (another word for circle, Sw. *cirkel*) when Sture pointed at a period in the periodic table of elements asking what that was (he had drawn the inner shell of the Lithium atom on the whiteboard and wanted the students to note that you probably needed a new shell for the third electron). Somewhat surprised, Sture picked up the word and went to the whiteboard to draw a new circle around the first shell. This is an example where a teacher makes it possible for the students to participate and to reify the use of the atomic model (Rogoff, 1995). In this case, it was achieved by using the periodic table to explain how the student's self-generated analogy of *rings* could be connected to the periods.

14.5.2 The Everyday Life Domain

The scientific content was quite frequently connected to the domain of everyday life. These analogies were mainly used to explain the atomic structure and its particles and they were based on everyday objects, such as *bicycle spikes* and *apples*. The analysis revealed that in teachers' expositions, these metaphors were often introduced as second participants, where the first participant was typically the student or an indefinite "you", while the process was mental (e.g., think), as in "if you think of an apple" or "if you think about the spikes in a bicycle wheel". By using such constructions, the teacher signals that some kind of analogy is used.

When talking about the atomic structure, the Swedish teachers used a big and well-known sports arena outside Stockholm, the Globe (shaped as a globe), to illustrate that atoms consist of a lot of empty space. This analogy typically involved other metaphors, as in the following sequence in Sture's classroom.

Sture started out talking about the fact that there are many different substances, and that they have similar structures. He told the students that he was going to present a simplified model, but before doing that, he showed the students a concrete wooden model of the Carbon atom (Fig. 14.2), asking the students what it "looks like inside". However, these wooden models are usually used to build molecular models. When Sture tossed the wooden ball in the air talking about the inside of an atom, the students appeared to feel uncertain. It took some time before they responded, but one student mentioned the different particles (protons, neutrons, and electrons) and another student added "and nothing".

With the probable intention of picking up "and nothing", Sture went on, introducing the Globe, making a drawing on the whiteboard while talking (Fig. 14.3). Sture asked a student whether he had been to the Globe and then commented: "it's pretty

Fig. 14.2 Concrete model of carbon atom, Sture's classroom

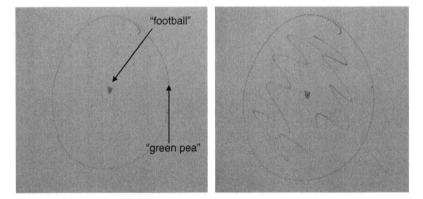

Fig. 14.3 "The Globe", two stages, Sture's classroom

big in between. I really do not know how big ... but it is amazingly large ... there is room for horses and stuff and hockey and everything". Then he went on using a football, size 5, as an analogy to the atomic nucleus: "now we are thinking like this that we hang up a football in size five like this ... ok?", drawing a dot in the middle of the circle. Then Sture asked a student: "What was it out here then?" pointing at the circle around the nucleus. One student responded: "Places to sit". This indicates the difficulties to follow the reasoning of the Globe as an analogy for the atom since the student refer to the Globe as a sports arena instead.

After that Sture drew a small dot on the circle, telling the students that it was "a small green pea". He emphasized that there is a lot of space in the arena, marking the area between the "football" and the "pea" with a green pen, saying that this is empty space. Interestingly, throughout the metaphoric explanation, Sture did not stress that the analogy was used to illustrate the atomic model. Also, throughout this analogous explanation, disciplinary language was more or less nonpresent (the nucleus and electrons were not mentioned). Further examples

Fig. 14.4 Blackboard note, Fredrika's classroom

Repetition of the atom:
An atom consists of:
protons   positive
                         in the nucleus
neutrons   neutral

electrons   negative – in the electronic cloud

below will reveal that this pattern was typical for Sture's classroom. As mentioned above, Sture introduced the Globe metaphor after the student had said "and nothing" as a response to what is inside an atom. However, the students in this example showed obvious difficulties participating in the reasoning about the Globe as an analogy for an atom, and they received no guidance from Sture for how to reify the model of the atom. Instead the discussion at times evolved around the actual sports arena. In other classrooms, teachers used the apple as an analogy for the basic atomic structure, stressing that the atom has a nucleus surrounded by electronic shells. What is important to note here is that in Swedish, the same word is used for "nucleus" and "core," or "seed" (Sw. *kärna*), and similarly, the Swedish word *skal* can be used both for (electron) "shell" and "peel", something which makes the apple a possible metaphor in that linguistic context. Here, we will use Fredrika as an illustration (Fig. 14.4).

Fredrika chalk-talked while going through the different parts of the atom. During her exposition, the students answered questions about the particles of the atom.

After having stated that the protons and neutrons are in the nucleus, Fredrika introduced the apple as a metaphor:

> if you have trouble understanding what kind of nucleus (Sw. *kärna*) this is or what is meant by it ... so it is in the middle ... the furthest inside you can get ... if you think of an *apple* it is like the *core* (Sw. *kärnhus*) inside ... where the *seeds* (Sw. *frön*) are ... here are the protons and neutrons ... while the electrons ... they move and are spinning around it [makes circular gestures with her arms] (Fredrika, lesson 1)

As opposed to Sture's exposition, we can note that Fredrika combined metaphoric language with subject-specific terminology (protons, neutrons, electrons). In her blackboard notes, only subject-specific terminology was used. Sometimes, as in this case, metaphors that are integrated into the scientific discourse, such as "electron cloud" were incorporated in the notes (Fig. 14.4).

As was commented on above, Fred, too, used the metaphor "electron cloud". However, when speaking about the electrons spinning around the nucleus like a cloud, Fred introduced two additional metaphors, taken from the domain of everyday life, *bicycle spikes* and a *thin film*, to explain the speed of the electrons. At the end of the transcription excerpt below, we can also note how Fred touched upon the subject-specific "electron cloud", but through an explicit simile, using the everyday word "cloud" without the subject-specific complement "electron":

> the electron [in the hydrogen atom] moves at the same distance from the nucleus but it spins around and around and around … quickly quickly … if you think about the spikes in a bicycle wheel ["draws" lines like bicycle spikes from a center point with his hands in the air] … you can see them when they are still … but when you spin the bicycle wheel [gestures as if giving speed to a wheel] you cannot distinguish the spikes and instead it is like a thin film when they spin [gestures with his hand in a two-dimensional circle] /… / this is similar . they spin around and you get the impression that they are everywhere . thus it is like a cloud. (Fred, lesson 3)

In the teaching episodes above, both Fredrika and Fred were very careful to name the elementary particles and making drawings. Similar to Sture, both Fredrika and Fred were careful to explain that the models were simplified, and in their attempts to explain the structure, they introduced analogies. Both Fredrika and Fred were explicit that they used an analogy (e.g., "if you have trouble /… / think of an apple", "it is like a cloud"), though neither of them asked the students how they interpreted the analogy, nor did they discuss the reach of the metaphors. The students in their classrooms did not use the metaphors in the discussions, nor did they participate in the use of metaphors for the atomic model (Rogoff, 1995). However, results from the overall project reported elsewhere (Danielsson, 2011b) revealed that in interviews some students talked about an "empty outer shell" when asked about the structure of, for instance, a Sodium ion in comparison with the atom. Such comments reveal that students had taken the shell metaphor literally and that the reach of it was not clear.

14.5.3 Anthropomorphic Metaphors

Numerous anthropomorphic metaphors were found in our data. Through these metaphors, chemical particles were described as having intentions and feelings. Examples are noble gases or ions being *content* or *lazy*, *jumping* electrons, or atoms *striving* to become stable, or *giving away* and *taking* electrons. Many of the less obvious metaphors in our data were anthropomorphic, and they were often expressed through processes (like the last two examples). In the data, these processes were quite often material, thus implying some kind of action from a participant being an actor with a clear intention, for example, "*jump* to another atom". But in some cases chemical particles were first participants involved in mental processes as in "atoms *want to have* a full outer shell". Other metaphors were attributive, as "the noble gases are *content*". Expressions like "Halogenes *gladly* (Sw. *gärna*) form chemical bonds" also imply feelings as well as intentions.

What is particularly interesting is that anthropomorphisms were seldom explicitly talked about as metaphors in the classroom discussions. This is especially true for the less obvious metaphors, with particles *striving*, *placing themselves* and *giving away* or *taking*. Such metaphors were used both in speaking and writing (Fig. 14.5).

Ionic bonds (see pictures p. 112)
• If a chemical substance wants to give away electrons and another one wants to take, then positive and negative ions are formed. • Different charges are drawn towards each other (attracts) and therefore ionic bonds are formed. • The ions are packed in crystalline structures.

Fig. 14.5 Blackboard note from Fredrika's classroom

As regards anthropomorphisms, Sture's classroom stood out. In the following, we will look a bit more closely at one of his expositions. When introducing valence electrons, Sture used Sodium as an example. Then he drew a model with three electronic shells with one electron placed on the outer shell. He pointed at the electron saying: "and this little *beggar* ... how to you think it *feels*?", with a student responding "*lonely*". Then another student clarified: "you said before that they *like* to be in pairs but this one is *lonely*". Sture commented that "he she or it does not have any feelings" but that they could try to imagine the situation for the electron, further elaborating a humanization of the particles: "does this *feel* good do you think? ... *this is really harassment* ... and what do you think the main office [points at the nucleus] *thinks* about it?".

The metaphoric way of explaining why some substances are more reactive than others was also filled with metaphoric language in Sture's explanations. Interestingly enough, after having stated that the atoms or particles have no feelings, Sture consistently used such metaphors during the rest of the teaching period, rather than using disciplinary language. Also, the fact that the expressions were metaphoric was seldom brought to the fore. Instead Sture talked about Noble gases as having "*reached their Nirvana*", "not *wanting to play* with others" or being "*drunk* (Sw. 'full') and *happy*"[4] and he stated that the reactive substances in the seventh group of the periodic table are "the *meanest* substances". The students picked up these metaphors and used them both in talking and writing in the classroom (Danielsson, 2011a).

14.6 Discussion and Conclusions

Through our analyses we discerned interesting patterns regarding the use of metaphors. Our first research question concerns the characteristics of metaphorical language used by teachers and students regarding the atomic structure and ion formation.

First, the metaphors used could be connected to the scientific or everyday life domains, or they were anthropomorphic. Connected to the scientific domain were

---

[4]This is a word game that works in Swedish, since Swedish *full* means both "drunk" and "full."

metaphors that have been integrated into the scientific discourse, such as *attraction*, *electron clouds*, and *atomic shells*, or metaphors connected to other scientific fields, like *planetary orbits*. Connected to the everyday domain were objects like *apples*, and *bicycle spikes*. A large number of metaphors were anthropomorphic, such as "they [noble gases] don't *want to play* with others".

Second, through the SFL-analysis (Halliday & Matthiessen, 2004) we discerned patterns as to how the metaphors were expressed linguistically, with metaphors being expressed as processes ("atoms *want to have* a full outer shell", "electrons *jump*"), first or second participants ("this *electronic cloud* swirls around", "think of an *apple*"), or attributes connected to the participants ("*drunk and happy*") and circumstances ("the electrons swirl *in the electronic cloud*"). Metaphors expressed in processes were often anthropomorphic, and what was especially interesting was the fact that a number of these were mental processes (*want, like*) with chemical particles or substances as the first participant having a role as an experiencer. When anthropomorphic metaphors involved material processes (*jump, give, take*), the processes imply a first participant with the role as an actor with an intention of doing something. An important finding was that metaphors or analogies were only seldom presented explicitly ("you can *think of the atom as an apple*") or through other linguistic choices indicating analogies ("as a kind of first explanation *we say that* they are in the electronic cloud"). Such expressions were mainly connected to relatively obvious metaphors such as *apples, clouds*, or *thin film*.

Our second research question concerns possibilities or hindrances for students' understanding of content, and their border-crossing into the field of science. In regard to the scientific content, different properties of the atom were foregrounded depending on the choice of metaphor, and at the same time other features were disregarded. This is as true for metaphors that have been integrated into the scientific discourse, such as *electron clouds, shells*, or *orbits*, or *attraction*, as it is for more provisional metaphors, such as *apples* or *bicycle spikes*. The cloud metaphor focuses on the fact that electrons do not move in exact circles. However, the everyday concept "cloud" does not imply any particular movement at all. The shell metaphor is useful for analyses of chemical reactions. However, it implies something quite concrete, and as mentioned, student comments about *empty shells*, previously reported from the overall project (Danielsson, 2011b) indicate that the border-crossing into the scientific domain can be challenging if metaphors are understood literally, or if the reach of them is unclear. The orbit metaphor, on the other hand, implies no concrete object and involves a movement. Yet, this orbital movement appears to be quite exact. Thus, each of these metaphors as resources to explain scientific aspects of the atom has different affordances (Gibson, 1977; Kress, 2009), or "gains and losses". In our data, we also noted that at times the use of metaphors could lead the focus away from the science content to the actual source of the metaphor, as when teacher and students started talking about horses, hockey, and places to sit, when a sports arena was used as an analogy to the atom.

Even though the teachers in our data clearly expressed that models are just simplified models, a striking result was the relative lack of discussions around metaphors. Neither did the teachers discuss the reach of the analogies or the affordances (gains and losses) that can be connected to the respective choices. Sometimes the teachers signaled that they were using analogies, but such signaling was mainly noted in relation to metaphors such as apples, or bicycle spikes. Conversely, in connection to anthropomorphic metaphors, no such signals or explanations were noted. Instead, chemical particles were presented as actors with intentions and feelings or with human attributes. Such metaphors were used to a great extent, something that indicates that this way of presenting the content is more or less a normal part of the discourse of school science. The ample use of such metaphoric language, in combination with the fact that previous research has indicated that students tend to consider anthropomorphic metaphors as true (Tibell & Rundgren, 2010), calls for a greater awareness among teachers about the challenges associated with such expressions.

We also noted that the tendency to use metaphoric expressions sometimes led to a situation where scientific language was avoided, thus hindering the students' border-crossing into the field of science with regard to the discourse of science. When using metaphors as pedagogic tools for explaining the scientific context through connections to the students' previous knowledge or everyday lives, there is a risk that the everyday expressions are used at the expense of disciplinary language. If you normally talk about noble gases as "drunk and happy" instead of stable, or reactive substances as "mean", students will not get the opportunity to gradually appropriate the disciplinary language (Lemke, 1990, 1998).

We believe that in order to enable students' border-crossing into the scientific arena, the metaphoric language used in science activities need to be made explicit to the students (Aikenhead, 1996; Tobin, 2006). A teacher's choice to use a metaphor is always based on an idea to foreground a specific aspect of a phenomenon. One example taken from our data is when the teacher used a large sports arena (the Globe) in analogy with the atom. The idea behind this analogy was to foreground that there is nothing in the space between the particles (Mortimer & Scott, 2003, p. 95). However, the teacher, in this specific situation, did not support the students in applying the scientific point of view by making explicit why to choose this particular analogy, and the reach of it. Therefore, the students' attention was drawn to misleading features of the analogy (in this case places to sit, instead of empty space). To encourage students to generate other metaphors (Haglund, 2013) (both disciplinary specific and more provisional) in order to highlight "empty space" could be another productive way of highlighting content in parallel with the gains and losses around metaphors.

To conclude, metaphoric language is part of the scientific discourse, therefore an essential characteristic of the scientific "story" (Mortimer & Scott, 2003, p. 18) and central to the development of scientific literacy. In addition, analogies are used as pedagogical tools to make visible or explain scientific content. However, in order for metaphors and analogies to be efficient tools in the science classroom, they need to be discussed explicitly or reflected upon, for example, in regard to their reach. The teachers in our study used an abundance of metaphoric language. At the same time, only seldom were such use discussed explicitly. Through such discussions, the scientific content is focused on, at the same time as the students get access to the scientific "story". For teachers to be able to accommodate such discussions, and to get the gains of using metaphors they need to be aware both of their own use of metaphoric language, and the gains and losses of the metaphors used. We hope that our study can be a contribution in raising such awareness among science educators and teachers.

References

Aikenhead, G. (1996). Science education: Border crossing into the subculture of science. *Studies in Science Education, 27*, 1–52.
Andersen, T., Petersen, U. H., & Smedegaard, F. (2001). *Sproget som resource. Dansk systemisk funktionel lingvistik i teori og praksis* [*The language as a resource. Danish systemic functional linguistics in theory and practice*]. Odense: Odense Universitetsforlag.
Artemeva, N., & Fox, J. (2011). The writing on the board: The global and the local in teaching undergraduate mathematics through chalk talk. *Written Communication, 28*(4), 345–379.
Askeland, N., & Aamotsbakken, B. (2010). Understandings and misunderstandings of metaphors and images in science textbooks among minority pupils in Norwegian primary school. *IARTEM e-journal, 3*, 62–80.
Aubusson, P.J., Harrison, A.G., & Ritchie, S.M. (2006). *Metaphor and analogy in science education*. Dordrecht: Springer.
Brown, B. A., & Kelly, G. J. (2007). When clarity and style meet substance. In W-M. Roth, & K. G. Tobin (Eds.), *Science, learning, identity*. Rotterdam: Sense.
Bryman, A. (2012). *Social research methods*. Oxford: Oxford University Press.
Cameron, L. (2002). Metaphors in the learning of science. A discourse focus. *British Educational Research Journal, 28*(5), 673–688.
Danielsson, K. (2011a). "Då blir de fulla och glada". Multimodala representationer av atommodellen i kemiklassrum. In B. Aamotsbakken, J. Smidt, & E. Tønnessen Seip (Eds.), *Tekst og tegn*. Trondheim: Tapir Akademisk Forlag. (in Swedish).
Danielsson, K. (2011b, 19 May). Texts in chemistry classrooms. Presentation at the conference *Oslo Visions*, Oslo.
Danielsson, K. (2016). Modes and meaning in the classroom – the role of different semiotic resources to convey meaning in science classrooms. *Linguistics & Education, 35*, 88–99.
Danielsson, K., & Selander, S. (2016). Reading multimodal texts: A model for cultivating students' multimodal literacy. *Designs for Learning, 8*(1), 25–36.
Eriksson, I. (Ed.). (2011). *Kemiundervisning, text och textbruk i finlandssvenska och svenska skolor: En komparativ tvärvetenskaplig studie*. Stockholm: Stockholms universitets förlag. (in Swedish).
Fang, Z. (2005). Scientific literacy: A systemic functional linguistics perspective. *Science Education, 89*(2), 335–347.
Gibson, J. (1977). The theory of affordances. In R. Shaw, & J. Bransford (Eds.), *Perceiving, acting and knowing: Toward an ecological psychology* (pp. 67–82). Hillsdale: Lawrence Earlbaum.
Golden, A. (2010). Grasping the point: A study of 15-year-old students' comprehension of metaphorical expressions in schoolbooks. In Low, G. (Ed.), *Researching and applying metaphor in the real world* (pp. 35–61). Amsterdam: John Benjamins.
Haglund, J. (2013). Collaborative and self-generated analogies in science education. *Studies in Science Education, 49*(1), 35–68.
Halliday, M. A. K. (1978). *Language as Social Semiotics: The Social Interpretation of Language and Meaning*. London: Edward Arnold.
Halliday, M. A. K., & Martin, J. R. (1993). *Writing science: Literacy and discursive power*. London: University of Pittsburgh Press.

Halliday, M. A. K., & Matthiessen, C. M. (2004). *An introduction to functional grammar*. 3rd ed. London: Arnold.
Holmberg, P., & Karlsson, A-M. (2006). *Grammatik med betydelse [Grammar with meaning]*. Uppsala: Hallgren and Fallgren.
Jewitt, C. (Ed.) (2017). *The Routledge Handbook of Multimodal Analysis*. London: Routledge.
Klein, P. D., & Kirkpatrick, L. C. (2010). Multimodal literacies in science: Currency, coherence and focus. *Research in Science Education, 40*, 87–92.
Kress, G. (2009). What is mode? In C. Jewitt (Ed.), *The Routledge handbook of multimodal analysis* (pp. 54–67). London: Routledge.
Kress, G. (2010). *Multimodality. A Social Semiotic Approach to Contemporary Communication*. London: Routledge.
Kress, G., Jewitt, C., Ogborn, J., & Tsatsarelis, C. (2001). *Multimodal teaching and learning: The rhetorics of the science classroom*. London: Continuum.
Kress, G., & van Leeuwen, T. (2006). *Reading images: The grammar of visual design*. 2nd ed. London: Routledge.
Lakoff, G., & Johnson, M. (1980). *Metaphors we live by*. Chicago: University of Chicago Press.
Lave, J., & Wenger, E. (1991). *Situated learning: Legitimate peripheral participation*. Cambridge: Cambridge University Press.
Lemke, J. L. (1990). *Talking science: Language, learning and values*. Norwood: Ablex Publishing Corporation.
Lemke, J. L. (1998). Multimedia literacy demands of the scientific curriculum. *Linguistics and Education, 10*, 247–271.
Martinec, R. (2004). Gestures that co-occur with speech as a systematic resource: The realization of experiential meanings in indexes. *Social Semiotics, 14*(2), 193–213.
Moje, E. B. (2007). Developing socially just subject-matter instruction: A review of the literature in disciplinary literacy teaching. *Review of Research in Education, 31*, 1–44.
Mortimer, E. F., & Scott, P. H. (2003). *Meaning making in secondary science classrooms*. Maidenhead: Open University Press.
Norris, S. P., & Phillips, L. M. (2003). How literacy in its fundamental sense is central to scientific literacy. *Science Education, 87*(2), 224–240.
Ogborn, J. (Ed.). (1996). *Explaining science in the classroom*. Bristol: Open University Press.
O'Halloran, K. L. (1999). Towards a systemic functional analysis of multisemiotic mathematics texts. *Semiotica, 124*(1–2), 1–29.
Rogoff, B. (1995). Observing sociocultural activity on three planes: Participatory appropriation, guided participation, and apprenticeship. In J. V. Wertsch, P. del Rio, & A. Alvarez (Eds.), *Sociocultural studies of mind*. Cambridge: Cambridge University Press.
Sutton, C. (1992). *Words, science and learning*. Buckingham: Open University Press.
Svanlund, J. (2007). Metaphor and convention. *Cognition and Linguistics, 18*(1), 47–89.
Swedish Research Council. (2016). *Humanistisk och samhällsvetenskaplig forskning*. Retrieved from http://www.codex.vr.se/forskninghumsam.shtml
Tan, E., Calabrese Barton, A., Kang, H., & O'Neill, T. (2013). Desiring a career in STEM-related fields: How middle school girls articulate and negotiate identities-in-practice in science. *Journal of Research in Science Teaching, 50*, 1143–1179.
Tibell, L., & Rundgren, C-J. (2010). Educational challenges of molecular life science: Characteristics and implications for education and research. *Life Sciences Education, 9*, 25–33.
Tobin, K. (2006). Why do science teachers teach the way they do and can they improve practice? In Aubusson, P. J., Harrison, A. G. & Ritchie, S. M (Eds.), *Metaphor and analogy in science education*. Dordrecht: Springer.
Wenger, E. (1998). *Communities of practice: Learning, meaning and identity*. Cambridge: Cambridge University Press.
Yore, L. D. & Treagust, D. F. (2006). Current realities and future possibilities: Language and science literacy – empowering research and informing instruction. *International Journal of Science Education 28*(2–3), 291–314.

Chapter 15
Image Design for Enhancing Science Learning: Helping Students Build Taxonomic Meanings with Salient Tree Structure Images

Yun-Ping Ge, Len Unsworth, Kuo-Hua Wang and Huey-Por Chang

Abstract Drawing on cognitive theories of graphic comprehension and on systemic functional semiotics, the intention of this study is twofold: first, to examine the effects of image design on reading comprehension of science texts; second, to investigate the process of meaning-making when reading image and verbal text. An experiment was conducted to test the hypothesis that image designs with salient tree structure can cue better reading comprehension about the concept of the biological classification system. A 5-phase interview was developed to investigate the reading comprehension in different textual conditions. 12 Taiwanese students from year 7 were assigned as the participants either in a control group to read the text with the textbook images or in a treatment group to read the same texts but with a salient tree structure image designed to be more coherent with the textual information. The participants are further identified in terms of low, medium, and high level of prior knowledge on the topic according to a pretest. The results support the hypothesis which shows the textbook image did not efficiently activate as many theme-related meanings as the tree-structure one. Moreover, there are many misunderstandings embedded in the design of the textbook image which might also be potential risks for the other readers. The influence of prior knowledge on the reading comprehension was negligible. Implications are drawn for the importance of image design in textbooks and biology pedagogy, and value of extended large-scale research in this area.

Y.-P. Ge (✉)
Department of Education and Human Potentials Development, National Dong-Hwa University, Hualien, Taiwan
e-mail: yunpingge@yahoo.com.tw

L. Unsworth
Learning Sciences Institute, Australian Catholic University, Sydney, Australia
e-mail: Len.Unsworth@acu.edu.au

K.-H. Wang
National Changhua University of Education, Changhua, Taiwan
e-mail: sukhua@cc.ncue.edu.tw

H.-P. Chang
Open University of Kaohsiung, Hualien, Taiwan
e-mail: president@ouk.edu.tw

Keywords Image design · reading comprehension · multiple representations · meaning-making · tree structure

15.1 Introduction

Images and words are the most common representations used in science textbooks to mediate modern scientific paradigms (Kuhn, 1972). The complementary use of images and language constrains the interpretation of each mode by reference to the other, facilitating deep understanding of the concepts (Ainsworth, 1999). However, not all images have the same facilitative effects for science learning and it has been shown that some image designs facilitate more efficient learning than the others (Canham & Hegarty, 2010; Schnotz & Bannert, 2003). In general, appropriate image designs can lead to better science learning whereas ill-designed images are possible causes of misconceptions (Catley, Novick, & Shade, 2010; Eilam, 2013).

The importance of image design has been articulated in many studies which are, however, limited in a number of ways. First, most studies focus on the images related to physics (Mayer & Gallini, 1990), chemistry (Seufert, 2003), meteorology (Canham & Hegarty, 2010), earth science (Lee, 2010), and geography (Schnotz & Bannert, 2003). Biological diagrams, particularly those depicting abstract concepts in science learning materials for younger students have received little attention. Second, many such studies have prescribed how images should be designed to be efficient but have not included any empirical validation (Fleming, 1987; Kosslyn, 2006). Third, most of these studies verify hypotheses through quantitative data analysis only and do not engage in qualitative analyses to determine why some image designs fail to facilitate efficient learning (Canham & Hegarty, 2010; Schnotz & Bannert, 2003).

Based on our previous finding, there are variant types of image designs which can represent taxonomic relations (Ge, Chung, Wang, Chang, & Unsworth, 2014). Some explicitly represent taxonomic relations by tables or tree structures whereas the others represent these relations implicitly by simply different positions. It has been claimed that taxonomic images with tree structure may better facilitate readers' understanding (Ifenthaler, 2010; Nesbit & Adesope, 2006). Furthermore, if visual cues are explicit enough to highlight the key concept, it is believed the cues can privilege the recognition of target ideas (Larkin & Simon, 1987). This study investigates the reading comprehension of taxonomic images by manipulating image design in order to reveal how different image designs affect meaning-making.

15.1.1 Research Questions

1. What are the effects of image design on junior high school students' reading comprehension of science texts with an accompanying taxonomic image?
2. Which image design do students judge most appropriate for use in science textbooks?
3. On what basis do students justify their choice of image?

15.2 Theoretical Framework

Our theoretical framework integrates both cognitive and semiotic perspectives to interpret how and why image design influences reading comprehension. The cognitive perspective focuses primarily on the design of an efficient taxonomic image in concert with words to build a coherent mental representation, while the semiotic perspective focuses more on the interplay of semiotic resources in image and language in communicating meaning. The integration of both perspectives can illuminate an understanding of the relationship between visual comprehension and semiotic construction of meaning through image–language relations in science texts.

15.2.1 Cognitive Theories About Image and Language Comprehension

When reading texts involving multiple representations, comprehension requires readers to build referential connections between corresponding elements in words and images (Seufert, 2003). Knowledge construction depends on whether readers could connect inter-related visual and verbal elements to construct a coherent mental representation. Readers would search for relevant elements across representations and identify the relevant relations in order to build meanings. If the mental representations built from a text and relevant image are not compatible with each other, meaning construction would be difficult due to the process of analogical structure mapping between different representations (Gentner & Markmann, 1997; Seufert, 2003). Therefore, the image display plays a crucial role in comprehension. If the visual representation generated from the image display maps onto the structure of a mental model then the comprehension will be reinforced. Otherwise, the meaning construction will be impeded by different representation configurations.

According to Gestalt principles, the visual perception is by no means the sum of the parts. The perception of what will be regarded as a whole and what will be as parts is determined by the functional relations of visual elements in an image (Wagemans et al., 2012). For example, common region is a tendency for the elements within the same bounded region to be perceived as a group (Palmer, 1992). If the elements within a given region share property similarity, such as two black dots circled by an oval, they are more likely to be perceived as a group (Wagemans et al., 2012).

Image structure is thus recommended to be designed to cue specific conceptions (Lee, 2010; Schnotz & Bannert, 2003; Seufert, 2003). For example, a world map representing time difference by time zone format is more efficient in cuing the understandings related to which kind of the task assessed (Schnotz & Bannert, 2003). It is believed that explicit cues can highlight and thus privilege the recognition of target ideas (Larkin & Simon, 1987). However, comprehension involves a variety of interdependencies among the visual elements (Lee, 2010). Some cues will orchestrate with the whole whereas others might counter the intended effects. Too many visual cues could overwhelm readers' cognitive load and lead to no benefits at all (Chandler & Sweller, 1991).

With regard to the representation of the classification system, two considerations are suggested. First, tree structure has been widely accepted as effective in representing hierarchical relationship through node-link assemblies (Hurley & Novick, 2010; Novick & Catley, 2007). Usually recognized as cladograms or concept maps, tree diagrams are powerful tools to represent evolutionary relationships in modern biology and externalize the structure of conceptual knowledge in a text (Nesbit & Adesope, 2006; Novak & Gowin, 1984). Second, it is advisable to avoid unnecessary details in order not to increase the cognitive load and lead readers astray (Cook, 2006; Paas & van Merrienboer, 1993). Third, prior knowledge is also essential in considering the effects of image design on meaning construction (Seufert, 2003). Readers with higher levels of prior knowledge tend to have better and deeper understanding of the image and text. For those with insufficient prior knowledge, comprehension of visual representation would be difficult and problematic (Cook, 2006).

When reading texts involving multiple representations, comprehension requires readers to build referential connections between corresponding elements in words and images (Mayer, 2003; Seufert, 2003). Knowledge construction depends on whether readers could connect inter-related visual and verbal elements to construct a coherent mental representation. Readers would search for relevant elements across representations and identify the relevant relations in order to build meanings. If the mental representations built from a text and relevant image are not compatible with each other, meaning construction would be difficult (Seufert, 2003).

15.2.2 A Semiotic Theory of Image and Language Relations

Based on the concept of language as a social semiotic resource (Halliday, 1978), our semantic framework extends the grammar of visual design (Kress & van Leeuwen, 2006) and intersemiotic identification (Unsworth & Cléirigh, 2009) to interpret image–text relations. Science texts are highly "packed" in meaning because the language configuration has been transformed to represent generalized and abstract ideas (Halliday, 1998). The process of linguistic unpacking of science texts could restore hidden meanings that are embedded in the disciplinary discourse.

The grammar of visual design proposes that the meaning-making in images can be described formally as with a verbal grammar (Kress & van Leeuwen, 2006). The visual components of a science image are independently self-organized and structured rather than being a merely duplication of verbal texts. Instead, each image has three embedded dimensions of meaning: ideational meaning referring to the meaning of subject matter; interpersonal meaning referring to the relationships between image and readers; textual meanings referring to the meaning of compositional arrangement of images.

The construction of meaning from the words and images depends on image–language relations (Unsworth, 2006). Unsworth and Cléirigh (2009) further elaborate how the meaning can be made by mutual identification. Either image or verbal text can be used as an "identifier" to identify the other mode. For example, if readers are not familiar with the image, the language elements can function as identifier to gloss the image. On the other hand, when readers are more familiar with the image, then the image will visualize the language. Therefore, we suggest that either image or text can be the mode of departure to identify the other. Though this intersemiotic identification is proposed from a semiotic perspective (Unsworth & Cléirigh, 2009), it matches the notion of structure mapping between corresponding elements in different representations (Gentner & Markmann, 1997; Seufert, 2003). When readers read an image and a text, the meaning construction requires referential connection from both resources which interrelate the corresponding elements and corresponding structures in different representations. However, if the image structure does not agree with the text, mapping might not be processed successfully and then comprehension will be impeded (Schnotz & Bannert, 2003).

15.3 Method

This study was mainly concerned with two variables that might affect students' reading comprehension. One was the impact of alternative image design on reading comprehension of the same text. To investigate this study, we designed an alternative image, the form of which we believed to be more facilitative in comprehending science concepts than the image in the textbook. Our concern was the effects of alternative image designs. As well as focusing on image design, we also accounted for prior knowledge as a variable affecting reading comprehension. The participants were differentiated into low, medium, and high level prior knowledge, based on the result of a pretest.

15.3.1 5-Phase Interview

Since readers construct meanings integratively from images and text and this process can be initiated either from image or text (Unsworth & Cléirigh, 2009), the investigation of visual comprehension was controlled in an interview contexts by first presenting the image isolated from the verbal information and then revealing the additional caption and finally the text in a 3-phase interview (Pozzer-Ardenghi & Roth, 2005). To investigate visual comprehension further, these novice learners were asked to evaluate synonymous images. Therefore, a 5-phase interview was developed with two additional phases (shown as Fig. 15.1). In the first phase, the image-only context, participants received no verbal information in textual format. The caption was provided in the second phase. In the third phase, the associated text was revealed. Then, in the fourth phase participants accessed the same text with the synonymous image. Finally, in the fifth phase participants were asked to judge one image that they believed to be more appropriate in science textbook and provide their reasons.

Rather than simply thinking aloud, our participants were invited to express the difficulties they experienced in relation to understanding particular parts of the image and text. Furthermore, they were prompted by asking "anything else?" to be exhaustive in

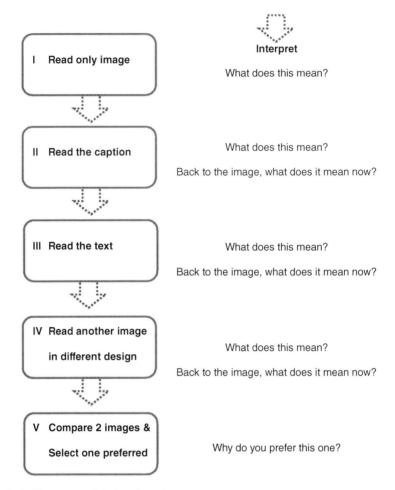

Fig. 15.1 The process of 5-phase interview

their interpretation of meanings. All the interviews were coded according to the themes of the reading material (shown in Fig. 15.3). If the underlying idea of the interpretation was not correct, the contents would be coded as a misunderstanding.

15.3.2 Reading Materials

The reading materials consisted of text and images about the biological classification system for defining groups of living things on the basis of shared characteristics and the name of those groups. The image used a laddered structure of classification labels to identify the grey wolf in seven sequential ranks (shown in Fig. 15.2). All verbal themes in the text were examined to check if corresponded to the visual elements. As shown in Fig. 15.3, in addition to the primary theme of naming the grey wolf in

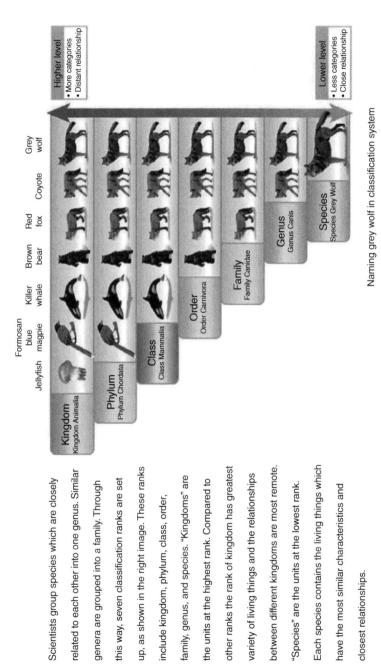

Scientists group species which are closely related to each other into one genus. Similar genera are grouped into a family. Through this way, seven classification ranks are set up, as shown in the right image. These ranks include kingdom, phylum, class, order, family, genus, and species. "Kingdoms" are the units at the highest rank. Compared to other ranks the rank of kingdom has greatest variety of living things and the relationships between different kingdoms are most remote. "Species" are the units at the lowest rank. Each species contains the living things which have the most similar characteristics and closest relationships.

Fig. 15.2 The reading text and image about classification system for control group (translated from Chen, Fang, Yao, Hsu, & Lee, 2010, p. 98, Figs. 15.3–15.4)

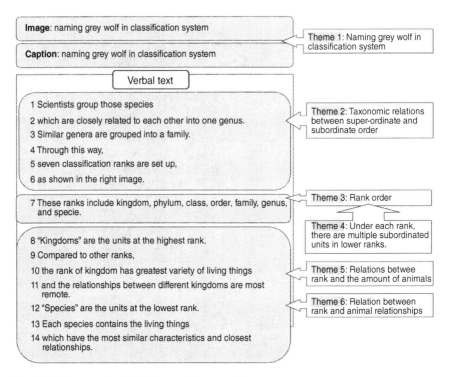

Fig. 15.3 Themes embedded in the reading material

the classification system, this image also represented rank order, taxonomic relations between superordinate and subordinate order, relations between ranks and the number of animals, and relations between ranks and animal relationships.

According to cognitive load theory, too many colors and locating the classification terms far from grey wolf could obscure the primary theme. Moreover, the result of examining the verbal text in Fig. 15.3 revealed that the textbook image was short of an embedded Theme 4 in which there should be other branches other than grey wolf to represent the real structure of classification system. Then, as suggested, an alternative image reformatted with tree structure which reduced the colors to make the grey wolf salient was created as a treatment image (shown in Fig. 15.4).

15.3.3 Participants

Our participants were 12 students in year 7 from two urban junior high schools in middle Taiwan who volunteered to take part in the study. These participants were randomly assigned into either control group or treatment group. The control group read the control image first and then the treatment image in the fourth phase. The treatment group had the opposite order of reading the images. According to the Taiwanese science curriculum, basic notions of classification in biology are

Scientists group species which are closely related to each other into one genus. Similar genera are grouped into a family. Through this way, seven classification ranks are set up, as shown in the right image. These ranks include kingdom, phylum, class, order, family, genus, and species. "Kingdoms" are the units at the highest rank. Compared to other ranks, the rank of kingdom has greatest variety of living things and the relationships between different kingdoms are most remote. "Species" are the units at the lowest rank. Each species contains the living things which have the most similar characteristics and closest relationships.

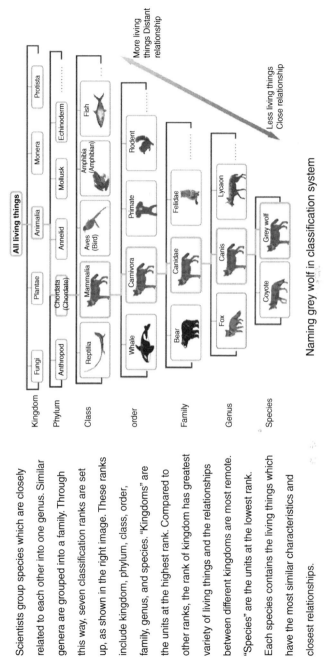

Fig. 15.4 The reading text and image for the treatment group

generally taught in the primary school. More advanced notions are located in the curriculum of year 7. None of our participants had received the relevant lesson on biological classification before the study.

15.4 Findings

All the interviews were transcribed in Mandarin and coded according to the themes of the reading material. If the underlying idea was not appropriate, the contents would be coded as a misunderstanding. The effects of image design are shown by Phases in the following:

15.4.1 Phase 1. Image Only

This phase isolates the participants from verbal information. It is not easy for the participants to identify the primary theme of the image. Naming the classification of grey wolf requires the participants to identify seven taxonomic terms in the ranks. Most participants were struggling in making meaning by picking up some of the visual elements. Only five could successfully name grey wolf in seven ranks and make this identification meaningfully. Four out of five are from the treatment group (shown in Table 15.1). Only one is from the control group.

According to the interview, these participants picked up the theme immediately without hesitation. For example, a participant, MT2 (the first code M indicating the level of prior knowledge; the second T indicating the group from), from the medium level of prior knowledge in the treatment group told how he identified the theme of the image as follows:

> 01 Interviewer: What does this image mean?
>
> 02 MT2: It (grey wolf) is chordate, mammal, carnivore, Order Canidae, Genus Canis, and Species grey wolf.
>
> 04 Interviewer: How do you know this?
>
> 05 MT2: I follow the ascending order (of the yellow boxes in this image).
>
> 06 Interviewer: Why don't you mention the other terms?
>
> 07 MT2: Oh, it's because these all in yellow boxes.

Table 15.1 Identification of the primary theme related to caption in phase 1

Identity prior knowledge	Control group		Treatment group	
	#1	#2	#1	#2
Low				
Medium			✓	✓
High	✓		✓	✓

According to Line 07, it was the yellow boxes framing the grey wolves which helped. Though the other participants from treatment group did not mention the yellow boxes, they all named the grey wolf by classification system. In contrast, the only participant from the control group, HC1, also did this but on a different basis.

> 11 Interviewer: What does this image mean?
>
> 12 HC1: This lets me know which type grey wolf is.
>
> 13 Interviewer: How do you know this?
>
> 14 HC1: Coyote looks like grey wolf. But it is not the same species with grey wolf. **If excluding** bear from Order Carnivore, **then the rest will be** Family Canidae. Jellyfish would be the species most distant from grey wolf. This is the least relationships between jellyfish and grey wolf.
>
> 18 Interviewer: How do you know this?
>
> 19 HC1: Because we can only find jellyfish and grey wolf in Kingdom Animal. Jellyfish is absent hereafter from this row (Kingdom Animal). It is likely that Formosan blue magpie is not mammal because mammals are listed in Order here. Formosan blue magpie appears (from top) until Phylum. It disappears in Class. Also killer whale is not Order Carnivore.
>
> 24 Interviewer: Why do you say so?
>
> 25 HC1: It (killer whale) only presents at mammal (box). It **disappears** in the rows down from Order Carnivore. Though red fox looks like coyote and grey wolf, it is not Genus Canis.
>
> 28 Interviewer: Why do you say so?
>
> 29 HC1: Because it (red fox) only presents (from top) until Family Canis. It is not Genus Canis.

Based on HC1, the classification is determined by excluding animals from the boxes. The approach, "if excluding … from …, then the rest would be …" (Lines 14 and 15) is different from that of the treatment group. However, this is not just a coincidence but shared by the other participants, HT1 and MT1, when they read the textbook image in Phase 4. As a later stage in the interview, the verbal text had been revealed but these two participants still made the identification in a manner similar to what HC1 has done in Phase 1. Neither did they use any words related to classification. Instead, many words with exclusive meaning were used. All these words are marked in bold face.

> 30 HT1: This is the result of **sieving** (the animals) many times. Jellyfish **is not** a vertebrate of animal kingdom. So it (this image) **separates** jellyfish from the group. And then mammals. Birds **are not** mammals. They are oviparous. Therefore, it (the bird here) is **excluded**. In Order Carnivore, whales **are not** carnivore. They eat algae. Then bears **are not** Family Canidae so it (this image) **picks** fox, coyote and grey wolf **out**.
>
> 40 MT1: Kingdom, then jellyfish disappears (from Phylum Chordata) because jellyfish is Mollusca, **not** vertebrate, therefore, it **is out of** phylum. The next, class, because Formosan blue magpie is oviparous, therefore (the image) **delete** it. Killer whale is omnivore so **delete** it. The next, only bear is bear family, the rest are all Family Canidae so keep them. The next is genus. Fox **is not** Genus Canis so **delete** it. The rest are coyote and grey wolf. Finally only the animal grey wolf is grey wolf species.

This alternative interpretation is realized like the process of sieving out unwanted things so that only what is wanted remains. All the participants interpreting via this alternative approach did not use words related to classify. Meanwhile, the verbs used by the participants who interpreted the tree-structure image were "differentiate," "tell from," "branch out," and "classify." The following is an example.

50 Interviewer: What does this diagram mean?

51 MT2: It is about classification. Carnivores are those differentiated from mammals. And then it keeps differentiating downward until Species Grey wolf and Species Coyote.

53 Interviewer: What else?

54 MT2: It is clearer to have these animals classified. And, that's it.

Actually the word "classify" in Chinese means first dividing things into several parts and then grouping those similar together. We were curious if the image design cued different ways of defining classification. If this was the case, then the word use should be different. An examination of the word use related to classification was therefore developed. The results in Tables 15.2 and 15.3 show that almost all the participants used classifying words when reading the tree-structure image but only two did when reading the textbook image.

In addition to the unusual way of defining classification, a misunderstanding was generated by reading the textbook image. MC1 identified "kingdom" as meaning the same as "kingdom animal."

Interviewer: What does this image mean?

MC1: The classification of all animals.

Interviewer: According to the diagram, how do you know this?

MC1: Kingdom is the category which includes the greatest number of animals. It includes all the animals.

Table 15.2 Word use of related to classification in interpreting the tree-structure image

Identity prior knowledge	Control group in phase 4		Treatment group in phase 1	
	#1	#2	#1	#2
Low	✓	✓	✓	✓
Medium	✓	✓	✓	✓
High		✓	✓	✓

Table 15.3 Word use of related to classification in interpreting the textbook image

Identity prior knowledge	Control group in phase 1		Control group in phase 1	
	#1	#2	#1	#2
Low				
Medium		✓		
High		✓		

Interviewer: What is kingdom? Have you learned it before?

MC1: Kingdom, mm, is animal kingdom. No, I haven't.

Interviewer: Can you say again what kingdom is?

MC1: Kingdom is animal kingdom.

We examined the textbook image. The reason MC1 responded in this way was probably due to these name tags in the same box. This is a design with potential risk of generating misunderstandings.

15.4.2 Phase 2. An Image with Caption

With the revealing of the caption, the number who could identify the primary theme increased slightly (shown in Table 15.4). The newly increased member is one from the treatment group with a low level of prior knowledge. The meaning-making in the control group involved little change.

Though the meaning-making did not increase markedly in this phase, a new misunderstanding appeared among the participants with low prior knowledge. LC2 could not understand why there were so many identical grey wolves as he misinterpreted these as evolved species.

LC2: This (Formosan blue magpie) is created by means of this jellyfish. And then, the next are whale and brown bear. One produces two and then more.

Interviewer: What do you mean "one produces two and then more"?

LC2: Finally there are many.

Interviewer: What do these grey wolves mean?

LC2: There are many kinds of grey wolves. This is a grey wolf belonging to Family Canis. This is another belonging to Genus Canis. This is Order Carnivore. And this is Family Canis. Grey wolf can be divided into these types.

Interviewer: What else?

LC2: We can tell different kinds of grey wolves. There are many kinds of grey wolves.

Interviewer: Do you think the grey wolf here (in Class Mammal) is the same as the grey wolf here (in Order Carnivore)?

LC2: No, they are different.

Table 15.4 Identification of the theme related to caption in Phase 2

Identity prior knowledge	Control group		Treatment group	
	#1	#2	#1	#2
Low				✓
Medium			✓	✓
High	✓		✓	✓

LC2 misunderstood that each grey wolf represented a distinct species. This misunderstanding is similar to the idea that Chinese have many food habits. Some are vegetarians. Some are omnivore. But all of them are classified as Chinese and looked very similar. The misunderstanding does not stop here. Both LC2 and LT1 demonstrated similar misunderstandings in Phase 4.

15.4.3 Phase 3. An Image with Caption and Verbal Text

With the revealing of verbal text, more thematic meanings were generated in this phase (shown in Table 15.5). The gap between low and high group, however, in this phase became larger rather than narrowing down.

In addition, the way of interpreting hierarchical order was reversed by three participants, LT2, MT2, and HT2. The original way was top-down from kingdom to species when provided with one of the images. Now the way was bottom-up from species to kingdom which was consistent with the verbal text. The following is an example.

> LT2: It seems that species grey wolf and species coyote belong to Genus Canis. Then Genus Canis, Genus Fox, and Genus Lycaon belong to Family Canidae. Family Bear, Family Canidae, and Family Felidae belong to Carnivora Order.
>
> Interviewer: Why is the direction of your explanation the opposite from what you have said a few minutes ago? At that time you explained the image from top to down.
>
> LT2: Because it (the verbal text) says that closely related species are combined into one Genus.

Though the verbal text causes a reversion of interpretation, it did not cancel misunderstandings. Both LC2 and LT1 maintained the misunderstanding about multiple grey wolves generated in Phase 2. Moreover, a missing Theme 4 was common among the control group. In contrast, almost all the participants from the treatment group could determine this embedded meaning. This is a very noteworthy difference between groups.

15.4.4 Phase 4. With Additional Image, Caption and Verbal Text

By reading an alternative synonymous image, most participants could directly derive multiple thematic meanings in this phase. In contrast, no participant demonstrated understanding of more than one thematic meaning in Phase 1. The difference in the number of participants identifying themes is shown in Table 15.6.

Table 15.5 Number of the themes identified in Phase 3

Identity prior knowledge	Control group		Treatment group	
	#1	#2	#1	#2
Low	0	0	0	2
Medium	2	4	3	5
High	3	4	5	4

In this phase, two synonymous images and all verbal information were revealed. Most participants made progress in identifying at least two themes. The rarely identified Theme 4 was now understood by the control group when they read the tree-structure image (shown in Table 15.7). Only LC1, LC2, and LT1 could not identify this implicit theme.

15.4.5 Phase 5. Image Judge and Justification

Here, the participants were asked to judge which image was more appropriate to use in textbook. The result shows that 8 out of 12 judged the tree-structure image as their preferred choice (shown in Table 15.8). The reasons after reorganization were represented in Table 15.9.

According to Table 15.9, all the reasons referred to ideational meanings. The missing Theme 4 was identified by LT2, MC1, MT2, HT1, and HC2. The reason to justify the textbook image by the other four participants was to confirm the helpful cues.

Table 15.6 Number of the themes increased from Phase 1 to Phase 4

Identity prior knowledge	Control group		Treatment group	
	#1	#2	#1	#2
Low	0→1	0→0	0→0	0→1
Medium	0→2	0→3	1→2	1→4
High	1→4	0→3	1→4	1→5

Table 15.7 The timing of identifying Theme 4 among all the phases

Identity prior knowledge	Control group		Treatment group	
	#1	#2	#1	#2
Low				Phase 1
Medium	Phase 4	Phase 4	Phase 1	Phase 1
High	Phase 5	Phase 4	Phase 4	Phase 3

Table 15.8 Image judge between the textbook image and tree-structure image

Identity prior knowledge	Control group		Treatment group	
	#1	#2	#1	#2
Low	♦	♦	♣	♣
Medium	♣	♣	♦	♣
High	♣	♣	♣	♦

♦ represents the textbook image.
♣ represents the tree-structure image.

Table 15.9 Reasons of selecting an image to be printed in textbook

Selection prior knowledge	Tree-structure image	Textbook image
Low	• [a]Classify by more detailed branches. List all the kingdoms. (LT2) • Highlight important places. (LT1)	• Easier to understand by the pictures and arrows. Label relationships clearly. (LC1) • Can see the living things more clearly. Easier to know the living things by labeling name tags. (LC2)
Medium	• [a]Classify by more detailed branches. (MC1 & MT2) • Provide more animals for us to know. (MC2)	• Represent more clearly by deleting animals which are not included. Help me to know which animals are included. (MT1)
High	• Can know more kinds of animals. (HT1) • Can easily tell the classification of grey wolf. (HT1) • [a]Represent classification by multiple and detailed branches. (HT1 & HC2) • [a]Find the difference between grey wolf and the other animals. (HC1)	• Easier to understand. Easier to tell the classification levels of grey wolf. The animals for comparison remained and labeled with names. The arrows and boxes marking the relations between relationships and categories are helpful. (HT2)

[a]represents the reason is related to the missing Theme 4.

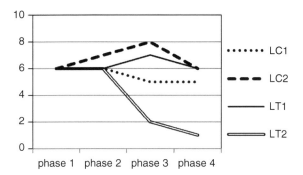

Fig. 15.5 Number of the misunderstandings with low prior knowledge

15.4.6 The Number of Misunderstandings in the Four Phases

As more information was revealed gradually, the number of misunderstandings fluctuated in different levels of prior knowledge groups (shown in Figs. 15.5–15.7). As the more verbal information was provided, it was expected that the misunderstanding would become fewer. In Fig. 15.5, the participants with low prior knowledge, however, had even more misunderstandings. LC2 and LT1 had the most misunderstandings in Phase 3.

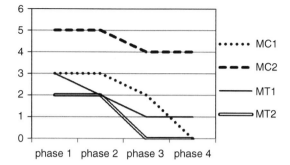

Fig. 15.6 Number of the misunderstandings with medium prior knowledge

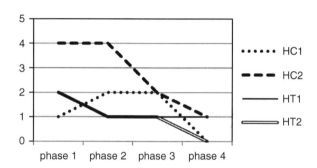

Fig. 15.7 Number of the misunderstandings with high prior knowledge

15.5 Discussion

15.5.1 *The Effects of Image Design*

The 5-phase interview helped us to investigate the effects of image design in different conditions. Particularly in Phase 1 when the image was isolated from the verbal information, the impacts between groups are clear. The tree structure was more successful in cuing the primary theme. Partly because the tree structure is so commonly used in representing classification that it quickly maps on to participants' mental model and retrieves the relevant concept. In contrast, the participants were not familiar with the textbook image so that the identification was not as easy as the tree-structure image. And the indented boxes cue another aspect of understanding classification by sorting out those that do not share similar characters. The left edge of the textbook image, regularly indented one by one, is like a v-shape sieve. Together with the common region formed by those animals repeated vertically, the tendency of grouping a sieve function mapping mental model so that the interpretations subconsciously used the verbs related, such as "separate," "exclude," "delete," and "is out of." In contrast, the salient tree structure in yellow frame successfully cues the understanding of classification by grouping those with similar characters. The words used by participants were "differentiate," "tell from," "branch out," and "classify." Their interpretations were highly related to the image design.

Actually both methods of interpretation are legitimate in biology. But there are possible risks for the textbook design. The regularity of these indented edges might suggest the classification system is neatly organized. Since the textbook image serves more like a metaphor to represent a rough idea about the classification of grey wolf rather than an authentic diagram to illustrate the system. If this motivation is not perceived by the novice learners, misunderstanding may well occur. Moreover, the boxes are framed by solid lines, instead of dotted ones, which also hint that the composition of each rank is real. For example, the animals in Genus Canis are definitely many more than those represented by the coyote and grey wolf. In addition, another risk is embedded in the arrangement of name tags. The coexisting classification ranks and units hint these tags are the same. Thus MC1 was guided to say "kingdom, mm, is animal kingdom."

One last problem of the textbook image is the most salient. The system lacks the other subordinated units other than the single hierarchical line represented by the name tags. That is, there should be other kingdoms in addition to the Kingdom animal. Therefore, all the participants from control group did not mention any other branches until the presence of the tree-structure image according to Table 15.7. For novice learners, their perceptions were limited by the image. An important question is why the verbal text could not recover the missing branches? Due to the nature of science texts, this theme is only embedded among the sentences. So that even after the text was revealed, the meanings remained sealed and not until the presence of the tree-structure image did the control group start to identify the existence of other branches.

From the linguistic perspective, mutual identification between representations provides a possibility to decode the meaning (Unsworth & Cléirigh, 2009). The already-made meaning construction from Phase 1 to 3 forms a basis for the control group to comprehend the tree-structure image. It is then easier to identify multiple themes and find out the missing theme unrepresented in their previous reading.

On the other hand, the cognitive perspective argues that comprehension requires readers to build referential connections between corresponding elements across representation modes (Seufert, 2003). If both words and image don't provide any referents as in the reading materials experienced by the control group, these novice learners could not possibly build Theme 4 by themselves. Only when the tree-structure image offers an alternative visual cue can the participants then recover the missing Theme. If meaning-making in reading comprehension is like solving a jigsaw puzzle, then the revelation of new information could offer clues to complete the original work.

15.5.2 The Effects of Verbal Text

The 5-phase interview helped us to see the effects of image design in different contexts of textual affordance. Many people might take for granted that readers would have better understanding about an image if additional verbal text were provided. To some extent, this is the case. However, if the image design is not appropriate, the text is limited in facilitating further comprehension.

The influence of the text is first shown by the shift of the meaning constructions after the text was revealed. Three participants reversed their way of interpreting the classification system and shifted to follow the bottom-up order of the verbal text. By reversing the direction, the structure mapping between representations and mental constructions can be resolved finally. If the image was read in association with the verbal text in the interview from the beginning, this change would not have been found. A second aspect of the addition of the text is that, although the text provided access to more understanding of the topic, the control group were still unable to identify Theme 4, suggesting that the image viewed by this group did not prompt interaction with the text to reveal this implicit theme to the participants.

15.5.3 The Influence of Prior Knowledge

The influence of prior knowledge is clear by comparing the results through phases. When those with medium or high prior knowledge could discern thematic theme in Phase 1, none of the participants from low level of prior knowledge were able to achieve this. Even when the majority of participants had successfully identified multiple themes in Phase 4, these low participants were still struggling. What they had identified remained as the surface features of the image, such as the animals or name tags. Their perception of the images was so disparate that no integrated meanings could be made to map with the verbal one. Moreover, as Fig. 15.3 reveals, even with the text, these participants still seemed unable to make more meanings and the text only offered more possibilities for them to mix their piecemeal perceptions into misunderstandings. In contrast, those who had more prior knowledge, as shown in Figs. 15.5 and 15.6, were able to use the text as a key to successfully decode the image. Therefore, their misunderstandings decrease in successive phases.

15.5.4 Participant Image Choices and Reasons

The ultimate theme identification in Phase 4 also resonates with the outcomes of the participant choices of images in Phase 5. The reason why the tree-structure image was judged as the more appropriate to use in textbooks relates to the evaluation of these two images. The representation of tree-structure image seemed more intelligible, plausible, and fruitful in interpreting the classification system, thus supporting it as a more efficacious image. In their reasons for preferring the tree-structure image, LT2, MC1, MT2, HT1, HC2, and HC1 mentioned their awareness of the missing Theme 4. These participants included those who had low and medium levels of prior knowledge, which is a further confirmation of the significance of the tree-structure image in cueing meaning construction for these students. Therefore, the judgment is significant.

15.6 Conclusion and Implications

The 5-phase interview helped us to see the effects of image design in different contexts of textual affordance. Two synonymous images represented the same topic but were associated with different reading comprehension results. As a powerful tool verified by the interview data, the salient tree-structure image design cues taxonomic relations more successfully than the original textbook image design. The factor of image design is crucial in meaning-making either from the image alone or from the image in combination with text.

The presence of verbal text could facilitate the comprehension. However, the facilitation is limited by two ways. First, it is limited by the nature of science text. Particularly when the image fails to represent a complete structure of the concept. The meaning-making of novice learners depends mostly on the nature of the image representation. The ill-designed image will result in either reading difficulty or misunderstandings. The representation of the tree-structure image is more consistent with the current scientific paradigm and compensates for the absence of missing theme in the text. Therefore, it facilitates better reading comprehension than the textbook image.

Second, a greater amount of semiotic information does not necessarily guarantee better understandings. The effect of the extent of semiotic information depends on readers' prior knowledge. For the readers with better prior knowledge, the caption and verbal text offer resources of mutual identification and enhance comprehension. However, for those readers with poor prior knowledge, the additional verbal text just cast more information to confuse them.

The tree-structure image is judged to be superior by the majority of participants. Through the comparison of two images, even as novice learners, our participants could discern the key differences between them. An effective image design is not difficult for learners to identify.

However, without effective pedagogic intervention, student reading of science materials might generate many misunderstandings. This suggests that teachers should develop explicit pedagogy to help students to comprehend the science materials. Particularly for those students with poor prior knowledge who struggle even when they are provided with ample semiotic resources.

Our findings inform the future research a deeper concern about image design. The results provide valuable information by arguing for textbook publishers, science teachers, and science educational researchers. The design of our treatment image offers empirical evidence to support the principles of image design (Fleming, 1987; Kosslyn, 2006). Though the findings from the interviews offer in-depth insight about the effects of image design, they were derived from the experience of a small number of participants. Nevertheless, the findings suggest the value of a large-scale study to investigate in greater breadth and depth the effects of variation in image design on student comprehension of science concepts from their interpretation of image–language relations in science texts.

Acknowledgements We are grateful to Azing Chen, Hsunfei Yang, Laxic Hsiao, Wikimedia, and all the publishers for the permission of copyright in this book.

References

Ainsworth, S. (1999). The functions of multiple representations. *Computers & Education, 33*, 131–152.

Canham, M., & Hegarty, M. (2010). Effects of knowledge and display design on comprehension of complex graphics. *Learning and Instruction, 20*, 155–166.

Catley, K. M., Novick, L. R., & Shade, C. K. (2010). Interpreting evolutionary diagrams: When topology and process conflict. *Journal of Research in Science Teaching, 47*(7), 861–882.

Chandler, P., & Sweller, J. (1991). Cognitive load theory and the format of instruction. *Cognition and Instruction, 8*, 293–332.

Chen, S.-H., Fang, C.-H., Yao, H., Hsu, K.-C., & Lee, T.-Y. (2010). *Science and Technology 2*. Taiwan: Han-Lin.

Cook, M. P. (2006). Visual representations in science education: The influence of prior knowledge and cognitive load theory on instructional design principles. *Science Education, 90*(6), 1073–1091.

Cox, R., & Brna, P. (1995). Supporting the use of external representations in problem solving: The need for flexible learning environments. *Journal of Artificial Intelligence in Education, 6*(2–3), 239–302.

Eilam, B. (2013). Possible constraints of visualization in biology: Challenges in learning with multiple representations. In D. F. Treagust & C.-Y. Tsui (Eds.), *Multiple representations in biological education* (pp. 55–74). London: Springer.

Fleming, M. L. (1987). Designing pictorial/verbal instruction: Some speculative extensions from research to practice. In D. A. Houghton & E. M. Willows (Eds.), *The psychology of illustration volume 2: Instructional issues* (Vol. 2, pp. 136–157). New York: Springer.

Ge, Y. P., Chung, C. H., Wang, K. H., Chang, H. P., & Unsworth, L. (2014). Comparing the images in Taiwanese and Australian science textbooks by grammar of visual design: An example of biological classification. *Chinese Journal of Science Education 22*, 109–134.

Gentner, D., & Markmann, A. B. (1997). Structure mapping in analogy and similarity. *American Psychologist, 52*(1), 45–56.

Halliday, M. A. K. (1978). *Language as social semiotic: The social interpretation of language and meaning*. London: Edward Arnold.

Halliday, M. A. K. (Ed.). (1998). *Language and knowledge: The 'unpacking' of text*. Beijing: Peking University Press.

Hurley, S. M., & Novick, L. R. (2010). Solving problems using matrix, network, and hierarchy diagrams: The consequences of violating construction conventions. *The Quarterly Journal of Experimental Psychology, 63*, 275–290.

Ifenthaler, D. (2010). Relational, structural, and semantic analysis of graphical representations and concept maps. *Educational Technology Research and Development, 58*, 81–97.

Kosslyn, S. M. (2006). *Graph design for the eye and mind*. New York: Oxford University Press.

Kress, G., & van Leeuwen, T. (2006). *Reading images: The grammar of visual design*. 2nd ed. New York: Routledge.

Kuhn, T. S. (1972). *The structure of scientific revolution*. Chicago: University of Chicago Press.

Larkin, J. H., & Simon, H. A. (1987). Why a diagrnm is (sometimes) worth ten thousand words. *Cognitive Science, 11*, 65–99.

Lee, V. R. (2010). How different variants of orbit diagrams influence student explanations of the seasons. *Science Education, 94*, 985–1007.

Mayer, R. E. (2003). The promise of multimedia learning: Using the same instructional design methods across different media. *Learning and Instruction, 13*, 125–139.

Mayer, R. E., & Gallini, J. K. (1990). When is an illustration worth ten thousand words? *Journal of educational psychology, 82*(4), 715–726.

Nesbit, J. C., & Adesope, O. O. (2006). Learning with concept and knowledge maps: A meta-analysis. *Review of Educational Research, 76*(3), 413–448.

Novak, J. D., & Gowin, D. B. (1984). *Learning how to learn*. New York: Cambridge University Press.

Novick, L. R., & Catley, K. M. (2007). Understanding phylogenies in biology: the influence of a gestalt perceptual principle. *Journal of Experimental Psychology: Applied, 13*(4), 197–223.

Paas, F., & van Merrienboer, J. J. G. (1993). The efficiency of instructional conditions: An approach to combine mental effort and performance measures. *Human Factors, 35*(4), 737–743.

Palmer, S. E. (1992). Common region: A new principle of perceptual organization. *Cognitive Psychology, 24*, 436–447. https://doi.org/10.1016/0010-0285(92)90014-S

Pozzer-Ardenghi, L., & Roth, W.-M. (2005). Making sense of photographs. *Science Education, 89*, 219–241.

Schnotz, W., & Bannert, M. (2003). Construction and interference in learning from multiple representation. *Learning and Instruction, 13*, 141–156.

Seufert, T. (2003). Supporting coherence formation in learning from multiple representations. *Learning and Instruction, 13*, 227–237.

Unsworth, L. (2006). Towards a metalanguage for multiliteracies education: Describing the meaning-making resources of language-image interaction. *English Teaching: Practice and Critique, 5*(1), 55–76.

Unsworth, L., & Cléirigh, C. (Eds.). (2009). *Multimodality and reading: The construction of meaning through image-text interaction*. London: Routledge.

Wagemans, J., Elder, J. H., Kubovy, M., Palmer, S. E., Peterson, M. A., Singh, M., et al. (2012). A century of Gestalt psychology in visual perception: I. Perceptual grouping and gigure–ground organization. *Psychological Bulletin, 138*(6), 1172–1217.

Part 5
Disciplinary Literacy and Science Inquiry

Chapter 16
Inquiry-Based Science and Literacy: Improving a Teaching Model Through Practice-Based Classroom Research

Marianne Ødegaard

Abstract This chapter is based on the outcomes of The Budding Science and Literacy research project in Norway. The project included several video-based classroom studies, which aimed to continually improve a teaching model of integrated science inquiry and literacy instruction in collaboration with practicing science teachers (six primary teachers and their students). The main aim of this chapter is to use these studies to improve and clarify the model so the essential features are easily communicated to users. The data sources for the studies were classroom video observations and interviews. The main analytical approaches were (i) variations in multiple learning modalities (read-it, write-it, do-it, talk-it) and (ii) the distribution of different phases of inquiry (preparation, data, discussion, communication). The results indicate that literacy activities embedded in science inquiry provide support for teaching and learning science. The greatest challenge for teachers is to find the time and courage to exploit opportunities for consolidating conceptual learning in the discussion and communication phases. Investigation on students while they communicated their inquiries revealed that students' word knowledge develops toward conceptual knowledge when they are required to apply the key concepts in their talk throughout all phases of inquiry. In interviews, students expressed that literacy and the role of text in science was not clear. Nevertheless, multiple literacies emerged, "schooled" and everyday literacy practices, when students connected literacy to science inquiry.

Keywords Inquiry-based science education · literacy · conceptual understanding · literacy practices · video study

16.1 Introduction

In this chapter, a teaching model based on the importance of inquiry and literacy in science is explored through several studies conducted as part of the Budding Science and Literacy project (Ødegaard, Haug, Mork, & Sørvik, 2014). All of the

---

M. Ødegaard (✉)
Department of Teacher Education and School Research, University of Oslo, Oslo, Norway
e-mail: marianne.odegaard@ils.uio.no

studies (Haug, 2015; Haug & Ødegaard, 2014, 2015; Melhus, 2015; Ødegaard et al., 2014; Sørvik, Blikstad-Balas, & Ødegaard, 2015; Sørvik & Mork, 2015) elaborate on different aspects of the Budding Science teaching model. The results are summed up, and implications for the teaching model are discussed. The focus is on improving and clarifying the model so the essential features are easily communicated to users, who are mainly science teachers in elementary and secondary schools.

Inquiry and literacy have a twofold role of providing structures that support science content learning and being important areas of content knowledge in the science curriculum (Knain & Kolstø, 2011; Norris & Phillips, 2003; Wellington & Osborne, 2001). Pearson, Moje, and Greenleaf (2010) claimed that science and literacy serve each other and that a curriculum based on the two will provide synergy effects. Science learning benefits from embedded literacy activities, as literacy learning benefits from being embedded within science inquiry.

16.2 Local Context

When the Norwegian national curriculum in science was changed in 2006 and then revised in 2013, the reform was in line with international trends in science education. The revised curriculum includes an increased focus on science inquiry, scientific practices, and the nature of science (Abd-El-Khalick et al., 2004; Ministry of Education and Research, 2006/2013; National Research Council, 2012; Rocard et al., 2007), with a new cross-curricular demand to integrate subject literacies, denoted as basic skills, in all subjects (reading, writing, arithmetics, oral, and digital competence). However, implementation studies showed that the demand to focus on basic skills did not seem to be understood and thus was not perceived as meaningful by teachers (Møller, Prøitz, & Aasen, 2009), which led to the adjustment in 2013 (Mork, 2013) (see Chap. 2, Knain and Ødegaard).

As a consequence of the new curriculum, many teachers expressed that teaching inquiry-based science was challenging, especially when it was combined with basic skills. This teaching style was new to them, and they lacked teaching resources and models. The promotion of the Budding Science teaching model was an attempt to meet the teachers' requests for resources and develop a framework for how inquiry and literacy together might foster good science teaching and learning.

16.3 Inquiry-Based Science, Scientific Literacy, and Literacy

Inquiry involves engaging students in critical thinking to deepen their understanding by using logic and evidence about the natural world (Crawford, 2014). Inquiry includes being curious, asking questions, designing and carrying out investigations, interpreting data as evidence, creating arguments, building models, and communicating findings (Barber, 2009). The Budding Science and Literacy project's understanding of science inquiry concurs with Crawford's (2014) and Barber's (2007) interpretations.

Inquiry-based science is often described as consisting of various inquiry features and is a "multifaceted activity" (National Research Council, 1996) that involves posing questions (Chinn & Malhotra, 2002), exploring (Bybee et al., 2006), testing hypotheses (Gyllenpalm Wickman, & Holmgren, 2010), designing and carrying out investigations (Crawford, 2014), analyzing data (Krajcik et al., 1998), making explanations based on evidence (Barber et al. 2007), and debating and communicating findings (Wu & Hsieh, 2006). Some present inquiry-based science as inquiry cycles, for example, the 5E learning cycle model (Bybee et al., 2006) that consists of the Engagement, Exploration, Explanation, Elaboration, and Evaluation phases. The Seeds of Science/Roots of Reading program describes several inquiry cycles with increasing sophistication in the ways students employ evidence to form logical explanations (Barber et al. 2007). However, Bell, Urhahne, Schanze, and Ploetzner (2010) emphasized that inquiry processes do not appear in a fixed order and should not be interpreted as steps in a linear fashion. In a recent review of inquiry-based learning studies, Pedaste et al. (2015) identified from various inquiry cycles five distinct general inquiry phases: orientation (stimulating curiosity, addressing a learning challenge), conceptualization (stating theory-based questions and/or hypotheses), investigation (planning exploration or experimentation, collecting, and analyzing data), conclusion (drawing conclusions from the data, comparing inferences with research questions), and discussion (presenting findings by communicating with others and/or engaging in reflection). These phases have several possible pathways and do not necessarily occur in cycles. The Budding Science and Literacy project describes inquiry-based science through four phases: preparation, establish data, discussion, and communication (Ødegaard et al., 2014).

An understanding of scientific inquiry and the nature of science is fundamental to the development of scientific knowledge and scientific literacy, and the notions of scientific literacy continue to be discussed. DeBoer (2000) suggested a broad conceptualization of scientific literacy that included the public's understanding and appreciation of science. Norris and Phillips (2015) noted that the concept of literacy in general has transitioned from being defined as the "ability to read and write" to include "having education and knowledge typically in a specific area" and that scientific literacy refers to being educated and possessing knowledge in and about science.

However, literacy, as the ability to read and write, is fundamental in order to engage in science inquiry and is a crucial part of scientific literacy. Literacy, in its fundamental sense, is based on the essential role of text in science and involves reading and writing and being fluid in the discourse patterns and communication systems of science (Norris & Phillips, 2003). Specific literacy practices that underlie science have often been seen as tools that help scientists do science, not as essential features of the nature of science. Norris and Phillips (2015) stated that scientific literacy practices should be incorporated into science teaching and learning as a natural part of science itself, just as observations are. Further, Norris and Phillips explained how science and reading involve inquiry by analyzing, interpreting, and taking a critical stance on information. Pearson et al. (2010) pointed out how doing inquiry-based science also has synergy effects on students' reading abilities because students use the same cognitive strategies.

Focusing on scientific literacy practices as a natural part of school science falls under the perspective of Vision I of scientific literacy defined by Roberts (2007); it looks inward at the products and processes of science itself. Vision II of scientific literacy is portrayed as related to situations with a scientific component that students are likely to encounter as future citizens. Roberts describes the relationship between the two visions as "these two visions [are] in a kind of mating dance wherein they complement one another" (2007, p. 730). Vision I forms the basis of Vision II, but Vision II can provide a context for making Vision I relevant to students (Roberts, 2007). Inquiry-based science in the Budding Science and Literacy project provides contexts that transcend the subject of science, bringing in the students' everyday lives and other socioscientific issues, thus relying on the broad perspective of scientific literacy included in Vision I and II.

16.4 Inquiry-Based Science and Literacy Studies

Several promising science and literacy projects have supported employing literacy tools with doing inquiry to acquire knowledge in science. In a review, Pearson et al. (2010) emphasized among many the Concept-Oriented Reading Instruction (CORI) program (Guthrie et al., 1996; Guthrie, Wigfield, & Perencevich, 2004). The CORI framework emphasized the role of science and science inquiry as a setting to provide students with various types of interaction that facilitate reading to promote reading engagement through content-area learning. Palincsar and Magnusson (2001) developed the Guided Inquiry Supporting Multiple Literacies research program, in which firsthand investigations (hands-on) and secondhand investigation (consulting the text to learn from others' interpretations) were combined to support teachers' and students' participation in science inquiry. The Seeds of Science/Roots of Reading program (Cervetti, Pearson, Bravo, & Barber, 2006) is described as an approach with multiple modalities in which inquiry-oriented reading, writing, and talking activities are embedded within hands-on scientific inquiry (see Fig. 16.1 from Pearson et al., 2010). In the Nordic action research project, Students as Researchers in School Science (Knain & Kolstø, 2011), language and literacy activities were developed as important support structures for science inquiry (Bjønness & Kolstø, 2015). Large-scale studies have shown that integrated inquiry-based science and literacy activities lead to improved learning outcomes (Cervetti, Barber, Dorph, Pearson, & Goldschmidt, 2012; Fang & Wei, 2010). However, there have been calls for more research to understand the challenges teachers encounter in the classroom when they integrate science and literacy (Howes, Lim, & Campos, 2009).

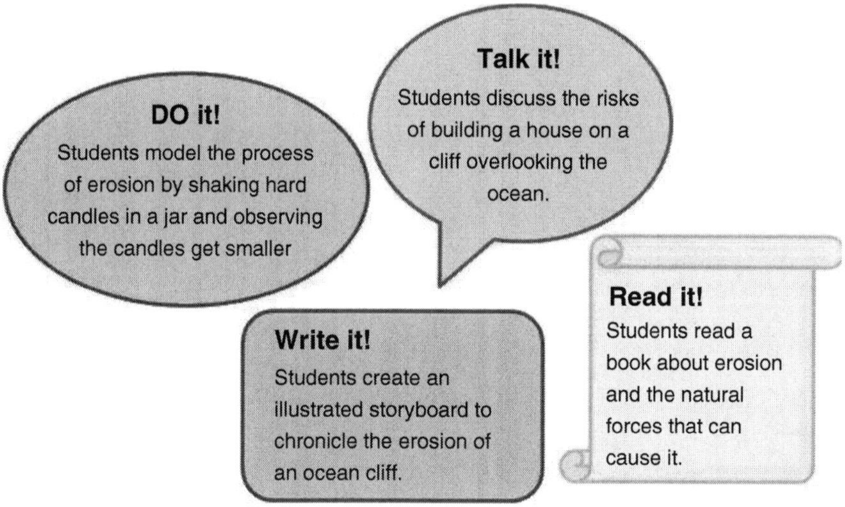

DO it!
Students model the process of erosion by shaking hard candles in a jar and observing the candles get smaller

Talk it!
Students discuss the risks of building a house on a cliff overlooking the ocean.

Write it!
Students create an illustrated storyboard to chronicle the erosion of an ocean cliff.

Read it!
Students read a book about erosion and the natural forces that can cause it.

Fig. 16.1 An approach with multiple modalities to learning about the concept of erosion (Barber et al., 2010) (adapted from Pearson et al. 2010)

16.5 The Budding Science Teaching Model

When the Budding Science teaching model was first developed, it was greatly influenced by other science and literacy projects. The main idea was that classroom teaching should consist of systematic variation of inquiry activities, which implies that students use multiple learning modalities (doing, reading, writing, talking) (see Fig. 16.2). This model was consistent with the Seeds of Science/ Roots of Reading teacher program (Cervetti et al., 2006). Science, as a knowledge-building enterprise, naturally includes inquiry and multiple activities, such as doing experiments, reading scientific articles, writing reports, and presenting and discussing scientific studies. The model proposed that by systematically varying the learning activities, students in science education were to learn more about the nature of science and gain multiple learning experiences connected to a science topic, and thus obtain more robust knowledge.

The students also systematically alternated between firsthand and secondhand investigations to make meaning of the world surrounding them. Firsthand investigations are practical "hands-on" investigations, and secondhand investigations are text-based investigations that involve texts written about scientific studies performed by others (Palincsar & Magnusson, 2001). In the program, text-based inquiries were integrated as a natural part of science inquiries as in academic science and showed how texts can enhance and support "hands-on" inquiries. The continuous shift between firsthand and secondhand inquiries was a crucial element of the model. The students practiced searching for evidence and arguments in experience-based information and text, and this activity supported the

Fig. 16.2 The first Budding Science teaching model (Ødegaard, 2011), inspired by Barber et al. (2007)

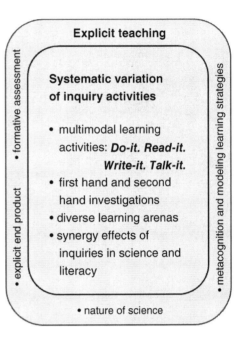

development of scientific language (Barber et al., 2007; Mork, 2005). The intention was for students to understand more about the dynamics between data and text in authentic science (Norris et al., 2008) and to recognize the synergy effects of inquiries in science and literacy (Cervetti et al., 2006; Pearson et al., 2010).

Another important feature of the initial Budding Science teaching model (Ødegaard, 2011) was that teachers explicitly focus on metacognition, modeling learning strategies and formative assessment, during the multiple classroom activities. The students were also systematically reminded to link their own work to the nature of science context and the end product of the inquiry. The teacher was to provide reasons for his or her pedagogical choices and make his or her teaching methods visible, so the students understood why they were doing what they did. This integrated model (not only teaching what the students should learn but also how they should do it; Weinstein, Bråten, & Andreassen, 2006) has also been used successfully in projects such as Communities of Learners (Brown, 1997) and CORI (Guthrie et al., 2004). Research conducted in Norwegian lower secondary reading and science classrooms indicate that explicit teaching strategies are not used systematically (Anmarkrud & Bråten, 2009; Ødegaard & Arnesen, 2010). The objective of explicit teaching strategies was that the students gained a metacognitive understanding and insight into their own learning strategies, and consequently the students' ability to transfer knowledge should increase. Lave and Wenger (1991) described this as transparency in the learning process: the use of tools (here: learning strategies) and understanding the use are integrated. In concrete situations, "transparency" makes knowledge and understanding more available to students. The nature of science was also made transparent. In this program, the students read narratives of scientists' work connected to the theme and performed practical inquiry activities that made the students see the links between learning and doing inquiry in science education and scientists' knowledge building.

16.6 The Budding Science and Literacy Research Project

To evaluate the Budding Science teaching model, a research project was designed. Several studies were completed during the research project using various cases with a range of perspectives. These studies used the same data material and constitute the basis for this review.

To develop a functional and valuable teaching model, there was a need for researchers to work closely with teachers and students. Interested and experienced elementary school teachers were invited to enroll in a professional development course based on inquiry-based science and literacy, operationalized through the Budding Science and Literacy teaching model. Thus, the teachers' professional classroom competence was combined with researchers' competence in science education. The teachers tried out classroom activities between course meetings and reflected together with researchers about the purpose of evaluating the practical experiences with the teaching model. To elaborate and increase students' knowledge about inquiry and literacy in science classrooms, a classroom research project was designed. The aim of the project was to study how the interaction between inquiry-based science and literacy could support the teaching of science and students' meaning making in science. The research findings from the various studies are presented here with a discussion of how they guided the improvement of the Budding Science teaching model.

16.7 Methodological Design

All studies in the Budding Science and Literacy project had the same methodological design: a combination of video observations and interviews. Video observations enabled analysis and comparison of different layers of activities in order to search for patterns (Ødegaard & Klette, 2012). The focus was to explore how an inquiry approach influenced literacy activities in science classrooms and vice versa. Analyzing videos of classroom activities provides an overview of the variation and succession in the inquiry and literacy activities (Ødegaard, Haug, Mork, & Sørvik, 2014). This video analysis formed the basis for several in-depth studies (Haug, 2014, 2015; Haug & Ødegaard, 2014, 2015; Sørvik, 2015; Sørvik et al., 2015) by providing a database of systematically coded activities that could underpin the selection of cases for further investigation.

Six teachers from four schools and their students were studied during a sequence of science lessons (five to ten lessons per teacher) (see Table 16.1 for more

Table 16.1 Overview of background information about participating teachers, schools, and recordings (adapted from Ødegaard et al., 2014)

Teacher	Years of teaching	Science credits[a]	Grade	No. of students	School location	Total video rec. (in min.)
Anna	0–5	16–30	5	14	S	343
Betsy	11–15	16–30	1	18	R	165
Birgit	11–15	16–30	4	24	R	426
Cecilia	20+	16–30	3	19	S	540
Ellinor	11–15	31–60	3	16	R	224
Emma	20+	16–30	3	21	R	269
					(Suburban/Rural)	Σ 1967

[a]Generalist teacher education includes 16–30 ECTS credits in science (60 credits is 1 year full-time study).

information). The grades ranged from first to fifth grade. The teachers were recruited from a professional development course that engaged them in the Budding Science and Literacy teaching model. All of the teachers used adapted versions of Seeds and Roots units with teacher guides (Barber et al., 2007) as curriculum resources. This included several short student textbooks in various genres, student investigation notebooks, and materials for hands-on activities. The teacher guides presented the do it, talk it, read it, and write it approach linked to special topics and urged teachers to expose students to these multiple learning modalities while learning central concepts (e.g., system, structure, and function in the "Body systems" unit and observation, evidence, and inference, included in all units). Although the teachers were encouraged to follow the teacher guide closely, this was not required. Each lesson was videotaped in a classroom overview angle. In addition, two students in each lesson wore a head-mounted camera to capture student dialogue and activities (Frøyland, Remmen, Mork, Ødegaard, & Christiansen, 2015). Teachers and a selection of students from each classroom were interviewed.

For the overview analyses, a coding scheme was developed for two layers of activities: learning modalities (oral, reading, writing, practical) and inquiry phases (preparation, data collection, discussion, communication; see Table 16.2). The video recordings were analyzed with Interact coding software, and the occurrence and duration of each code were recorded (Ødegaard et al., 2014). The in-depth project studies analyzed the video observations and teacher and student interviews (Haug, 2014; Sørvik, 2015).

16.8 Review of Studies: Findings

All studies in the Budding Science and Literacy project used the coding as the starting point for further analyses. An overview of this coding was published in the article "Challenges and Support in an Integrated Inquiry and Literacy

Table 16.2 Coding scheme for video analysis, based on visible classroom activities (adapted from Ødegaard, Mork, Haug, & Sørvik, 2012, Ødegaard et al., 2014)

	Category	Specific codes
Inquiry	Preparation	Background knowledge/wondering/researchable questions/prediction/hypothesis/planning
	Data	Collection/registration/analysis
	Discussion	Discussing interpretations/inferences/implications/connecting theory and practice
	Communication	Orally/in writing/assessing their work
Multiple learning modalities	Oral activities	Whole class/group/pair/individual
	Writing activities	Whole class/group/pair/individual
	Reading activities	Whole class/group/pair/individual
	Practical activities	Whole class/group/pair/individual
	Focus on key concepts	

Teaching Model" (Ødegaard et al., 2014). In addition to providing an outline of the Budding Science and Literacy data material, this study explored how the relationship between multiple learning modalities and science inquiry might challenge and support the teaching and learning of science. The video observations were analyzed using the coding scheme presented in Table 16.2, and variations in multiple learning modalities and the distribution of different phases of inquiry were scrutinized. The inquiry phases were identified in an iterative process as the researchers reflected on theory and watched video examples of classroom activities. Four overarching inquiry phases were distinguished: preparation, data, discussion, and communication, which were operationalized by what were identified as central inquiry processes (Table 16.2). A graphical representation of the main findings is presented in Fig. 16.3. A summary of the coding for all six classrooms revealed that the inquiry phases had separate patterns. For example, the multiple learning modalities were well distributed in the preparation and data phases, whereas the discussion and communication phases were characterized mainly by oral activities. Four of the six teachers spent less time in the discussion phases than the teacher guides suggested. This result implies that the teaching resources for an integrated inquiry and literacy approach in science could include increased support for teachers to provide their students more time in the discussion phase of their inquiries. Designing a greater range of learning modalities to scaffold the discussion phase is suggested.

The study "From Words to Concepts. Focusing on Word Knowledge When Teaching for Conceptual Understanding in an Inquiry-Based Science Setting" (Haug & Ødegaard, 2014) aimed to investigate how two teachers' instruction supported the development of students' conceptual understanding from the preparation phase to the communication phase of an inquiry. The two teachers were selected based on the criteria frequent use of science concepts and incidents of students communicating their understanding based on hands-on activities (Figs. 16.4 and 16.5). The video observations were used to analyze the classroom discourse

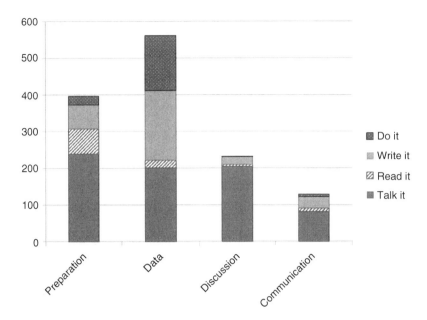

Fig. 16.3 Variation in multiple learning modalities during the inquiry phases summarized for all teachers and displayed in coded minutes (adapted from Ødegaard et al., 2014)

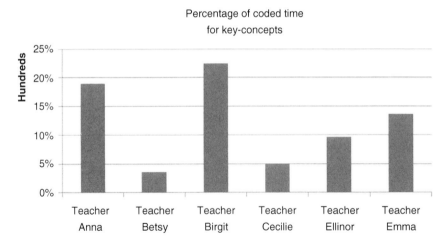

Fig. 16.4 Results for the first criterion of selecting participants for the study. The figure shows the frequency the six volunteer teachers focused the classroom dialogue on key concepts. Teacher 1, Anna, and teacher 3, Birgit, had the highest percentages (adapted from Haug & Ødegaard, 2014)

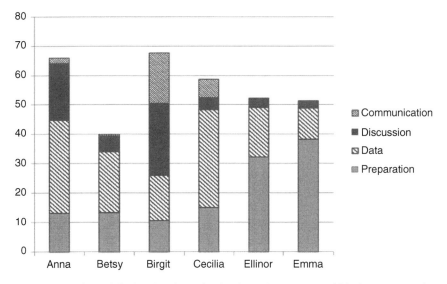

Fig. 16.5 Durations of the inquiry phases for the six teachers. Anna and Birgit spent more time discussing data (Ødegaard et al., 2014)

Table 16.3 Framework for word-knowledge (Haug & Ødegaard, 2014). Based on Bravo et al. (2008)

Level of word-knowledge	Cognitive processes	Explanation
Low	Recognition	Knowing how the word sounds or looks when it is written.
Passive	Definition	Being able to recite a word's definition but has little understanding of the meaning of the word or its implications.
Active word knowledge / Conceptual knowledge	Relationship	Knowing the word's relationship to other words and concepts.
	Context	Knowing how to use the word in context. Understanding how the word fits into different sentences.
	Application	Knowing how to apply the word in context when engaging in inquiry about the phenomena being taught. Linking the word to the empirical data.
	Synthesis	Knowing how to employ the word when communicating the emerging knowledge about the phenomena under study. Solving problems in new situations by applying acquired knowledge.

and talk actions, using a combination of students' conceptual understanding (Bravo, Cervetti, Hiebert, & Pearson, 2008) and the teachers' use of link-making strategies (Scott, Mortimer, & Ametller, 2011) as the analytical framework. Conceptual understanding involves understanding words in context and in relation to other words within the discipline. There is a continuum from low and passive word knowledge to active word knowledge in context; considered to be conceptual knowledge (Bravo et al., 2008) (see Table 16.3).

The study showed that there were clear differences between the two classrooms when students communicated their inquiry results. In one classroom, the students demonstrated a low passive level of conceptual understanding, whereas in the other classroom, the students displayed an active level of conceptual understanding. This difference was linked to the observations that in the first classroom the teacher reformulated the students' answers, and she was the one who used the key concepts. In the second classroom, the students were requested to use the concepts through all inquiry phases, and the students actively used everyday language as the starting point in discussions. The transformation from passive to active conceptual understanding detected in the second classroom is a quality the Budding Science teaching model attempts to pursue. Conceptual understanding is closely linked to the practice of formative assessment (Haug & Ødegaard, 2015). Although the Budding Science and Literacy project emphasized key concepts and the teachers expressed great commitment to teaching the key concepts, the analyses of the classroom practices disclosed few observations of teachers adapting their teaching when students revealed a lack of understanding (Haug & Ødegaard, 2015). Haug (2014) distinguished between planned and spontaneous teachable moments that can foster conceptual understanding and suggested using consolidating phases of inquiry (discussion and communication) for planning teachable moments of conceptual understanding.

Some of the published studies in the Budding Science and Literacy project focused especially on how literacy was connected to science (Sørvik, 2015; Sørvik et al., 2015; Sørvik & Mork, 2015). In the article "Do Books Like These Have Authors? New Roles For Text and New Demands on Students in Integrated Science-Literacy Instruction in Primary School" (Sørvik et al., 2015), the aim was to investigate students' emerging situated literacy practices in science classroom inquiries and identify how texts are actually used in science. In this study, student interviews were used in addition to video observations. The students' encounters with and use of text in specific literacy events (activity where text plays a role) were analyzed by drawing on the New Literacies Studies perspective (Barton, 2007)—seeing literacy as a situated social practice. Students' views and experiences related to science and science texts articulated in the interviews were also analyzed to enrich the picture. Literacy events constituted more than 50% of the video-recorded time, indicating that text played an important role in the science classrooms involved. Multiple literacy practices emerged, within the contexts of science and school and the students' everyday life. Inquiry processes opened up to include students' everyday literacy practices in the science classroom. However, the student interviews revealed that literacy and the role of text in science was not clear to the students. Thus, the utterance: "Do books like these have authors?" The study indicated that explicit attention is needed for how science texts have a sender and a receiver and are written for a purpose, and awareness of representational and communicative aspects of science and school science. Sørvik and Mork (2015) asserted that there will always be multiple school science literacies because of the changing nature of learners' lived lives in the digital information age.

In addition to studying conceptual understanding and students' literacy practices in science a small study explored on how creativity was fostered in the Budding Science and Literacy project (Melhus, 2015). Inquiry is closely linked to creative thinking (Kind & Kind, 2007; Vygotsky, 1930/2004) and one of the features of the Budding Science and Literacy model was to link students' creative ways of doing inquiry to the creative ways scientists build knowledge. Analyses of four parts of a teaching sequence were conducted. The focus was on how the teacher promoted creativity in inquiry-based science teaching. Melhus (2015) found that by explicitly modeling, emphasizing, and encouraging the use of imagination and diversity of ideas during inquiry, the teacher stimulated the students to use their creativity during inquiry tasks. The teacher gave positive or neutral feedback to students and in that way created a pleasant and trustworthy environment for the students, where it was acceptable to make mistakes.

16.9 The Improved Budding Science Model: Discussion

The improved Budding Science teaching model represented in Fig. 16.6 maintains the central principles from the first model. However, the model has been simplified; some elements are emphasized, and others are de-emphasized. In addition, some new elements are included. The model still has a core of multiple learning modalities. *Do it! Read it! Write it! Talk it!* appeared to be an important and viable slogan and was kept as the central idea of the model. Teachers expressed that the slogan helped them remember to diversify their teaching with different learning activities and in that way automatically practice basic literacy skills. For each topic within a content area, the students are to do something hands-on, talk about it, read about it, and write about it. The order of the activities attempted to be authentic, determined by how the activities support the investigations in which the students are engaged. For instance, before investigating how to make the best glue, the students read about materials and properties, observe possible glue ingredients,

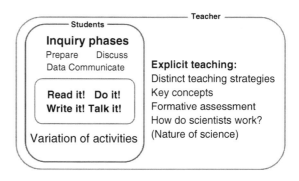

Fig. 16.6 The new improved Budding Science teaching model

and write down their predictions of what makes the best glue, do a glue testing experiment, register their data in a table, and discuss their results, before they compare their findings to other glue tests in the literature. Eventually students also design their own experiment based on the previous experiments and new ideas. When teachers create their own teaching sequences, the teachers should be aware of which role text should play in the inquiry and think of where to place text in the teaching sequence (e.g., being introduced to new concepts; learning to make a data table). The authenticity links the investigation to the way scientists naturally alter their working mode to continue their research, but as indicated by Sørvik et al. (2015), the link between literacy practices performed in everyday life (e.g., searching the Internet for information on a new topic), school science (e.g., searching for information in a science textbook), and scientific literacy practices (searching for and reading science articles to prepare a new study) must be explicitly expressed. In this way, literacy practices become a natural part of science itself as asserted by Norris and Phillips (2015). However, Phillips and Norris (2009) reminded us that authentic scientific journal articles use argument to convince the reader of the study's value. Science textbooks and trade books primarily use exposition; thus, the language is not argumentative. Phillips and Norris suggested adapting primary literature texts for school science purposes. Although the Budding Science teaching model does not include adapted texts, through student inquiries, with naturally integrated literacy practices, the need for expressing evidence-based arguments will increase.

16.10 Four Inquiry Phases

Several of the studies in the Budding Science and Literacy project showed the importance of the different inquiry phases (Haug, 2015; Haug & Ødegaard, 2014; Ødegaard et al., 2014). Including the four inquiry phases (preparation, data, discussion, and communication) in the Budding Science model became imperative (see Table 16.2).

The preparation phase is characterized by activities such as wondering, making questions, activating preknowledge, making predictions and hypotheses, planning. In this phase, the teacher has a crucial and supportive function in helping the students find interesting inquiries and find suitable and informative texts. It is important that the students feel ownership for their inquiries.

In the data phase, the students collect empirical data. They also register and analyze the data that are collected. How to register and analyze is crucial and requires critical thinking. The inquiry questions to be answered determine how the data should be analyzed, which, in turn, influences how to register the data.

In the discussion phase, the students discuss the meaning of their collected data. This phase includes discussing different interpretations, making inferences, discussing how the findings connect with theory, and making implications. Did the students' findings coincide with their predictions and hypotheses? In the discussion phase, the students try to explain and make meaning of their results by exchanging experiences, discussing interpretations, and drawing conclusions.

In the communication phase, it is essential that those who make inquiries—scientists or students—communicate their work to others. It can be done in an oral presentation, a written report, or a combination (a research poster they have written).

Although the phases are presented and often appear in this order, they did not always occur in that sequence. Discussing results occurs all the time, and it is not unusual that a communication phase and a discussion phase are entangled, and that a discussion might appear when a group of students communicate their results to their classmates. The focus of the discussion phase is continuously on interpretations and inferences based on data, which are findings that are presented during the communication phase. Scientists ensure the quality of their work by presenting and discussing it with other scientists and, thus, agreeing on what is reliable new knowledge. Likewise, when students are requested to present and discuss their work, they build and consolidate knowledge. This entanglement of discussion and communication resembles the discussion phase in Pedaste and colleagues' (2015) review of inquiry cycles. Table discussions can also appear in smaller sub-inquiries that arise during the main inquiry, for instance, while planning experiments or registering data when students try out different alternatives before they make a decision. In the Pedaste et al. framework, these discussions are part of the main inquiry phases (e.g., investigation or conclusion), and in the 5E inquiry cycle (Bybee et al., 2006), the act of discussion is not emphasized in any of the phases.

As described in the Methodology section, the four phases in the Budding Science and Literacy model were first developed as empirical video codes to provide an overview of the inquiry activities involved in the project. Thus, the phases are based on visible classroom activities, unlike the 5E instructional inquiry cycle (Bybee et al., 2006) and the phases of inquiry-based learning identified by Pedaste et al. (2015) that are based on analyses to find the core features of inquiry-based learning. The concrete, visible activities are recognizable to teachers and connect easily to the Budding Science phases. Consequently, the Budding Science phases were retained in the model, but the phases may be reconsidered in future improvements. However, activities involving learner discussions in which students make evidence-based arguments, draw inferences, and debate implications are important consolidating activities for knowledge building in science and reading (Klette, 2013; Norris & Phillips, 2003); therefore, it is crucial that they are highlighted through the discussion phase.

16.11 Explicit Teaching

The significance of the link between formative assessment and key concepts became evident through the research project. For students to understand feedback during a teaching sequence, they had to be acquainted with the aim of the lesson.

Relating the aim to central concepts made teachers and students more focused during the learning process. The teachers expressed increased confidence in teaching science topics when the teaching focused on central key concepts. The teachers could then design learning activities in which students used the key concepts to expose their understanding (Haug & Ødegaard, 2014, 2015).

Thus, the improved Budding Science Model includes a focus on a small number of key concepts for each science topic that the students meet repeatedly through writing, doing, reading, and oral activities. Key concepts are defined as words that are central for students to learn when they investigate a certain science topic. They could be scientific concepts linked to distinct science topics, like *habitat* or *digestion*, or science inquiry concepts that are necessary for planning and conducting investigations in science topics, such as *observation, prediction, evidence, hypothesis*, or *conclusion*. These concepts convey how scientific knowledge comes into existence and are important tools in classroom dialogue. Scientific and science inquiry concepts are required to do inquiries into scientific issues (Haug & Ødegaard, 2014).

The Budding Science model includes a strong emphasis on students seeing links between how they might find answers to their own questions and how scientists do it. By systematically asking the following questions when students do inquiry, they gain awareness of what characterizes scientists' work and knowledge of science: how do we resemble scientists? What would a scientist do? How would a scientist find a solution? Students also get to know how their own creative endeavor and hard work in the science classroom resemble what scientists do (Melhus, 2015; Ødegaard, Haug, Mork, & Sørvik, in press). Although the school science inquiries are mostly guided, they are based on students' curiosity and students finding answers to their own questions. Several of the metacognitive strategies the students use in doing so resemble the ones scientists use (Norris & Phillips, 2003; Pearson et al., 2010).

16.12 Conclusions

The studies in the Budding Science and Literacy project demonstrated that the improved Budding Science model should put more emphasis on students using their own data in reflective discussions and in literacy activities. It is imperative to recognize the different phases of science inquiry. The model should stress the importance of students using a small selection of key concepts actively together with everyday language. The model should encourage students' own literacy practices and link them to science practices and encourage modeling the use of imagination, not the "correct" scientific answer.

The Budding Science and Literacy studies support that ownership of data seems to be an important force of engagement. Literacy activities might provide scaffolding for discussing inquiry data and thus help teachers pay more attention to the challenging discussion phase when teaching science. Inquiry creates engagement that can be taken advantage of in reading, writing, and oral activities. In return, reading, writing, and oral activities offer structure to inquiry in science. And most importantly, most teachers who have used the Budding Science model express deep satisfaction with the multiple learning modalities. Hopefully, the teaching model will influence the future learning environments the teachers create for their students.

References

Abd-El-Khalick, F., BouJaoude, S., Duschl, R., Lederman, N. G., Mamlok-Naaman, R., Hofstein, A., et al. (2004). Inquiry in science education: International perspectives. *Science Education, 88*, 397–419.

Anmarkrud, Ø., & Bråten, I. (2009). Motivation for reading comprehension. *Learning and Individual Differences, 19*(2), 252–256.

Barber, J. (2009). *The seeds of science/roots of reading inquiry framework*. Retrieved from www.scienceandliteracy.org

Barber, J., Pearson, P. D., Cervetti, G., Bravo, M., Hiebert, E. H., Baker, J., et al. (2007). *An integrated science and literacy unit. Seeds of science. Roots of reading*. Nashville: Delta Education.

Barton, D. (2007). *Literacy: An introduction to the ecology of written language*. Malden: Blackwell.

Bell, T., Urhahne, D., Schanze, S., & Ploetzner, R. (2010). Collaborative inquiry learning: Models, tools, and challenges. *International Journal of Science Education, 32*(3), 349–377.

Bjønness, B., & Kolstø, S. D. (2015). Scaffolding open inquiry: How a teacher provides students with structure and space. *NorDiNa, 11*(3), 223–237.

Bravo, M. A., Cervetti, G. N., Hiebert, E. H., & Pearson, P. D. (2008). From passive to active control of science vocabulary. In *56th yearbook of the National Reading Conference* (pp. 122–135). Chicago: National Reading Conference.

Brown, A. (1997). Transforming schools into communities of thinking and learning about serious matters. *American Psychologist, 52*(4), 399–413.

Bybee, R., Taylor, J. A., Gardner, A., van Scotter, P., Carlson, J., Westbrook, A., et al. (2006). *The BSCS 5E instructional model: Origins and effectiveness*. Colorado Springs: BSCS.

Cervetti, G. N., Barber, J., Dorph, R., Pearson, P. D., & Goldschmidt, P. G. (2012). The impact of an integrated approach to science and literacy in elementary school classrooms. *Journal of Research in Science Teaching, 49*(5), 631–658.

Cervetti, G., Pearson, P. D., Bravo, M. A., & Barber, J. (2006). Reading and writing in the service of inquiry-based science. In R. Douglas, M. Klentschy, & K. Worth (Eds.), *Linking science and literacy* (pp. 221–244). Arlington: NSTA Press.

Chinn, C. A., & Malhotra, B. A. (2002). Epistemologically authentic inquiry in schools: A theoretical framework for evaluating inquiry tasks. *Science Education, 86*, 175–218.

Crawford, B. A. (2014). From inquiry to scientific practices in the science classroom. In N. Lederman & S. Abell (Eds.), *Handbook of research on science education*. (Vol. II, pp. 515–541). New York: Routledge.

DeBoer, G. E. (2000). Scientific Literacy: Another look at its historical and contemporary meanings and its relationship to science education reform. *Journal of Research in Science Teaching, 37*, 583–601.

Fang, Z., & Wei, Y. (2010). Improving middle school students' science literacy through reading infusion. *Journal of Educational Research, 103*(4), 262–273.

Frøyland, M., Remmen, K. B., Mork, S. M., Ødegaard, M., & Christiansen, T. (2015). Researching science learning from students' view – the potential of headcam. *NorDiNa, 11*(3), 249–267.

Guthrie, J. G., Van Meter, P., McCann, A., Wigfield, A., Bennett, L., Poundstone, C., et al. (1996). Growth in literacy engagement: Changes in motivations and strategies during concept-oriented reading instruction. *Reading Research Quarterly, 31*, 306–333.

Guthrie, J. T., Wigfield, A., & Perencevich, K. C. (2004). *Motivating reading comprehension: Concept-oriented reading instruction*. Mahwah: Erlbaum.

Gyllenpalm, J., Wickman, P.-O., & Holmgren, S. O. (2010). Teachers' language on scientific inquiry: Methods of teaching or methods of inquiry? *International Journal of Science Education, 32*(9), 1151–1172.

Haug, B. S. (2014). Teaching for conceptual understanding in science within an integrated inquiry-based science and literacy setting. Unpublished doctoral dissertation. University of Oslo, Norway.

Haug, B. S. (2015). Inquiry-based science: Turning teachable moments into learnable moments. *Journal of Science Teacher Education, 25*(1), 79–96.

Haug, B. S., & Ødegaard, M. (2014). From words to concepts: Focusing on word-knowledge when teaching for conceptual understanding within an inquiry-based setting. *Research in Science Education, 44*(5), 777–800.

Haug, B. S., & Ødegaard, M. (2015). Formative assessment and teachers' sensitivity to student responses. *International Journal of Science Education, 37*(4), 629–654.

Howes, E. V., Lim, M., & Campos, J. (2009). Journeys into inquiry-based elementary science: Literacy practices, questioning, and empirical study. *Science Education, 93*(2), 189–217.

Kind, P. M., & Kind, V. (2007). Creativity in science education: Perspectives and challenges for developing school science. *Studies in Science Education, 43*(1), 1–37.

Klette, K. (2013). Hva vet vi om god undervisning? Rapport fra klasseromsforskningen [What do know about good teaching? Report from classroom studies]. In R. J. Krumsvik & R. Säljö (Eds.), *Praktisk pedagogisk utdanning. En antologi [Practical pedagogical education: An anthology]* (pp. 173–200). Bergen: Fagbokforlaget.

Knain, E., Bjønness, B., & Kolstø, S. D. (2011). Rammer og støttestrukturer i utforskende arbeidsmåter [Support structures in inquiry]. In E. Knain & S. D. Kolstø (Eds.), *Elever som forskere i naturfag [Students as researchers in school science]* (pp. 85–126). Oslo: Universitetsforlaget.

Knain, E., & Kolstø, S. D. (2011). *Elever som forskere i naturfag [Students as researchers in school science]*. Oslo: Universitetsforlaget.

Krajcik, J. S., Blumenfeld, P. C., Marx, R. W., Bass, K. M., Fredricks, J., & Soloway, E. (1998). Inquiry in project based science classrooms: Initial attempts by middle school students. *Journal of the Learning Sciences, 7*(3–4), 313–350.

Lave, J., & Wenger, E. (1991). *Situated learning: Legitimate peripheral participation. Learning in doing: Social, cognitive, and computational perspectives*. New York: Cambridge University Press.

Melhus, F. A. (2015). *Kreative spirer. En kvalitativ studie hvor utforskende arbeidsmåter i naturfag kobles mot kreativ tenkning [Creative Buds. A qualitative study of inquiry-based science and creative thinking]*. Unpublished master's thesis. University of Oslo, Norway.

Ministry of Education and Research. (2006). *The (LK06) national curriculum for knowledge promotion in primary and secondary education and training*. Oslo: Ministry of Education and Research. Retrieved from http://www.udir.no/Stottemeny/English/Curriculum-in-English/

Ministry of Education and Research. (2006/2013). *The (LK06) national curriculum for knowledge promotion in primary and secondary education and training (revised 2013)*. Oslo: Ministry of Education and Research. Retrieved from http://www.udir.no/kl06/NAT1-03/

Møller, J., Prøitz, T. S., & Aasen, P. (2009). Kunnskapsløftet - tung bør å bære? Underveisanalyse av styringsreformen i skjæringspunktet mellom politikk, administrasjon og profesjon [Knowledge promotion: Analysis]. *NIFU STEP Rapport* (Vol. 42). Oslo.

Mork, S. M. (2005). Argumentation in science lessons: Focusing on the teacher's role. *Nordic Studies in Science Education, 1*(1), 17–30.

Mork, S. M. (2013). Revidert læreplan i naturfag – Økt fokus på grunnleggende ferdigheter og forskerspiren [Revised curriculum in science – Increased focus on basic skills and the budding scientist]. *Nordina, 9*(2), 206–210.

National Research Council. (1996). *National science education standards*. Washington: National Academy Press.

National Research Council. (2012). *A framework for K-12 science education: Practices, cross-cutting concepts, and core ideas. Committee on a conceptual framework for new K–12 science education standards*. Washington: The National Academies Press.

Norris, S. P., & Phillips, L. M. (2003). How literacy in its fundamental sense is central to scientific literacy. *Science Education, 87*(2), 224–240.

Norris, S. P., & Phillips, L. M. (2015). Scientific literacy: Its relationship to "Literacy". *Encyclopedia of science education* (pp. 947–950). Dordrecht: Springer.

Norris, S. P., Phillips, L. M., Smith, M. L., Guilbert, S. M., Stange, D. M., Baker, J. J., et al. (2008). Learning to read scientific text: Do elementary school commercial reading programs help? *Science Education, 92*(5), 765–798.

Ødegaard, M. (2011). Forskerføtter og leserøtter – et tilpasningsdyktig prosjekt i naturfag. [Budding science and literacy – an adaptable project in science]. *Bedre skole, 4*(11), 38–42.

Ødegaard, M., & Arnesen, N. E. (2010). Hva skjer i naturfagklasserommet? Resultater fra en videobasert klasseromsstudie; PISA+ [What happens in a science classroom? Results from a video-based classroom study: PISA+]. *Nordic Studies in Science Education, 1*, 16–32.

Ødegaard, M., Haug, B. S., Mork, S. M., & Sørvik, G.O. (2014). Challenges and support when teaching science through an integrated inquiry and literacy approach. *International Journal of Science Education, 36*(18), 2997–3020.

Ødegaard, M., Haug, B. S., Mork, S. M., & Sørvik, G. O. (2016). *På forskerføtter i naturfag [On scientists' feet in school science]*. Oslo: Universitetsforlaget.

Ødegaard, M., & Klette, K. (2012). Teaching activities and language use in science classrooms. In D. Jorde & J. Dillon (Eds.), *Science education research and practice in Europe* (pp. 181–202). Rotterdam: Sense.

Ødegaard, M., Mork, S. M., Haug, B., & Sørvik, G. O. (2012). *Categories for video analysis of science lessons*. Retrieved from http://www.naturfagsenteret.no/binfil/download2.php?tid=1995769

Palincsar, A. S., & Magnusson, S. J. (2001). The interplay of firsthand and text-based investigations. In S. Carver & D. Klahr (Eds.), *Cognition and instruction* (pp. 151–194). Mahwah: Erlbaum.

Pearson, P. D., Moje, E., & Greenleaf, C. (2010). Literacy and science: Each in the service of the other. *Science, 328*, 459–463.

Pedaste, M., Mäeots, M., Siiman, L.A., de Jong, T., van Riesen, S. A. N., Kamp, E. T., et al. (2015). Phases of inquiry-based learning: Definitions and the inquiry cycle. *Educational Research Review, 14*, 47–61.

Phillips, L. M., & Norris, S. P. (2009). Bridging the gap between the language of science and the language of school science through the use of adapted primary literature. *Research in Science Education, 39*(3), 313–319.

Roberts, D. A. (2007). Scientific literacy/science literacy. In S. K. Abell & N. G. Lederman (Eds.), *Handbook of research on science education* (pp. 729–780) New Jersey: Lawrence Erlbaum Associates.

Rocard, M., Csermely, P., Jorde, D., Lenzen, D., Wallberg-Henriksson, H., & Hemmo, V. (2007). *Science education NOW: A renewed pedagogy for the future of Europe*. Brussels: European Commission.

Scott, P., Mortimer, E., & Ametller, J. (2011). Pedagogical link-making: A fundamental aspect of teaching and learning scientific conceptual knowledge. *Studies in Science Education, 47*(1), 3–36.

Sørvik, G. O. (2015). *Multiple school science literacies. Exploring the role of text during integrated inquiry-based science and literacy instruction*. Unpublished doctoral dissertation. University of Oslo, Norway.

Sørvik, G. O., Blikstad-Balas, M., & Ødegaard, M. (2015). "Do books like these have authors?" New roles for text and new demands on students in integrated science-literacy instruction. *Science Education, 99*(1), 39–69.

Sørvik, G. O., & Mork, S. M. (2015). Scientific literacy as social practice: Implications for reading and writing in science classrooms. *Nordina, 11*(3), 268–281.

Vygotsky, L. S. (1930/2004). Imagination and creativity in childhood. *Journal of Russian & East European Psychology, 42*(1), 7–97.

Weinstein, X., Bråten, X., & Andreassen, X. (2006). Læringsstrategier og selvregulert læring: Teoretisk beskrivelse, kartlegging og undervisning [Learning strategies and self regulated learning: Theoretical description, mapping and teaching]. In E. Elstad & A. Turmo (Ed.), *Læringsstrategier. Søkelys på lærernes praksis [Learning strategies. focus on teacher practice]*. (pp. 27–54). Oslo: Universitetsforlaget.

Wellington, J., & Osborne, J. (2001). *Language and literacy in science education*. Buckingham: Open University Press.

Wu, H.-K., & Hsieh, C.-E. (2006). Developing sixth graders' inquiry skills to construct explanations in inquiry based learning environments. *International Journal of Science Education, 28*(11), 1289–1313.

Chapter 17
Infusing Literacy into an Inquiry Instructional Model to Support Students' Construction of Scientific Explanations

Kok-Sing Tang and Gde Buana Sandila Putra

Abstract Disciplinary literacy is increasingly emphasised as an important enabler for students to learn science inquiry. However, the nature of literacy instruction and how it supports inquiry-based science in practice still remains unclear. This chapter reports on the design and enactment of an integrated literacy-inquiry instructional model aimed to support students' development of disciplinary literacy in science. With the goal of understanding how literacy instruction supports inquiry-based science in practice, the study reported in this chapter utilised a design research to develop, enact, and test the literacy-inquiry model in four secondary school science (physics and chemistry) classrooms in Singapore. Analytical cases are shown to illustrate the nature of the literacy activities involved in the classrooms, and how they supported science inquiry in terms of: (a) framing driving question, (b) conducting experiments and collecting evidence, (c) constructing explanations and (d) communicating and evaluating explanations. The cases also illustrate how the participating teachers utilised and integrated literacy activities to support inquiry in their classrooms in order to enable the students to construct and communicate scientific explanations.

Keywords Disciplinary literacy · science inquiry · scientific explanation

17.1 Introduction

Recent curriculum reforms and standards around the world have increasingly emphasised the role of inquiry and literacy in the teaching and learning of science (see this book volume). This has resulted in a growing attention to the

K.-S. Tang (✉)
School of Education, Curtin University, Perth, Australia
e-mail: kok-sing.tang@curtin.edu.au

G.B.S. Putra
National Institute of Education, Nanyang Technological University, Singapore, Singapore
e-mail: buana.sandila@gmail.com

synergy between inquiry and literacy. In the USA for example, researchers have explored the intersection between the Next Generation Science Standards (NGSS; National Research Council, 2012) for science education and the Common Core State Standards (CCSS; Council of Chief State School Officers, 2010) for literacy education, and arrived at the consensus that three scientific practices outlined in NGSS are highly relevant to literacy. In particular, the practices of constructing scientific explanations, engaging in argumentation and communicating information are central to literacy and science inquiry (National Research Council, 2014).

Various studies on the role of literacy in inquiry-based science have highlighted that literacy and inquiry are complementary (Fang & Wei, 2010; Hand et al., 2003; Ødegaard, Haug, Mork, & Sørvik, 2014; Pearson, Moje, & Greenleaf, 2010). Pearson, Moje, and Greenleaf (2010, p. 459) argued that 'when science literacy is conceptualised as a form of inquiry, reading and writing activities can be used to advance scientific inquiry, rather than substitute for it'. Fang and Wei (2010) found that students from an inquiry-based science curriculum infused with an explicit reading strategy instruction performed better than those exposed to only an inquiry-based science curriculum. Overall, these studies raise questions of how literacy can be used as a tool to support inquiry-based science, and conversely, how literacy instruction can benefit when embedded within an inquiry-based learning setting.

Although the need to integrate literacy and inquiry is apparent, much remains unclear about how exactly literacy instruction supports inquiry-based science in practice. Moreover, there is a need for more research to understand how science teachers integrate literacy instruction with inquiry-based learning, and the challenges they face in classrooms (Howes, Lim, & Campos, 2009). For instance, some studies revealed that teachers spent significantly more time on the preparation and investigation phases of the inquiry and less time on discussion, writing and student presentation of their understanding (Ødegaard et al., 2014; Poon, Lee, Tan, & Lim, 2012). In particular, Tan, Talaue, and Kim (2014) highlighted that in Singapore, inquiry-based science often takes the form of guided inquiry with little emphasis on student discussion and exploration. Thus, there is a need to find ways to support teachers in using literacy instruction to support their inquiry-based lessons.

With these challenges in mind, the purpose of this study was to examine the role of infused literacy in supporting scientific inquiry within a design research carried out to develop, enact and test an integrated literacy-inquiry instructional approach in Singapore. Through the design research, four teachers enacted multiple lesson units designed to integrate literacy in inquiry-based lessons. For this study, we focus on the construction of scientific explanations as the targeted scientific practice to be learned by the students through the lesson units. Our choice in foregrounding scientific explanation does not imply the neglect of other scientific practices listed in NGSS. Instead, we see these other scientific practices, such as engaging in argumentation and communicating information, as necessary and leading to students' ability to construct scientific explanations. The research questions that guided this study were:

1. How is scientific inquiry supported through the literacy activities (e.g. reading, writing, talking) enacted by the teachers in the classrooms?
2. How do these inquiry-literacy activities support students in constructing scientific explanations?

17.2 Theoretical and Instructional Frameworks

17.2.1 Scientific Inquiry and Inquiry-Based Teaching

Scientific inquiry generally refers to the activities and processes scientists engage in to study the natural world. As many researchers have noted (Crawford, 2000; Keys & Bryan, 2001), inquiry in general is not a specific teaching method nor curriculum model that dictates how students should learn. Instead, inquiry should be considered as a set of disciplinary-specific skills to be developed among science students (Settlage, 2007). As such, besides an approach to science instruction (i.e. teaching *by* inquiry), science as inquiry (i.e. teaching *for* inquiry) is also a goal in itself in which students engage in the skills of inquiry as a way to learn about scientific practices as well as develop conceptual understanding in science. For two decades, inquiry has been a key focus in many science curricula around the world (National Research Council, 1996). However, with the recent development in NGSS, there is now a notable shift to use the term 'practice' instead of 'inquiry' in order to highlight the awareness that engaging in scientific inquiry require a range of practices specific to science (Ford, 2015). These practices, as documented in NGSS, are: asking questions, developing and using models, planning and carrying out investigations, analysing and interpreting data, using mathematics and computational thinking, constructing explanations, engaging in argument from evidence, and obtaining, evaluating, and communicating information.

There are currently a number of inquiry-based pedagogical approaches that incorporate these essential features of inquiry. A popular pedagogical approach among science teachers in Singapore (Lau, Wong, Chew, & Ong, 2011) is the 5E Inquiry Model (Bybee et al., 2006), which consists of five phases of inquiry: Engage, Explore, Explain, Elaborate and Evaluate. At the beginning of the cycle, the Engage phase gets learners engaged with a puzzling natural phenomenon through the use of short activities that promote curiosity and elicit prior knowledge. Next, the Explore phase provides learners with first-hand experiences or real-world contexts to investigate the phenomenon further. They may complete laboratory activities or conduct preliminary investigations. The Explain phase then involves learners in building scientific ideas and generating explanations of the phenomenon. This phase also provides opportunities for teachers to directly introduce scientific concepts involved in the explanation. In the Elaborate phase, learners are provided with new questions and contexts for them to apply their new knowledge so as to deepen and extend their conceptual understanding and explanation skill. Finally, the Evaluate phase encourages learners to assess their own understanding and teachers to evaluate learners' progress towards the learning objectives.

17.2.2 Disciplinary Literacy

This study is informed by the research literature on disciplinary literacy. Disciplinary literacy refers to the specific ways of talking, reading, writing and thinking valued and used by people in a discipline in order to successfully access and construct knowledge in that discipline (Moje, 2007). The argument for teaching disciplinary literacy in schools stems from the awareness of adolescents finding it more difficult to comprehend and produce complex texts in the subject areas (Shanahan & Shanahan, 2008). Research has also shown that different disciplines have specialised ways of communication which students need to master in order to be successful in the discipline (Lemke, 1990; Schleppegrell, 2004). Thus, the higher level literacy skills demanded in the discipline are not something that students can easily learn on their own as they differ significantly from everyday language and communication. A common literacy skill that is challenging for many science students is the construction of written scientific explanation (Wellington & Osborne, 2001).

A new development in disciplinary literacy is the increasing recognition that literacy is inherently multimodal (Kress, Jewitt, Ogborn, & Tsatsarelis, 2001). Thus, literacy is not confined to reading, writing and the use of print media, but also extends to specific ways of drawing, graphing, doing, acting and gesturing. Many literacy and science education researchers (Cervetti, Pearson, Bravo, & Barber, 2006; Ødegaard et al., 2014; Tang, Ho, & Putra, 2016; Tytler, Prain, Hubber, & Waldrip, 2013) have incorporated a multimodal approach into their conceptual frameworks to foster and analyse the development of disciplinary literacy. For instance, the Seeds of Science/Roots of Reading program by Cervetti et al. (2006) uses the do-it, talk-it, read-it and write-it literacy approach to expose students to multiple learning modalities during the inquiry process. The do-it component involves hands-on manipulation of apparatus, objects or modelling tools. The talk-it component involves students discussing with one another or their teachers. Lastly, the read-it and write-it components involve reading and writing activities respectively.

17.2.3 Scientific Explanation and PRO Strategy

The construction of scientific explanation is both an important literacy skill outlined in CCSS and one of the scientific practices outlined in NGSS. As such, one of the goals in our study was to help students develop the skill of constructing scientific explanations. Informed by work in the philosophy of science and systemic functional linguistics (SFL), we developed a literacy strategy called PRO (Premise-Reasoning-Outcome) to explicitly support students in constructing oral and written explanations. This strategy was conceptualised based on our understanding of the structure of a scientific explanation, which comprises three primary components: (a) premise, (b) reasoning and (c) outcome (Putra & Tang, 2016; Tang, 2015, 2016a).

From the philosophy of science, the *premise* provides the basis of the explanation. A premise can be a natural law or 'law-like' statement that is well established and accepted in the scientific community, or it can be a general theory or big idea that connects multiple phenomena with an overarching framework (Braaten & Windschitl, 2011). Once a law or theory is established – and until they are invalidated by the scientific community, scientists seek to use them as the basis to account for the phenomenon in the explanation. As the basis or 'first cause' of an explanation, the premise therefore does not require further elaboration or justification in the context of the explanation. (However, this does not mean that students should not question the source of their knowledge for the premise.)

In SFL, Halliday and Martin (1993) identified an explanation as one of the four major genres commonly found in scientific texts, besides information report, experimental report and argument. A genre has distinct functional stages which can be identified on the basis of lexical and grammatical shifts in the text (Martin, 1992). The functional stages of an explanation genre are phenomenon identification (what is being explained), implication sequences (series of logical clauses) and closure (Unsworth, 2001; Veel, 1997). The implication sequences, which are grammatically joined by the use of conjunctions (e.g. because, subsequently, although, if), comprise the *reasoning* part of the PRO structure. Unsworth (1998) calls the patterns of logical relations formed by conjunctions the 'language of reasoning' within an explanation and they are responsible for building up the causation of the explanation and leading to the *outcome* of the PRO structure. (For examples of scientific explanation using a PRO structure, see Tang, 2015.)

17.2.4 Literacy-Inquiry Instructional Model to Foster Scientific Explanation

As Singapore teachers are familiar with 5E Inquiry Model, we decided to adopt it as a starting model and modify it by infusing several literacy strategies. While the essence of the pedagogical model is still inquiry, there is also an additional focus on literacy, specifically on writing. The literacy-infused inquiry model features literacy activities to support fostering both conceptual knowledge and disciplinary literacy skills. The revamped model is summarised in Table 17.1. A key difference in this model lies in the Explain stage where the epistemic practices of writing scientific explanation are being taught through the PRO instructional strategy,

Table 17.1 The literacy-infused inquiry model *(Italics denote emphasis on literacy)*

Stage	Activity description	
	5E Model	Literacy-Inquiry Model
Engage	• Engaging students in experiments that trigger their prior knowledge • Engaging students in problems or situations that they need to solve	• Engaging students in experiments, video demonstration and/or *reading activity to explore a phenomenon* • Students or teacher introduce a problem to explore and solve
Explore	• Students exploring the earlier activity further through hands-on activities	• Students exploring and *discussing the activity/problem further* • *Students presenting their initial understanding*
Explain	• Students providing explanation • Teacher providing the necessary vocabulary, concepts and explanation	• Students *learning the necessary vocabulary, concepts and epistemic practices involved in constructing explanation (using the PRO strategy)* • *Students writing explanation*
Elaborate	• Students applying their new knowledge to related but new situations	• Students applying their new knowledge to related but new situations • *Students presenting their elaboration*
Evaluate	• Teachers evaluating students' conceptual understanding	• Teachers or students evaluating conceptual understanding

such as understanding how an explanation works, differentiating the functions among the premise, reasoning and outcome, and identifying relevant laws or theories as premises. This emphasis on explanation is in response to research that found explanation construction is not widely implemented in most inquiry-based instruction (Ruiz-Primo, Li, Tsai, & Schneider, 2010). To support the PRO instructional strategy, the Engage and Elaborate stages also provide students with the necessary phenomenon and context to formulate their explanation. Additionally in the Explore stage, students are required to discuss and come up with an initial explanation which may elicit students' prior knowledge and potential misconceptions.

17.3 Methodology

17.3.1 Research Context and Design

The data for this chapter were taken from a 3-year design research project aimed at developing disciplinary literacy instruction in science. The project involved two government secondary schools and four physics and chemistry teachers: John, Anne, Kathryn and Derrick. These teachers were recommended by their school leaders to participate in the project on the basis of their teaching experience and willingness to implement new teaching strategies. There were 106 student participants, with 86 students in 9th grade and 20 students in 10th grade.

Prior to the intervention phase of the research, most of the teachers were observed to focus on conceptual understanding and teach their students mostly through a traditional way of lecturing and giving practice questions. During the research intervention, all the teachers attended three workshops on disciplinary literacy conducted by the research team where they were introduced to the literacy-inquiry instructional model and PRO instructional strategy. They also regularly brainstormed and codeveloped lesson plans based on the literacy-inquiry instructional model on various topics with the research team. Although the lesson plans and instructional approaches were codeveloped, all the teachers had the liberty to enact the lessons on the ground according to their teaching styles and preferences as well as the environmental constraints. Thus, part of the value of this study was to examine how the various teachers interpreted the literacy-inquiry instructional model and integrated the model (with the PRO strategy) into their existing teaching practices.

17.3.2 Data Sources and Analysis

The data used in this study included video records of classroom observations and students' artefacts. A total of 68 hours of video records were collected during the intervention research phase. The videos were then viewed and segmented using Transana software into discrete units according to discernible boundaries in order to facilitate coding and annotation. Episodes of literacy activities were identified and transcribed for further analysis using discourse analysis (see Tang, 2016b for further elaboration of the segmentation and coding process). Students' artefacts collected included notes, worksheets and laboratory manuals. These artefacts were primarily used as secondary data to corroborate our findings. This study adheres to the ethical principles outlined by the Institutional Review Board concerning research involving human participants. All names used in this chapter are pseudonyms to ensure confidentiality.

17.4 Analyses and Findings

From the analysis of the classroom videos for features of inquiry and literacy, we identified four features of inquiry: (1) framing driving question, (2) conducting experiments and collecting evidence, (3) constructing explanations and (4) communicating and evaluating explanations, commonly found in various stages of the 5E inquiry model enacted by the teachers. We also identified the nature of the literacy activities that supported each feature of inquiry. In the following, we show several analytical cases to illustrate how the teachers utilised and integrated literacy activities to support inquiry in their classrooms in order to enable the students to construct scientific explanations.

17.4.1 Framing Driving Question

A driving question forms the core of scientific inquiry, serving as a starting point of inquiry to shape investigation. According to the National Research Council's (2000) recommendation, the level of inquiry can vary from an open inquiry where students formulate their own questions to a guided or structured inquiry where students investigate questions provided by the teachers. In Singapore where students tend to be more reserved, it is more common for teachers to provide the driving question to guide students through the inquiry process (Tan et al., 2014). However, in this study, we observed that the teachers did not simply state the inquiry question but carefully framed the question through literacy activities. These literacy activities support the inquiry process by providing the context for understanding the problem space set by the question.

For example, in an inquiry lesson unit on Electrolysis, the teacher Anne began her lesson with a reading activity consisting of an article for her students to read and follow-up questions for discussion in groups. As the concept of electrolysis could not be easily encountered in daily life, the article provided the students a preliminary idea of electrolysis to ground their subsequent inquiry experience. The follow-up questions, which included asking students to write down their predictions of the outcome of the electrolysis of water, also elicited the students' initial understanding. At the end of the reading activity, Anne gathered the students' responses by projecting some of the groups' written answers onto a screen.

Some of the groups had different initial understanding of electrolysis despite having read the same article. One group of students understood electrolysis as 'electrical decomposition' while another as '[using] electricity to analyse something'. One group predicted that water would produce 'hydrogen and oxygen' (unclear if they meant gases or atoms) while another predicted that water would be split into its atomic constituents, 'hydrogen and oxygen atoms'. Through this reading activity, Anne elicited competing points of view and engaged in a dialogic discussion (Mortimer & Scott, 2003) with her students. In her feedback to the students, Anne frequently had to reiterate there was 'no right or wrong' answer as this was only their 'first understanding'. This assurance was necessary as the students were used to having the teachers provide the right answers. In fact, the initial uncertainty and puzzlement served as the basis for the subsequent inquiry tasks to, in Anne's words, 'refine [their understanding] towards the end'. It also set the context for Anne to frame the driving question or objective for the topic, 'what is electrolysis and its outcome?' as a preamble to the next activity of conducting experiments.

In another example, Kathryn adopted a different approach to frame the driving question for her inquiry lessons on Qualitative Analysis. Instead of reading and predicting, she immersed her students in a real-world problem and engaged in a conversation with them to frame the question for the inquiry collaboratively. She began her lesson by holding a test tube filled with a colourless solution that she claimed to be a water sample from Ang Mo Kio (a housing estate in Singapore) that had been suspected to cause stomach ache. She asked her students to smell and visually inspect the sample to confirm if it was indeed water. They agreed that the sample looked like water but refused to drink it. She then questioned them why, but nobody gave a reason, and she continued asking if they would want to know what was inside the sample, as shown here:

Speaker	Utterance
Kathryn	**Why are you so keen to find out what's inside?**
Students	(inaudible)
Kathryn	Why are you so keen to find out what's inside? Do you want to drink it?
Student 1	Yes (softly) …
Kathryn	Then what do you think you want to do, if you don't want to drink it? You say that no, maybe you think it is not safe. **But how do you know it's not safe? The many questions that I've asked you, you keep referring me back to water. You say it does resemble water. So why are you not drinking it?**
Student 2	Because of ions (inaudible)
Kathryn	Okay, I see somebody is trying to experiment. Miranda, hold up higher.

Unlike Anne, Kathryn did not elicit different points of view to create 'conflict' and uncertainty among her students in order to frame the driving question. Rather, she established a consensus among the students and led them to the element of 'unknown'. She managed to initially convince her students that the sample she brought was water by appealing through their everyday knowledge of water – that it is colourless and odourless. Despite their suspicion that the water could contain something, as reflected by their refusal to drink it, they could not disprove Kathryn's claim. Their limited ability to disprove the claim created the problem space of inquiry that they lacked the chemistry knowledge to be able to collect relevant evidence and also drove them to be 'keen to find out what's inside' the water. As a matter of fact, out of curiosity, a student Miranda quickly tested the water using blue litmus paper to see if the water was acidic. Unlike typical classroom dialogue where a driving question is readily given by teachers without much context or rationale, here Kathryn skilfully challenged students to the problem, and this allowed her to frame a driving question for the topic, 'how could we know what is inside the solution?'

The two examples above illustrate how literacy activities of reading, writing and whole-class discussion could be utilised to support the framing of a driving question in an inquiry-based pedagogy. In both examples, the driving question surfaced through the various literacy activities in a manner that made the students more motivated to find out more in subsequent activities. In addition to piquing and maintaining their curiosity and interest, the integration of literacy activity also enabled the students to construct scientific explanations at the later stage of the lessons by providing them with the context or problem space that they had to address in their explanations.

17.4.2 Conducting Experiments and Collecting Evidence

The next inquiry feature identified was conducting experiments and collecting evidence to address the driving question framed at the earlier stage. Literacy activities that were utilised by the teachers to support students in conducting experiments and collecting evidence included providing guidance for multimodal observation, reading and translating experimental procedures, and discussion.

In the earlier example of the 'mystery water' problem, Kathryn had framed the inquiry question for the topic of Qualitative Analysis. To support the subsequent experiments and data collection, she proceeded to guide her students on how they could observe and record their observations. She asked them to 'look at [her] action' and demonstrated the scientific procedures of collecting data. These included adding dilute aqueous hydrochloric acid into a test-tube of water sample drop-by-drop, and writing the action taken as accurate as possible right after performing the action, including the amount and manner of adding chemicals. Furthermore, Kathryn provided the scientific terms that the students should use to accurately describe their actions such as 'adding dropwise' and 'pouring in excess', and their observations such as 'white precipitate is observed' and 'no visible reaction'. These activities supported the students on how to record their experiments properly so that they could get accurate and precise data to be interpreted and used as evidence in the subsequent explanations (Shanahan, Shanahan, & Misischia, 2011).

Additionally, reading experimental procedures and carrying them out methodologically is an important aspect of scientific inquiry. This requires the ability to 'translate' inscriptions (Latour & Woolgar, 1979) from one form to another (e.g. from written instructions to adding acid to mixture, from the colour of litmus paper to a coded table) and underscores the literacy tasks undertaken by scientists to produce data and evidence (Tang, Tighe, & Moje, 2014). In this respect, literacy activities can be designed to help students translate inscriptions in the process of conducting experiments and collecting evidence. For instance, Kathryn taught her students how to translate an experimental procedure into a flowchart. This included creating 'a flowchart to help [students] see what [was] going on' in the experiments by using boxes and arrows, as illustrated in Fig. 17.1. The procedural information from the text was decoded into a visual representation that was more functional and helpful in the visualisation of the experiment (Shanahan, Shanahan, & Misischia, 2011). Consequently, the translation of the experimental procedure into the flowchart allowed the students to conduct the experiment methodologically following the norms and conventions of experimental chemistry. Furthermore, the students were asked to record their findings in the flowchart for a reference, which would later be used in the next stage of constructing scientific explanations.

17 Infusing Literacy into an Inquiry Instructional Model ... 291

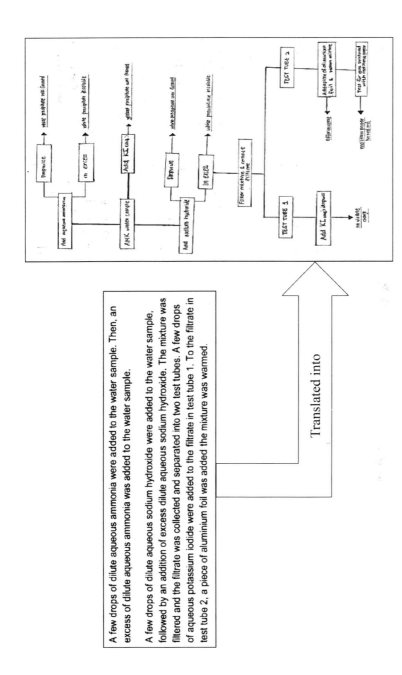

Fig. 17.1 A student's work of translating a procedure into a flowchart

17.4.3 Constructing Explanations

Constructing scientific explanations of the phenomenon based on accepted theories and collected evidence is an important process in scientific inquiry. However, many students have difficulty in this area even though they may have the necessary scientific knowledge or concepts related to the phenomenon. Traditionally, teachers tend to provide the articulation of scientific explanation and emphasise the necessary facts and information contained in the explanation, rather than getting students to think about the epistemic processes of constructing an explanation such as the nature of an explanation, how an explanation works and how scientists construct explanations. What was different in this study was that the teachers incorporated the PRO literacy strategy to support their students in learning these epistemic processes involved in scientific explanations.

For example, in a physics lesson on forces, John introduced the PRO strategy to his students after several students expressed their difficulty in writing an explanation:

> Janice tells me that, 'I don't know how to phrase the answer', or, 'I don't know how to phrase the explanation'. Well, what you can do that is, in science when we craft our answer, you could actually **follow this model, P-R-O, PRO**. So basically, you **start with your Principle**, any **science principle that you have or you know of**. Obviously I haven't taught you any science principle on Forces. You can use any principle that you know, that you have learnt before. And **then what is the Reason for it happening and how does it explain the Outcome**. Obviously **your Outcome has already come out here** (pointing to the screen). Alright, so try to use a Principle that you have known and some Reasons to explain your Outcome.

In this instruction, John briefly stated the steps needed to construct explanations and elaborated the elements of PRO. As the lesson progressed, he modelled to the students how to craft the explanation using the PRO in a step-by-step manner and supplied them with a writing template as a form of scaffolding (Fig. 17.2). The

Principles (What do you know? What laws/principles involve? What's the concept behind?)	When the skater was pushing _____ , a force _____
	According to Newton's _____ , a _____ force _____
Reasoning (What follows from the principles?)	Therefore, _____
Outcome (What is your conclusion?)	Thus, _____

Fig. 17.2 Written scaffolding on John's students worksheets

scaffolds consisted of a table to organise each element of PRO, elaboration of each element of PRO as reminder, adverbs such as *according to* and conjunctions such as *when, therefore*, and *thus* to signal and connect the clauses together into a logical and coherent sequence. The written scaffolds were gradually removed in order to let the students practise writing without any explicit support.

The form of scaffolding was not restricted only to a written mode such as the table in Fig. 17.2. In another lesson on Density where the inquiry question was: 'why does a raisin in soda water sink and float again numerous times?', John co-constructed the explanation orally with his students using an Initiation-Response-Feedback (IRF) classroom discourse pattern (Mortimer & Scott, 2003). John integrated his questioning techniques with the PRO strategy and asked specific questions to elicit particular element about the epistemic structure of the explanation. He asked 'what did you observe?' to elicit the Outcome, and 'why?' and 'what happens?' to prompt a chain of Reasoning from his students. In sum, John used the PRO structure to help the students explain the dancing raisins phenomenon systematically, as shown here:

Speaker	Utterance
John	Usually **by principle** … you can state the formula or definition of density…. Density is mass per unit volume. **What did you first observe**? Let's write down the observation. The raisin, the moment you put it in, actually most of them actually?
Student	Sink
John	Outcome is the raisin sink. So **why do you think the raisin will sink**?
Student	Due to fact that they are more denser than (inaudible)
John	Due to the density of the raisin is higher …. **And so your raisin sink**. After that, what **do you observe**? The raisin actually?
Student	Float
John	**Why did the raisin float**?
Student	Carbon dioxide …
John	Now **what actually happens** to the carbon dioxide gas bubbles?
Student	Attach itself to the raisins
John	Yes, it attached itself to the raisins. The gas bubbles attached themselves to the raisins. And when it is attached to the raisin, **what happens to it**? What increases?
Student	Volume
John	The volume increases. The volume increases. **So what happen** when the volume increases?
Student	Average density decreases
John	The average density decreases. The average density decreases. So, **so what's the outcome**? The raisin with the air bubble?
Student	Floats

The teaching of the PRO structure is crucial in enabling the construction of scientific explanation in the inquiry process. The PRO structure serves as a literacy tool for the students to synthesise their earlier inquiry activities and findings. The context, ideas and concepts that students had gathered from the earlier activities

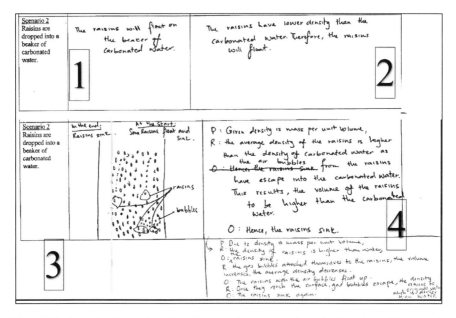

Fig. 17.3 A student's work of predicting, observing and explaining the raisin experiment (divided into 4 segments that show the progression of explanation construction)

were assembled into a coherent logical sequence to answer the initial driving question. For example, as illustrated in Fig. 17.3, segment 1 and 2 showed a student's initial prediction and explanation (although inaccurate) respectively prior to the experiment which shaped the context (i.e. about the sinking or floating of a raisin) of the problem. In segment 3, which was written after the experiment, the student wrote a more accurate version of the phenomenon through a visual representation of the experiment and, together with John's lecture on density, the relevant concept (i.e. density) for the explanation. Finally, in segment 4, the ideas from the first three sections were integrated using the PRO structure, which provide a more comprehensive and coherent explanation of the phenomenon observed.

17.4.4 Communicating and Evaluating Explanation

In the pursuit of scientific inquiry, communication and evaluation of ideas and explanations are crucial as scientific ideas and explanations must reach a consensus within the scientific community. Reflecting this idea of a consensus, science students should therefore communicate their findings to their peers and subject them to peer critiques. There are various ways to do so and in this study, we observed that oral presentation of the students' works (by the teachers or the students) was the preferred literacy activity taken by the teachers to support this communication aspect of the inquiry.

Kathryn, for example, tasked her students to present their explanations in front of the class so that they could practise communicating and evaluating each other's explanations. In the following excerpt, a group of students presented their explanation of the electrolysis of dilute aqueous copper (II) sulphate solution. Throughout the presentation, Kathryn evaluated their explanation continually.

Speaker	Utterance
Pauline	My group did the dilute aqueous copper (II) sulphate. There we observed that effervescence occurs at the anode and reddish-brown coating is formed on the cathode.
Kathryn	On the cathode. That piece of pencil lead, is it? Okay.
Pauline	Then our group explanation is, at the anode, OH and SO_4 …
Pauline	At the anode, OH^- and SO_2, erm, SO_4^{2-} ions are attracted to the anode.
Kathryn	Good. **Pause there. Alright, did you hear what she said? Do you agree?**
Students	(nodding heads)
Kathryn	Yes, ah. All your negative ions will be attracted to the anode. Okay, continue.
Gaby	SO_4^{2-} ions are not discharged.
Kathryn	**Are not discharged? Okay. Instead …?**
Gaby	Instead, OH^- ions are preferably discharged as oxygen gas.
Kathryn	**How do you know?**
Pauline	Because effervescence is observed at the anode.
	(…)
Gaby	At the cathode, H^+ and Cu^{2+} ions are attracted to the cathode.
Kathryn	Yes
Gaby	Copper is less reactive than hydrogen, that's why it is discharged as copper metal.
Kathryn	Copper metal. Yes. And?
Gaby	Which forms the reddish-brown coating.
Kathryn	Okay.
Gaby	Reddish brown coating on the cathode.
Kathryn	Yes. Okay. **Girls, how many of you saw this?** How many passed by this station and you saw that the pencil lead, there was actually pink coating on it? Erm, reddish brown coating?
Students	*raise hands*
Kathryn	**All managed to see that? Okay. Good**. Right. Okay. Continue. Is there anything else that you want to share with us with regard to your observation?
Zhiwen	The remaining solution is acidic because H^+ ions SO_4^{2-} are left.
Kathryn	Okay. Because H^+ ions and SO_4^{2-} ions are left. **How do you know that it's acidic? You just guessed it?**
Pauline	It forms sulphuric acid.
Kathryn	Yeah. But **how do you know when you were carrying out the experiment**?
Zhiwen	Test using litmus paper.

(continued)

(continued)

Kathryn	Okay. You tested using the litmus paper. Alright. What happens to the litmus paper? To prove that is acidic solution?
Zhiwen	Blue litmus paper turns red.
Kathryn	Alright. Thank you very much.

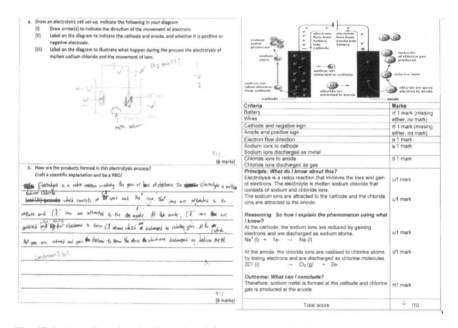

Fig. 17.4 An evaluated student's work and the assessment guideline

We can see from the above excerpt that Kathryn sought the students' agreement to the presenters' explanations by continually pausing the presentation and asking the rest of the students if they had the same or agree with the observation and explanation. This action ensured that there was a consensus among the students that the explanation presented was conceptually sound. In addition, Kathryn also assessed the explanation by probing with 'how do you know?' to get the presenters further justify their explanation. By doing so, she ensured that the explanation was completely based on evidence gathered, such as observation of effervescence or the changing colour of litmus paper, and not by guessing.

On the other hand, Anne had her students to be peer reviewers. They were exposed to their peers' written explanations and had to evaluate the explanations for accuracy. Furthermore, Anne provided the students with an assessment guideline and had them use it to review the explanation. As illustrated in Fig. 17.4, a student reviewed his peer's explanation by ticking the relevant points as suggested by the assessment guideline and correcting mistakes by crossing and adding missing points, using a red pen. Although some guidelines were given, this approach trained the students to be sharp in evaluating explanation and learn the difference between a good and bad explanation.

17.5 Discussion

From the analytical cases, we saw how the teachers enacted literacy activities to support scientific inquiry and how the literacy-infused inquiry pedagogical model enabled the process of constructing scientific explanations by first framing the context and problem to be explained, eliciting ideas and concepts through discussion and experiments, and lastly assembling the context and concepts in a logical structure through the PRO writing framework. Table 17.2 summarises the connection between the features of inquiry and literacy for every stage of the inquiry pedagogical model.

Overall, this study provides several insights into the nature of literacy instruction in enabling inquiry-based science and further supports the argument that literacy skills necessary for scientific inquiry need to be explicitly taught (Moje, Collazo, Carrillo, & Marx, 2001). First, making predictions of an experimental outcome while reading is a common way of reading among scientists (Shanahan & Shanahan, 2008), and should thus be integrated into an inquiry lesson similar to the way Anne enacted. Furthermore, as this is a rather disciplinary-specific way of reading, students have to be supported with some forms of written scaffolds and whole-class discussion led by teachers. Second, the ability to make observations and translate inscriptions from one form to another in the process of conducting experiments and collecting data is also another literacy skill that needs to be pedagogically supported, as illustrated by an example of Kathryn modelling experimental procedure and translating a procedure into a flowchart. Lastly,

Table 17.2 Features of inquiry and literacy enacted by the teachers

5E Stage	Inquiry features	Literacy features
Engage	• Framing driving questions • Carrying out experiments	• Reading articles • Writing prediction and initial explanation • Having class discussion
Explore	• Carrying out experiments • Collecting evidence	• Writing observations and hypothesis • Translating inscriptions across multiple modes
Explain	• Explaining observed phenomena • Applying content knowledge	• Teaching/learning structure of explanations • Writing explanations
Elaborate	• Explaining new but related phenomena	• Writing explanations • Presenting explanations
Evaluate	• Evaluating explanations	• Presenting explanations • Critiquing explanations

writing and evaluating scientific explanations is another important literacy skill to be developed, and this study provides several examples on how to support student development in this area, such as using the PRO scaffolds for oral discussion and written practice, and presenting explanations for consensus and peer critique.

Although we saw some success in infusing literacy elements into an inquiry pedagogical model, which resulted in some changes in the teachers' instructional practices, there are still many challenges and improvements ahead. For instance, we felt that many of the group discussions during the Explore stage can be more argumentative rather than consensus building. More can be done to get students to propose alternative claims and support them with evidence and rebuttals. These are also literacy skills that we hope to implement in the next stage of our research. One of the constraints faced in this study was the readiness of the students to participate in literacy-infused inquiry-based pedagogy as well as argumentative talk involving multiple 'answers'. There is also a need to further differentiate between an explanation (which can be facilitated using the PRO strategy) and an argument (which may require a different strategy). For more discussion on their differences, see Tang (2016a) and Osborne and Patterson (2011).

Another challenge that must be considered for future research is teacher training and preparation. In this study, much of the professional development was short and limited to 'just in time' feedback as the researchers and teachers codeveloped the instructional model and lesson plans. More extensive teacher training and preparation will be needed in order to sustain and scale up the pedagogical move towards an integrated literacy-inquiry instructional model.

References

Braaten, M., & Windschitl, M. (2011). Working toward a stronger conceptualization of scientific explanation for science education. *Science Education, 95*, 639–669. https://doi.org/10.1002/sce.20449

Bybee, R. W., Taylor, J. A., Gardner, A., Van Scotter, P., Powell, J. C., Westbrook, A., et al. (2006). *The BSCS 5E instructional model: Origins and effectiveness*. Colorado Springs: BSCS.

Cervetti, G., Pearson, P. D., Bravo, M. A., & Barber, J. (2006). Reading and writing in the service of inquiry-based science. In R. Douglas, M. Klentschy, & K. Worth (Eds.), *Linking science and literacy in the K-8 classroom* (pp. 221–244). Arlington: NSTA Press.

Council of Chief State School Officers. (2010). *Common core state standards*. Washington: National Governors Association Center for Best Practices, Council of Chief State School Officers.

Crawford, B. A. (2000). Embracing the essence of inquiry: New roles for science teachers. *Journal of Research in Science Teaching, 37*(9), 916–937.

Fang, Z., & Wei, Y. (2010). Improving middle school students' science literacy through reading infusion. *The Journal of Educational Research, 103*(4), 262–273.

Ford, M. J. (2015). Educational implications of choosing "practice" to describe science in the next generation science standards. Science Education, 99(6), 1041–1048. https://doi.org/10.1002/sce.21188

Halliday, M. A. K., & Martin, J. R. (1993). *Writing science: Literacy and discursive power*. Pittsburgh: University of Pittsburgh Press.

Hand, B., Alvermann, D. E., Gee, J., Guzzetti, B. J., Norris, S. P., Phillips, L. M., et al. (2003). Message from the "Island group": What is literacy in science literacy? *Journal of Research in Science Teaching, 40*(7), 607–615.

Howes, E. V., Lim, M., & Campos, J. (2009). Journeys into inquiry-based elementary science: Literacy practices, questioning, and empirical study. Science Education, 93(2), 189–217. https://doi.org/10.1002/sce.20297

Keys, C. W., & Bryan, L. A. (2001). Co-constructing inquiry-based science with teachers: Essential research for lasting reform. Journal of Research in Science Teaching, 38(6), 631–645. https://doi.org/10.1002/tea.1023

Kress, G., Jewitt, C., Ogborn, J., & Tsatsarelis, C. (2001). *Multimodal teaching and learning: The rhetorics of the science classroom*. London: Continuum.

Latour, B., & Woolgar, S. (1979). *Laboratory life: The construction of scientific facts*. Princeton: Princeton University Press.

Lau, C. Y., Wong, D., Chew, C., & Ong, K. S. (Eds.). (2011). *Handbook for teaching secondary physics*. Singapore: Ministry of Education.

Lemke, J. L. (1990). *Talking science: Language, learning and values*. Norwood: Ablex.

Martin, J. R. (1992). *English text: System and structure*. Amsterdam: Benjamins.

Moje, E. B. (2007). Developing socially just subject-matter instruction: A review of the literature on disciplinary literacy teaching. *Review of Research in Education, 31*, 1–44.

Moje, E. B., Collazo, T., Carrillo, R., & Marx, R. W. (2001). "Maestro, what is 'quality'?": Language, literacy, and discourse in project-based science. *Journal of Research in Science Teaching, 38*(4), 469–496.

Mortimer, E. F., & Scott, P. (2003). *Meaning making in secondary science classrooms*. Buckingham: Open University Press.

National Research Council. (1996). *National science education standards*. Washington: National Academy Press.

National Research Council. (2000). *Inquiry and the national science education standards: A guide for teaching and learning*. Washington: National Academy Press.

National Research Council. (2012). *A framework for K-12 science education: Practices, cross-cutting concepts, and core ideas*. Washington: The National Academies Press.

National Research Council. (2014). *Literacy for science: Exploring the intersection of the next generation science standards and common core for ELA standards. A workshop summary*. Washington: The National Academies Press.

Ødegaard, M., Haug, B., Mork, S. M., & Sørvik, G. O. (2014). Challenges and support when teaching science through an integrated inquiry and literacy approach. *International Journal of Science Education, 36*(18), 2997–3020. https://doi.org/10.1080/09500693.2014.942719

Osborne, J. F., & Patterson, A. (2011). Scientific argument and explanation: A necessary distinction? *Science Education, 95*, 627–638. https://doi.org/10.1002/sce.20438

Pearson, P. D., Moje, E., & Greenleaf, C. (2010). Literacy and science: Each in the service of the other. *Science, 328*(5977), 459–463. https://doi.org/10.1126/science.1182595

Poon, C. L., Lee, Y. J., Tan, A. L., & Lim, S. S. L. (2012). Knowing inquiry as practice and theory: Developing a pedagogical framework with elementary science teachers. *Research in Science Education, 42*(2), 303–327.

Putra, G. B. S., & Tang, K.-S. (2016). Disciplinary literacy instructions on writing scientific explanations: A case study from a chemistry classroom in an all-girls school. *Chemistry Education Research and Practice, 17*(3), 569–579. https://doi.org/10.1039/c6rp00022c

Ruiz-Primo, M. A., Li, M., Tsai, S.-P., & Schneider, J. (2010). Testing one premise of scientific inquiry in science classrooms: Examining students' scientific explanations and student learning. *Journal of Research in Science Teaching, 47*, 583–608. https://doi.org/10.1002/tea.20356

Schleppegrell, M. (2004). *The language of schooling: A functional linguistics perspective*. Mahwah: Lawrence Erlbaum Associates.

Settlage, J. (2007). Moving past a belief in inquiry as a pedagogy: Implications for teacher knowledge. In E. Abrams, S. A. Southerland, & P. Silva (Eds.), *Inquiry in the classroom: Realities and opportunities* (pp. 204–215). Charlotte: Information Age.

Shanahan, C., Shanahan, T., & Misischia, C. (2011). Analysis of Expert Readers in Three Disciplines. *Journal of Literacy Research, 43*(4), 393–429. https://doi.org/10.1177/1086296X11424071

Shanahan, T., & Shanahan, C. (2008). Teaching disciplinary literacy to adolescents: Rethinking content-area literacy. *Harvard Educational Review, 78*(1), 40–59.

Tan, A. L., Talaue, F., & Kim, M. (2014). From transmission to inquiry: Influence of curriculum demands on in-service teachers' perception of science as inquiry. In A. L. Tan, C. L. Poon, & S. S. L. Lim (Eds.), *Inquiry into the Singapore science classrooms*. Singapore: Springer.

Tang, K. S. (2015). The PRO instructional strategy in the construction of scientific explanations. *Teaching Science, 61*(4), 14–21.

Tang, K. S. (2016a). Constructing scientific explanations through premise – reasoning – outcome (PRO): An exploratory study to scaffold students in structuring written explanations. *International Journal of Science Education, 38*(9), 1415–1440. https://doi.org/10.1080/09500693.2016.1192309

Tang, K. S. (2016b). How is disciplinary literacy addressed in the science classrooms? A Singaporean case study. *Australian Journal of Language and Literacy, 39*(3), 220–232.

Tang, K. S., Ho, C., & Putra, G. B. S. (2016). Developing multimodal communication competencies: A case of disciplinary literacy focus in Singapore. In M. Mcdermott & B. Hand (Eds.), *Using multimodal representations to support learning in the science classroom* (pp. 135–158). New York: Springer.

Tang, K. S., Tighe, S. C., & Moje, E. B. (2014). Literacy in the science classroom. In P. Smagorinsky (Ed.), *Teaching dilemmas and solutions in content-area literacy* (pp. 57–80). Thousand Oaks: Sage.

Tytler, R., Prain, V., Hubber, P., & Waldrip, B. (2013). *Constructing representations to learn in science*. Rotterdam: Sense.

Unsworth, L. (1998). "Sound" explanations in school science: A functional linguistic perspective on effective apprenticing texts. *Linguistics and Education, 9*(2), 199–226.

Unsworth, L. (2001). Evaluating the language of different types of explanations in junior high school science texts. *International Journal of Science Education, 23*(6), 585–609.

Veel, R. (1997). Learning how to mean - scientifically speaking: Apprenticeship into scientific discourse in the secondary school. In C. Frances & J. Martin (Eds.), *Genre and institutions: Social processes in the workplace and school* (pp. 161–195). London: Cassell.

Wellington, J., & Osborne, J. (2001). *Language and literacy in science education*. Philadelphia: Open University Press.

Chapter 18
Representation Construction as a Core Science Disciplinary Literacy

Russell Tytler, Vaughan Prain and Peter Hubber

Abstract There is growing interest in and understanding of the material basis of epistemic practices in science, and consequently of the role of multimodal representation construction in reasoning and learning in science classrooms. From this perspective learning in science crucially involves induction into the interplay between experimental exploration and construction and coordination of representations as a core element of scientific disciplinary literacy. In this chapter we argue that learning to explain and problem-solve effectively in science involves students actively generating and coordinating multiple, multimodal representations and material artifacts in exploring material phenomena, in a guided inquiry process. We describe the development of a 'representation construction' approach to inquiry in science classrooms that is grounded in pragmatist perspectives on learning and knowing, which engages students in active experimental exploration and generation and refinement of core representations underpinning science concepts. We provide evidence of the success of the approach in supporting quality learning and reasoning. We propose that the construction of representations such as drawings, animations, role-plays or mathematical/symbolic systems works to support learning and knowing through the affordances of different modes to productively constrain exploration and explanation of the material world. We conclude that induction into multimodal representation construction processes in response to grappling with real world problems is central to the development of scientific disciplinary literacy, and that this approach represents a significant innovation in its use of authentic inquiry to serve a serious conceptual learning agenda in science.

Keywords Literacy · representation · affordances · concepts · inquiry

---

R. Tytler (✉) · V. Prain · P. Hubber
Deakin University, Victoria, Australia
e-mail: russell.tytler@deakin.edu.au

V. Prain
e-mail: vaughan.prain@deakin.edu.au

P. Hubber
e-mail: peter.hubber@deakin.edu.au

18.1 Introduction

Increasingly science education researchers accept the sociocultural insight that learning in science, as with any discipline, entails students being inducted into the particular processes through which knowledge is generated, validated and communicated in this discipline. By implication, in learning science, students are acquiring a distinctive disciplinary literacy (Linder, Ostman, & Wickman, 2007; Moje, 2007; Norris & Phillips, 2003; Shanahan & Shanahan, 2008). Norris and Phillips (2003) argue that rather than being knowledgeable about science content, with a declarative focus, really understanding science needs to involve students becoming literate in the sense of being able to interpret, assess and represent scientific claims. This moves the focus from science as knowledge in the abstract to science as a discourse, as a set of practices for thinking, acting and representing claims scientifically. From this perspective, science disciplinary literacy entails both meaningful immersion in the epistemic processes of science inquiry and knowledge-generation (Duschl, 2008), as well as the ability to generate science texts to represent and communicate scientific claims arising from these processes. More broadly, this literacy also entails understanding and valuing the rationale for this disciplinary enterprise (Hurd, 1998). In this chapter we focus mainly on the role of representations in science learning processes, but also consider their relation to text interpretation and production.

There is increasing recognition of the role of material and representational tools in framing how the world is perceived and how theory is constructed (Amin, Jeppsson, & Haglund, 2015). Latour (1986, p. 3) argued that the emergence of scientific thought depended on the development of representational tools or 'inscriptions' that can be combined, transformed across modes including being turned into figures or supported by writing, and reproduced. His study of two scientists working together on soil profiles in the Amazon basin, at the boundary between rainforest and savannah, traced the process by which they generated data and progressively transformed it into the theory reported in scientific papers, representing abstracted and transportable knowledge, through a series of representational redescriptions. The raw soil was assembled into an ordered box arrangement, analysed and represented through a colour chart and numbering system, to a table which was the form in which they carried the information back with them to Paris to be further transformed into a scientific paper. The relation between the theoretical scientific claims made in papers, and raw data, is not unitary as imagined in much of the writing on the epistemic processes of science, but rather distributed across these representational redescription pathways. Drawing on these insights, we argue that the process of induction into scientific disciplinary literacies needs to include an appreciation, gained through practical problem solving, of the way data is generated and shaped, and progressively transformed through representation construction and redescription across modes.

A substantial body of work now exists that confirms the central role of representational generation and manipulation in the process of scientific discovery.

Gooding's research into Faraday's work on the relation between magnetism, electricity and motion, realised through his detailed diary accounts, demonstrated the central role of representational generation and refinement and improvisation in developing 'plausible explanations or realisations of the observed patterns' (Gooding, 2005, p. 15). Gooding identified a recurring pattern in Faraday's work, whereby he would generate chains of diagrams moving from 2D to 3D to 4D (involving representation of temporal change) and back to 2D as a general principle was established (Tytler, Prain, Hubber, & Waldrip, 2013).

This recognition of the key role of visual representation and reasoning is reflected in a strand of research in science education that investigates effective pedagogies to develop modeling competence aimed at the capacity for visualisation (Gilbert, 2005; Gilbert, Reiner, & Nakhleh, 2008, p. 3). Researchers working within a conceptual change tradition, such as Vosniadou (2008a, b), diSessa (2004), Duit and Treagust (2012), have incorporated representational work as a feature of pedagogies aimed at student conceptual growth. Researchers within a socio-semiotic tradition have investigated the challenges for this new literacy of harnessing the resources of a scientific multimodal discourse (linguistic, mathematical and visual) to identify the challenges of learning this new literacy (Gee, 2004; Kress & van Leeuwen, 2006; Lemke, 2003). Our own research sits within a sociocultural tradition that has focused on the meaning-making practices of scientists to provide the major lead for developing classroom pedagogies that align with these (Greeno & Hall, 1997; Hubber, Tytler, & Haslam, 2010; Lehrer & Schauble, 2006a, b; Manz, 2012; Tytler & Prain, 2010). Each orientation foregrounds representational competence as crucial to learning science.

Socio-semiotic research represents a diverse range of approaches to formal analyses of meaning-making processes and practices in science discourse and activity. They include genrist approaches (Halliday & Martin, 1993; Parkinson & Adendorff, 2004) focusing on textual features that affect interpretation, taxonomic structuralist accounts of visual language (Kress, Jewitt, Ogborn, & Tsatsarelis, 2001), post-structural multimedia semiotics and discourse analysis (Lemke, 2004), and sociocultural perspectives on science discourse (Gee, 2004; Moje, 2007) that seek to foreground the effects of situational factors on different learner cohorts' engagement with science. These perspectives are broadly united by the view that students must learn primarily to understand and reproduce the meaning-making practices of the science community if they are to become scientifically literate (Bazerman, 2009; Klein & Boscolo, 2016; Unsworth, 2001). Prescribed genres to achieve this end include formal laboratory reports, posters and science workbooks. However, the issue of which writing types will best facilitate disciplinary learning remains an open question.

In our own approach to the development of scientific disciplinary literacy, we take as a starting point that classroom work should involve induction into scientific disciplinary norms through enacting pedagogical processes parallel to those of practicing scientists. We draw on the work of Vygotsky (1978, 1981), and researchers such as Keys, Hand, Prain and Collins (1999), Moje (2007), Lehrer and Schauble (2006a, b), Duschl (2008) for this focus. While we recognise that classroom teaching and learning practices differ from those of practising scientists in purposes, knowledge bases, resources and rewards, we argue that they can parallel in productive ways the processes of inquiry of the research laboratory through engaging with experimental exploration and representational generation, refinement and validation. The classroom community can be configured to parallel the research team community, where students use practical workbooks to engage in experimental design, observations, reflections, and representing scientific reasoning and claim-making. This approach not only focuses on developing applied representational competence, but also includes formal genres such as posters and reports.

In developing scientific literacy, students need to learn to switch between material, verbal, written, visual, mathematical and 3D modeling modes, including digital form, and coordinate these in generating and justifying scientific explanations. They need to participate in authentic knowledge-producing activities that require the use of these culturally specific resources to develop competence in the diverse reasoning practices of science (Ford & Forman, 2006). In this, the classroom operates as a learning community in which their representations are shared, discussed and justified to arrive at a reasoned consensus that is consistent with accepted scientific understandings (Greeno, 2009; Kozma & Russell, 2005).

In our own approach to engaging students in disciplinary literacy practices, we acknowledge that teachers and students need to know the form and function of both generic and discipline-specific representational conventions. We argue that students, in learning to use these, are advantaged by having first-hand experience of the affordances of the different representational modes as they generate and use them to solve problems and construct explanations. Representations and their use perform active conceptual work in shaping how phenomena are perceived and understood, and this is true for learning in classrooms (Kozma & Russell, 2005) as it is for science (Gooding, 2006). They are the reasoning resources through which we know, and cannot be seen as simply tools for understanding some higher, abstracted form of knowledge that evades representation. We have argued that concepts must be understood through the representational practices through which they are performed (Tytler, Haslam, Prain, & Hubber, 2009). From our perspective, student learning proceeds through the active engagement with and coordination of representational resources, with different representations and modes having specific affordances that offer insight into a phenomenon through productively constraining attention (Prain & Tytler, 2012). Thus, for instance, as students construct drawings of invertebrates in response to a challenge to explain their movement, they select key features needing representation, notice and make claims about relations between structures, and abstract as they refine and coordinate the spatial and temporal features of the animals' structures relating to movement. Such drawings represent a claim, and can involve substantive reasoning. Similarly, role plays can focus attention on key spatial and temporal features of phenomena, and again productively constrain attention to provide embodied engagement with, in this same example, the animal's movement mechanisms. It is our contention that actively engaging with the construction of material and symbolic representations offers gains through this process of productive constraint, and that understanding of a phenomenon entails the coordination of multiple representations each offering partial explanatory insight.

18.2 Describing the Representation Construction Inquiry Approach

Over 3 years of an Australian Research Council-funded project – The Role of Representation in Learning Science (RiLS) – we worked with a small number of teachers of science, both primary and secondary, to develop and refine an approach to guided inquiry teaching. The project used a design experiment methodology (Cobb, Confrey, diSessa, Lehrer, & Schauble, 2003) where an iterative process of development and trialing, and evaluating outcomes was conducted with teachers as partners in the process. The team suggested activities and activity sequences that involved challenging and supporting students to generate representational responses to explicit material problems and challenges, which were then refined and embellished, and further developed by the teachers. This process involved regular planning meetings with the teachers, analysis of video records of teaching sequences including records of student groups' discussion and artefacts, feeding back into further discussion. The research team brought to the planning process a detailed knowledge of the literature around student conceptions and learning challenges in significant topics such as force and motion, adaptation, or changes to substance, and ongoing analyses of the key representational resources that underpinned these major conceptual topics. The teachers brought knowledge of their students' capacities and experience of the practicalities of establishing productive classroom investigations and processes. As the teachers become more confident and self-generating in the approach, they took increasing control of the process of planning and implementation. Investigation of the development of the teaching approach, and teachers' experience, was based on video capture and analysis, and teacher interviews (Hubber, Tytler, & Haslam, 2010). Documentation and analysis of student learning occurred through analysis of class discussion through whole class and small group video capture, collection of student artefacts, pre- and post-tests, and student stimulated recall interviews (Tytler, Prain, Hubber, & Waldrip, 2013), where ethical considerations, such as voluntary participation, informed consent, and use of pseudonyms were followed.

In a series of research workshops involving both the research team, critical friends and the partner teachers, the major principles of the approach were identified, and progressively refined. That process has been continued over subsequent projects, described below, so that the major features of the approach are:

1. Students construct representations in response to explicit challenges. This process involves strategic scaffolding so that students' representational work is focused and productive. The challenge involves a shared practical problem that is meaningful to students.
2. The representation work is underpinned by experimental exploration or appeal to evidence based in experience.
3. Teachers orchestrate shared discussion/evaluation of representation work.
4. There is explicit discussion of representations and representational adequacy and their role in science knowledge building.
5. Assessment is ongoing and a core aspect of learning.

18.3 The Nature of Representational Work in the Inquiry Approach

The nature of a representation challenge is diverse, and how a challenge is orchestrated is a core skill in the teaching and learning process. In some cases a challenge or series of challenges might begin a topic, for instance in introducing the arrow convention of force through a series of tasks in which students struggle to communicate the action of force on a piece of plasticine (Hubber, Tytler, & Haslam, 2010), in representing the imagined relations between particles in a solid to explain specific properties such as elasticity of rubber, or expansion of metal on heating (Hubber & Tytler, 2013), or in planning and constructing a 3D model of an invertebrate to explain its movement (Tytler, Haslam, Prain, & Hubber, 2009). In other cases teachers might plan a sequence of challenges involving representational redescription across modes, such as a sequence of activities in which students develop their understandings of particle models of evaporation using role-play, drawing, discussion of a 3D demonstration, and a cartoon representation of a single particle's history (Prain & Tytler, 2012). In cases where the scientific model is more complex, students may begin by redescribing an existing model in response to a specific challenge, such as taking digital simulations of the rotation of the earth and constructing drawings to explain how the sun moves around the horizon when seen from above the arctic circle in the northern summer (Hubber, 2010).

The following examples of students' representational work to illustrate the approach occurred within junior secondary classrooms from an Australian metropolitan school. The teachers were teaching the nationally set curriculum which mandated that students learn 'sedimentary, igneous and metamorphic rocks contain minerals and are formed by processes that occur within Earth over a variety of timescales' (ACARA, 2012). The initial exploration of rock types occurred by students in small groups creating a dichotomous key from a chosen boxed set of several rocks from a collection of sets. The evaluations of the keys were undertaken at the small group level whereby each group was to self-assess their own key in addition to evaluating another group's key by testing it with an unknown rock. Students as part of the sequence also explored the modeling of the earth's internal structure, critiquing models based on a boiled egg, and an orange, in terms of features that were and were not represented. Central literacy features of these activities were student representational construction of multimodal text, critique of models and understanding of the purpose of models.

A main learning outcome of the teaching sequence was for students to gain an understanding of the rock cycle whereby students get insights into the nature of the main rock types in addition to the processes by which they are individually formed and the processes by which one rock type can transform into another. There was not one canonical rock cycle that was advanced by the teachers for the students to study. Rather, students were to critique different diagrammatic forms of the rock cycle to then construct their own rock cycle. In the

Fig. 18.1 Small group critique of diagrammatic forms of the rock cycle

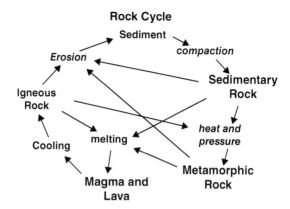

following example, the teacher laid out seven different diagrammatic forms of the rock cycle in different locations in the classroom. In groups of three students they were to move around the room critiquing each rock cycle in terms of addressing the questions, 'What does it show well?' and, 'What does it not show well?'. Figure 18.1 shows a particular rock cycle with a transcript of a discussion between the group and their teacher following the group's critique of the rock cycle representation.

T: *Looking at the cycle what can you tell me about it?*

S1: *It shows how everything is formed and connected*

T: *When you say everything what do you mean?*

S1: *The types of rocks*

S2: *And it is colour coded too*

T: *Does that help?*

S2: *Yes because if you follow the arrows you find what you are looking for.*

S1: *For example, both sedimentary and igneous rocks have similar processes that they can through heat and pressure form the metamorphic rocks* [pointing to the dark red arrows] ... *it shows how they are connected to the metamorphic rock*

S3: *... it gives you options about where to go*

S1: *The second example is sedimentary rocks can melt to form magma, which when it cools becomes igneous rocks; the igneous though can become a sedimentary rock once again through erosion* [tracing the path with an pen]

T: *So erosion is leading from that one* [pointing at igneous]

S1: *Connected to sediments to sedimentary ...*

S2: *its like a never ending cycle* [point out various cycle on the diagram]

T: *Does it show weathering?*

S1: *it shows erosion but doesn't show weathering*

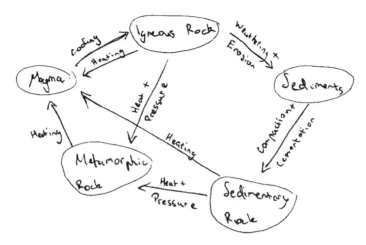

Fig. 18.2 Student constructed rock cycle from an end of sequence test

T: *So does this help explain the ideas?*

S2: *Looking at it first it was kind of confusing but once you had time to look at it and follow the arrows it makes a lot of sense.*

From the discussion students were challenged to construct their own rock cycle. None of the students chose a rock cycle from the critique challenge in its entirety but chose to take various features from several rock cycles to construct their own. Tests at the end of the sequence then a subsequent formal exam showed a consistency and high quality in students' representations over time. Figure 18.2 shows one example of a rock cycle constructed during the end of sequence test, illustrating engagement with diagrammatic claim-making and with the rock cycle concept.

As part of the approach, students engage with complex forms of reporting, including posters of extended investigations, group constructed models with explanatory digital text, or reports of investigations. Figure 18.3 shows a Grade 5/6 students' report of a group investigation into the dissolving of food colouring in hot and cold water, with explanatory text supported by diagrammatic particle representations. The class had discussed particle ideas and the group explorations were accompanied by a class brainstorm of ideas about dissolving, with the report instructions emphasising explanation and visual representation. The coordination of diagrams and text had been modeled consistently on the whiteboard and in reporting on teacher-scaffolded investigations.

This work is in some respects similar to a formal template of the type traditionally used for practical reports, except that the emphasis is on explanation rather than stepping out prediction, method etc. The students here have clearly engaged with the problem and the text and drawings represent complex claims related to experimental evidence. We argue, acknowledging Lemke's (2002) point that the science community does not follow the genre norms often promoted

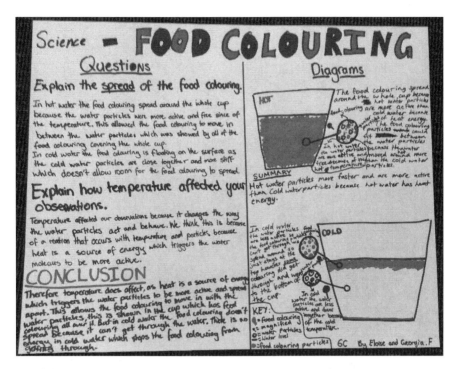

Fig. 18.3 Student report on a dissolving investigation, focused on explanatory text

as central to scientific disciplinary literacy, that such productions are important and generative examples of engagement with scientific literacy practices, using representational resources including text as tools to engage with significant reasoning. The tools achieve potency and meaning through their bending to interpretive, explanatory purposes that are both fresh, and shared within the classroom community.

Schools we have been working with on this inquiry approach have increasingly seen the value of text production within student workbooks that are lined on one side and blank on the other, encouraging diagrammatic exploration and presentation of ideas. Figure 18.4 shows an excerpt from a workbook in which a student, following class discussion on gravity, the moon and tides, plays with different ways of representing gravity on different objects. A subsequent entry represents how tides form and also explains why the moon doesn't fall to earth. Teachers have reported how students take great pride in these workbooks as evidence of their developing ideas. The workbooks sit within a strong tradition in science of field note taking and journal writing, both genres that play to informal and formal reasoning in developing science knowledge, and that capture important aspects of the interplay of evidence and idea generation in the representation circulation processes leading from data to knowledge production (Latour, 1999).

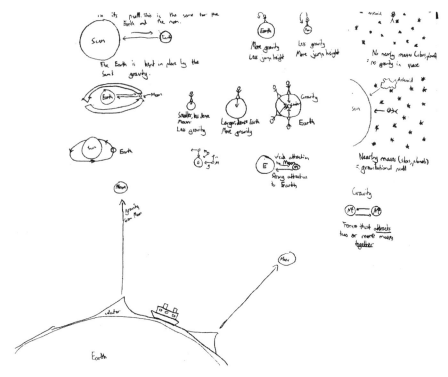

Fig. 18.4 Student workbook entry in response to a challenge to represent how gravity affects astronomical objects, and tides

18.4 Student Learning Outcomes: Building Disciplinary Literacy

In arguing for the authenticity and effectiveness of the approach for building students' disciplinary literacy, we argue that scientific disciplinary literacy involves a number of facets that are attended to by the approach:

1. Genuine engagement with classroom practices that parallel the epistemic practices of science;
2. Representational work that indicates commitment to explanation and problem solving through creating non-standard representational resources;
3. Evidence of high level reasoning through engagement with representational practices;
4. Mastery of science conceptual understanding of key concepts;
5. Productive disposition demonstrated by motivation to pursue investigation and problem solving;
6. Meta representational competence demonstrated by understanding of the role of representations and models in knowledge production and dissemination;

7. Flexible adaptation of traditional science genres to engage and extend student learning;
8. Explicit discussion of representational form and function and modeling of the integration of different modes.

To evaluate the effectiveness of the approach in building disciplinary literacy we draw on a number of sources of evidence to deal with these aspects of literacy in turn. First, in our original research developing the approach, video evidence of classroom activities and discussions shows students involved with high level problem solving as they develop individual representational practices to investigate and communicate populations of invertebrates in the school ground (Tytler, Haslam, Prain, & Hubber, 2009), build representations of animal movement, or develop and critique representations of force in explaining motion (Hubber, Tytler, & Haslam, 2010). Students work in groups and whole class discussions to construct and refine representations, drawing on empirical observation and experimentation in ways that parallel the operation of research laboratories (Tytler, Haslam, Prain, & Hubber, 2009). There is evidence also of increased student engagement with ideas and motivation, across the spectrum from advanced classes being challenged with high level problem solving, to lower level classes where teachers report students becoming more engaged with the active participation implied by the approach in contrast to more teacher delivered material.

Second, the level of student representational work in solving challenges is evidenced by the examples above. Similarly, examples from a range of topics and year levels show imaginative and individual engagement with representational work in solving problems, such as particle representations of evaporation from wet handprints or puddles demonstrating representational flexibility and conceptual ideas beyond expected for the grade level (Ainsworth, Prain, & Tytler, 2011; Tytler, Prain, Hubber, & Haslam, 2013), or engaging with astronomical problems using multiple and multimodal, sophisticated representations (Hubber, 2010) that show detailed command of astronomical perspective through diagrammatic work.

Third, with regard to high level reasoning through representational work, teachers have consistently attested to the liveliness and depth of classroom discussions around representational practices, more so than with traditional pedagogies. Again, evidence from video and student artefacts show significant reasoning occurring at multiple points in the representational work, from data generation structured by representational framing, to interpretation of observations and data and argumentation around representation construction, to analysis and argumentation around representation evaluation (Tytler, Prain, Hubber, & Haslam, 2013).

Fourth, while there has been no formal comparative research carried out, comparing the approach with other approaches, pre- and post-test data has consistently shown a significant gain in understanding as measured on multiple choice items. We have used, for instance, a recognised astronomy test instrument as part of an astronomy sequence in the RiLS project, to track outcomes. The test was used by Kalkan and Kiroglu (2007) in a study that involved 100 pre-service primary and secondary education teachers who participated in a semester length course in astronomy. This allows us to compare results with those obtained by Kalkan and Kioglu, who used the normalised gain index, $<g>$, as a measure of comparison of pre- and post-test results (Zeilik, Schau, & Mattern, 1998). $<g>$ is a measure of the ratio of the actual average student gain to the maximum possible average gain: $<g> = (\text{post}\% - \text{pre}\%)/(100 - \text{pre}\%)$. Gain index values can thus range from 0 (no gain achieved) to 1 (all possible gain achieved). For multiple choice questions, a gain index of 0.4 for an item indicates that for instance if 50% of students in the pre-test answered the question correctly, 70% answer correctly in the post test, being 0.4 of the possible gain from 50% to 100%. Kalkan and Kiroglu (2007, p. 17) reported a mean gain of a 'respectable 0.3'.

In our original study we worked, at secondary level, mainly with two teachers, Lyn and Sally, who were biology majors. For our second sequence, a four-week year 8 astronomy unit they expressed lack of confidence in astronomy concepts. A third teacher, Ben, who was a physics major and confident with astronomy, initially joined the project but shortly after the planning sessions he declined to continue on the grounds he preferred to teach astronomy as he had previously done. During the unit, Lyn and Sally progressively increased in confidence. The pre- and post-test data was collected for all three classes, which were not streamed, and the results for the gain index are shown in Table 18.1 for Lyn and Sally, and Ben, compared with the Kalkan and Kioglu results. The gain index shows clearly that the two classes using a representation construction inquiry approach outperformed by a wide margin both the class taught by the physics specialist, Ben, and the pre-service teachers undertaking a semester length course. Comparison using a two-tailed t-test showed difference at significant levels of 0.013 against Ben, and

Table 18.1 Normalised gain indices for Sally and Lyn's classes compared to Ben, those reported by Kalkan and Kioglu, and a later set from Sutton school using the approach

	Astronomy Context	Sally & Lyn	Ben	K&K	Sutton
1	Day and night	0.785	0.83	0.22	0.8
2	Phases of the moon	0.605	0.38	0.09	0.36
3	Sun Earth distance scale	0.4	0.13	0.05	0.44
4	Altitude of midday Sun	0.635	−.31	0.14	0.53
5	Earth's diameter estimate	0.415	0.23	0.09	0.44
6	Seasons	0.59	0.13	0.61	0.23
7	Sequence of objects from Sun	0.5	0.38	0.46	0.49
8	Time for Moon's orbit of Earth	0.75	0.71	0.22	0.72
9	Time for Earth's orbit of Sun	0.875	−1	0.41	0.7
10	Eclipse and phase of moon	0.795	0.42	0.22	0.32
11	Moon's motion around Earth	0.5	0.23	0.17	0.48
12	Centre of universe location	0.5	0.33	0.66	0.48
13	Seasons	0.9	0.5	0.64	0.81
	Mean gain index	0.63	0.23	0.31	0.52

0.00033 against the K&K result. This gain has been repeated for a number of classes since this initial investigation, for astronomy (results for Sutton school, Year 7, are shown as the final column) and also for other challenging topics such as the particle nature of matter (Tytler, Prain, Hubber, & Waldrip, 2013, p. 47). This comparison should be taken as indicative rather than a formal experimental proof, since there are unaccounted-for, possibly confounding factors present in the comparisons, and we do not know in detail what Ben's approach entailed. Nevertheless, the consistent strength of the gain across classes and topics does indicate a strong conceptual outcome attributable to this inquiry approach, on measures that target acknowledged high level concepts in difficult topics.

Fifth, students interviewed concerning their response to the approach, and teacher perceptions of student engagement with learning, show consistently increased motivation to become involved in pursuing representational practices and high level ideas, through group work and in classroom discussion. Teachers have reported being surprised by high levels of student competence and commitment to problem solving.

Sixth, a key feature of the approach is explicit discussion of the nature and role of representations in learning and reasoning about phenomena. Test items have been constructed that explore students' understanding of the nature of models in scientific explanation (Tytler et al., 2013, p. 45). Teachers report that students who have been exposed to the approach for a year or more become sophisticated consumers of text book representations, offering critique as a matter of course. As Lyn described (Tytler et al., 2013, p. 48):

> ... we're not teaching the particle model as in, this is the model and see how it relates to real life. It's more, this is real life and we have a model and does it actually explain real life, and does it explain this and that? And particularly ... how good is the representation?

A year 8 student, in interview, described the relationship between representations thus:

> Through many representations you can come to an understanding. So many representations help you get an understanding ... but then, through your understanding you can give many representations. So it works both ways. (Tytler et al., 2013, p. 48)

Thus, we argue that through this guided inquiry approach students can achieve a meta-level competence in the disciplinary literacies of science, through explicit attention to the nature of representations and their role in reasoning, learning and knowledge building.

Seventh, and finally, student production with the method is varied and primarily associated with the construction of multimodal text to generate and justify ideas. Traditional disciplinary genres are positioned in this production as resources to support reasoning, advancing claims and supporting these with evidence. These practices are positioned within a classroom community of inquiry with a focus on the construction, critique and refinement of representational forms. We argue that in this way, the scientific literacies being developed engage students in meaningful epistemic processes and text production that are an important adjunct to the more formal literacy genres foregrounded in much of the disciplinary literacy literature.

We also acknowledge that our guided inquiry approach required refinements over time and also posed some significant challenges for participant teachers. These refinements included the need to develop a range of challenges, tasks and learning processes that (a) catered for mixed ability classes, (b) offered generative scope for diverse students' responses and (c) could be broadly aligned with prescribed learning outcomes in the national curriculum. The challenges for teachers included the development of skills in interpreting, guiding and consolidating progress as the students responded to sequences of tasks, made divergent claims, and raised unscripted challenges to the teachers' own conceptual and representational understandings. The teachers also had to manage time spent on this deeper learning against the content demands of the curriculum. However, as noted in the preceding paragraphs, there were many overriding learning gains, as noted by both teachers and students.

18.5 Conclusion

In this chapter we have argued that a view that learning science involves induction into scientific disciplinary literacies implies a need for the promotion of classroom practices that more authentically parallel the epistemic practices of the discipline. Contemporary perspectives on processes of scientific discovery foreground the crucial role of representations and representational work in framing, building and sharing new knowledge, and this therefore needs to be a driving consideration in framing classroom inquiry approaches.

We have further argued that our view of learning and knowing offers a powerful perspective on the importance of active inquiry in which students engage with experimental exploration and the creation and critique of representations in the pursuit of knowledge. This is supported by our account of representational affordance as productive constraint, as a way of understanding the way representation construction within guided inquiry can productively mirror science epistemic processes.

Our account of student work engendered by the approach emphasises both the nature of student experimental exploration and generation, evaluation and refinement of representations as they grapple with conceptual challenges, and the quality of the representational work that can ensue. We argue that the approach reveals important aspects of what it means to develop scientific disciplinary literacies, such as engagement in classroom processes that mirror scientific epistemic processes, reasoning through construction and coordination of representations that results in deepening conceptual knowledge, a disposition and capacity to engage with scientific problems, and the development of meta-representational competence and awareness. The approach shows promise of supporting students to develop these scientific literacies to a high level, as evidenced by the quality of student engagement with reasoning illustrated in our examples, and elsewhere in our writing (Tytler, Prain, Hubber, & Waldrip, 2013), and evidence from pre- and post-test results and teacher accounts.

While we acknowledge the importance of a focus on formal scientific genres in supporting literacy development, we argue that if we are to engage students in thinking and working scientifically, these need to be positioned as resources for reasoning within contexts in which students explore, make claims and reason about material phenomena through imaginative, multimodal text production that draws on diverse, often informal scientific practices and genres.

References

ACARA (Australian Curriculum, Assessment and Reporting Authority) (2012). Foundation to Australian Curriculum: Science. ACARA. http://www.acara.edu.au/curriculum/learning-areas-subjects/science.

Ainsworth, S., Prain, V., & Tytler, R. (2011). Drawing to learn in science. *Science, 333* (26 August), 1096–1097.

Amin, T., Jeppsson, F., & Haglund, J. (2015). Conceptual metaphor and embodied cognition in science learning: Introduction to special issue. *International Journal of Science Education, 37* (5–6), 745–758.

Bazerman, C. (2009). Genre and cognitive development: Beyond writing to learn. In C. Bazerman, A. Bonini, & D. Figueiredo (Eds.), *Genre in a changing world* (pp. 279–294). Fort Collins: The WAC Clearinghouse.

Cobb, P., Confrey, J., diSessa, A., Lehrer, R., & Schauble, L. (2003). Design experiments in educational research. *Educational Researcher, 32*(1), 9–13.

diSessa, A. (2004). Metarepresentation: Native competence and targets for instruction. *Cognition and Instruction, 22*(3), 293–331.

Duit, R., & Treagust, D. (2012). Conceptual change: Still a powerful framework for improving the practice of science education. In K. Tan & M. Kim (Eds.), *Issues and challenges in science education research moving forward*. Dordrecht: Springer.

Duschl, R. (2008). Science education in three-part harmony: Balancing conceptual, epistemic, and social learning goals. *Review of research in education, 32*(1), 268–291.

Ford, M., & Forman, E. A. (2006). Refining disciplinary learning in classroom contexts. *Review of Research in Education, 30*, 1–33.

Gee, J. P. (2004). Language in the science classroom: Academic social languages as the heart of school-based literacy. In E. W. Saul (Ed.), *Crossing borders in literacy and science instruction: Perspectives in theory and practice* (pp. 13–32). Newark: International Reading Association/National Science Teachers Association.

Gilbert, J. K. (2005). *Visualization in science education*. New York: Springer.

Gilbert, J., Reiner, M., & Nakhlel, M. (2008). *Visualization: Theory and practice in science education*. New York: Springer.

Gooding, D. (2005). 1 Visualisation, inference and explanation in the sciences. *Studies in multi-disciplinarity, 2*, 1–25.

Gooding, D. (2006). From phenomenology to field theory: Faraday's visual reasoning. *Perspectives on Science, 14*(1), 40–65.

Greeno, J. (2009). A framework bite on contextualizing, framing, and positioning: A companion to Son and Goldstone. *Cognition and Instruction, 27*(3), 269–275.

Greeno, J. G., & Hall, R. P. (1997). Practicing representation: Learning with and about representational forms. *Phi Delta Kappan, 78* (5), 361–368.

Halliday, M., & Martin, J. (1993). *Writing science: Literacy and discursive power*. London: Falmer Press.

Hubber, P. (2010). Year 8 students' understanding of astronomy as a representational issue: Insights from a classroom video study. In D. Raine, L. Rogers, & C. Hurkett (Eds.), *Physics community and cooperation: Selected contributions from the GIREP-EPEC & PHEC, 2009 International Conference* (pp. 45–64). Leicester: Lulu, the Centre for Interdisciplinary Science, University of Leicester.

Hubber, P., & Tytler, R. (2013). Models and learning science. In *Constructing representations to learn in science* (pp. 109–133). Rotterdam: Sense Publishers.

Hubber, P, Tytler, R., & Haslam, F. (2010). Teaching and learning about force with a representational focus: Pedagogy and teacher change. *Research in Science Education, 40*(1), 5–28.

Hurd, P. D. (1998). Scientific literacy: New minds for a changing world. *Science Education, 82*, 40–416.

Kalkan, H., & Kiroglu, K. (2007). Science and nonscience students' ideas about basic Astronomy concepts in preservice training for elementary school teachers. *The Astronomy Education Review*, 1(6), 15–24.

Kelly, G. J., & Chen, C. (1999). The sound of music: Constructing science as sociocultural practices through oral and written discourse. *Journal of Research in Science Teaching, 36*, 883–915.

Keys, C., Hand, B., Prain, V., & Collins, S. (1999). Using the science writing heuristic as a tool for learning from laboratory investigations in secondary school. *Journal of Research in Science Teaching, 36* (10), 1065–1084.

Klein, P. D., & Boscolo, P. (2016). Trends in research on writing as a learning activity. *Journal of Writing Research, 7*(3), 311–350. https://doi.org/10.17239/jowr-2016.07.3.01

Kozma, R., & Russell, J. (2005). Students becoming chemists: Developing representational competence. In J. Gilbert (Ed.), *Visualization in science education* (pp. 121–145). Dordrecht: Springer.

Kress, G., Jewitt, C., Ogborn, J., & Tsatsarelis, C. (Eds.). (2001). *Multimodal teaching and learning: The rhetorics of the science classroom*. London: Continuum

Kress, G., & van Leeuwen, T. (2006). *Reading images: The grammar of visual design* (2nd ed.). London: Routledge.

Latour, B. (1986). Visualization and cognition: Drawing things together. *Knowledge and Society, 6*, 1–40.

Latour, B. (1999). *Pandora's hope: Essays on the reality of science studies*. Cambridge: Harvard University Press.

Lehrer, R., & Schauble, L. (2006a). Cultivating model-based reasoning in science education. In K. Sawyer (Ed.), *Cambridge handbook of the learning sciences* (pp. 371–388). Cambridge: Cambridge University Press.

Lehrer, R., & Schauble, L. (2006b). Scientific thinking and science literacy. In W. Damon & R. Lerner (Eds.), *Handbook of child psychology* (6th ed., Vol. 4). New York: Wiley.

Lemke, J. (2002). *Multimedia genres for science education and scientific literacy*. Retrieved from http://static1.1.sqspcdn.com/static/f/694454/12422300/1306520367633/Multimedia Genres-Science-2002.pdf?token=bunzpX1zcgDpSmDWzwuAnauCo%2Bw%3D

Lemke, J. L. (2003). Mathematics in the middle: Measure, picture, gesture, sign, and word. In M. Anderson, A. Sàenz-Ludlow, S. Zellweger & V.V. Cifarelli (Eds.), *Educational perspectives on mathematics as semiosis: From thinking to interpreting to knowing* (pp. 215–234). Ottawa: Legas.

Lemke, J. (2004). The literacies of science. In E. W. Saul (Ed.), *Crossing borders in literacy and science instruction: Perspectives on theory and practice*. Newark: International Reading Association/National Science Teachers Association.

Linder, C., Ostman, L., & Wickman, P-O (Eds.). (2007). Promoting scientific literacy: Science education research in transition. *Proceedings of the Linnaeus Tercentenary symposium*, Uppsala University, Uppsala, May 28–29.

Manz, E. (2012). Understanding the codevelopment of modeling practice and ecological knowledge. *Science Education, 96*(6), 1071–1105.

Moje, E. B. (2007). Developing socially just subject-matter instruction: A review of the literature on disciplinary literacy teaching. *Review of Research in Education, 31*, 1–44.

Mortimer, E. F., & Scott, P. H. (2003). *Meaning making in secondary science classrooms.* Maidenhead: Open University Press.

Norris, S. & Phillips, L. (2003). How literacy in its fundamental sense is central to scientific literacy. *Science Education, 87,* 224–240.

Parkinson, J., & Adendoff, R. (2004). The use of popular science articles in teaching science literacy. *English for Specific Purposes, 23,* 379–396.

Posner, G. J., Strike, K. A., Hewson, P. W., & Gertzog, W. A. (1982). Accommodation of a scientific conception: Toward a theory of conceptual change. *Science & Education, 66* (2), 211–227.

Prain, V., & Tytler, R. (2012). Learning through constructing representations in science: A framework of representational construction affordances. *International Journal of Science Education. available online.* https://doi.org/10.1080/09500693.2011.626462

Shanahan, T., & Shanahan, C. (2008). Teaching disciplinary literacy to adolescents: Rethinking content area literacy. *Harvard Educational Review, 78,* 40–61.

Tytler, R., Haslam, F., Prain, V., & Hubber, P. (2009). An explicit representational focus for teaching and learning about animals in the environment. *Teaching Science, 55*(4), 21–27.

Tytler, R., Hubber, P., Prain, V., & Waldrip, B. (2013). A representation construction approach. In R. Tytler, V. Prain, P. Hubber, & B. Waldrip (Eds.), *Constructing representations to learn in science* (pp. 31–50). Rotterdam: Sense.

Tytler, R., & Prain, V. (2010). A framework for re-thinking learning in science from recent cognitive science perspectives. *International Journal of Science Education, 32*(15), 2055–2078.

Tytler, R., & Prain, V. (2013). Representation construction to support conceptual change. In S. Vosniadou (Ed.), *Handbook of research on conceptual change* (pp. 560–579). New York: Routledge.

Tytler, R., Prain, V., Hubber, P., & Haslam, F. (2013). Reasoning in science through representation. In R. Tytler, V. Prain, P. Hubber, & B. Waldrip (Eds.), *Constructing representations to learn in science* (pp. 83–108). Rotterdam: Sense.

Unsworth, L. (2001). *Teaching multiliteracies across the curriculum: Changing contexts of text and image in classroom practice.* Buckingham: Open University Press.

Veel, R. (1997). Learning how to mean—Scientifically speaking. In F. Christie (Ed.), *Genre and institutions: The language of work and schooling* (pp. 161–195). London: Cassell Academic.

Vosniadou, S. (2008a). Bridging culture with cognition: A commentary on "culturing" conceptions: From first principles. *Cultural Studies of Science Education, 3,* 277–282.

Vosniadou, S. (2008b). Conceptual change research: An introduction. In S. Vosniadou (Ed.), *Handbook of research on conceptual change.* New York: Routledge.

Vygotsky, L. S. (1978). *Mind in society.* Cambridge: Harvard University Press.

Vygotsky, L. (1981). *Thought and language* (revised and edited by A. Kozulin). Cambridge: MIT Press.

Waldrip, B., Prain, V. & Carolan, J. (2010). Using multimodal representations to improve learning in junior secondary science. *Research in Science Education, 40*(1), 65–80.

Zeilik, M., Schau, C., & Mattern, N. (1998). Misconceptions and their change in university-level Astronomy courses. *The Physics Teacher, 36*(1), 104–107.

Part 6
Science Teacher Development

Chapter 19
Science and Language Experience Narratives of Pre-Service Primary Teachers Learning to Teach Science in Multilingual Contexts

Mariona Espinet, Laura Valdés-Sanchez and Maria Isabel Hernández

Abstract The command of at least three languages is considered one of the most important basic educational competences in Europe. In response to this demand, new teaching approaches have been promoted, such as Content and Language Integrated Learning (CLIL). A CLIL approach to science education implies the need of teachers capable of teaching both science and foreign language. This is a study on a specific group of pre-service primary teachers who are enrolled in an English-mediated primary education degree in Catalonia. The aim is to characterize and compare pre-service primary teachers' science and language experiences enacted in their science and language experience narratives (SN and LN) that might have shaped their beliefs, practices, and expectations about science and languages teaching. A content analysis is performed on these narratives grounded on a two dimensional space narrative structure defining the Fields of Experience. The results indicate that SN are mostly rooted in school contexts, and associated to learning difficulties. This is in strong contrast with LN which appear to be anchored in a wide variety of contexts, for purposes that go beyond school, and associated to progressive and positive learning trajectories. The question for science teacher education lies in how to help pre-service primary teachers make connections between both life trajectories and how to increase experiences of a social nature emphasizing the presence, value and utility of science, and science education.

Keywords Science experience narratives · language experience narratives · science teacher education · CLIL · multilingualism

---

M. Espinet (✉) · L. Valdés-Sanchez · M.I. Hernández
Departament de Didàctica de la Matemàtica i de les Ciències Experimentals, Universitat Autònoma de Barcelona, Catalonia, Spain
e-mail: mariona.espinet@uab.cat

L. Valdés-Sanchez
e-mail: lauravaldessanchez@gmail.com

M.I. Hernández
e-mail: mariaisabel.hernandez@uab.cat

19.1 Introduction

One of the demands that a global society places on most European education systems is related to the fact that our society and schools are multilingual contexts and that language diversity is a cultural heritage in need of conservation. The command of at least three languages is considered one of the most important basic competences that every European citizen should acquire through compulsory education (European Commission, 2007). However, the repertoire of language used in Europe can be seen as divided into two types of multilingualism (Guasch & Nussbaum, 2007): a first order multilingualism constituted by the big European languages which are strongly valued and worth learning, and the second order multilingualism constituted by the minority languages present as a consequence of immigration which can be tolerated but have a lower status.

The multilingual context experienced in Catalonia adds a third factor making multilingual education more difficult. In fact, Catalonia is an autonomous region of Spain considering itself a nation without a state. The Catalan social and cultural identity is built around the core element of its particular language Catalan which has been suppressed throughout the history of the country in several occasions and particularly during the time of General Franco's dictatorship. Since Franco's death, language policies promulgated by the Catalan government have had an important role in ensuring that Catalan is now commonly used in many aspects of daily life including education. The region has officially two languages: Catalan and Spanish although the school system has adopted a compulsory immersion model in Catalan as a tool for social cohesion.

19.2 A Content and Language Integrated Learning Approach (CLIL) to Primary Science Education

In response to this demand, European educational institutions at all levels are developing new teaching approaches which could be included under the broad umbrella of bilingual education. Cummins (2008), one of the fathers of multilingualism in education, defined bilingual education as the "use of two or more languages of instruction at some point in a student's school career" (p. xii). One of the bilingual education approaches recently promoted in Europe has been the "Content and Language Integrated Learning" (CLIL) (Coyle, Hood, & Marsh, 2010; Dalton-Puffer, Nikula, & Smit, 2010; Escobar Urmeneta, Evnitskaya, Moore, & Patiño, 2011). This approach advocates the need to design learning environments in which both specific content and a specific foreign language can be taught and learned together: "The acronym CLIL is used as a generic term to describe all types of provision in which a second language (a foreign, regional or minority language and/or another official state language) is used to teach certain subjects in the curriculum other than languages lessons themselves" (Eurydice,

2006, p. 8). A CLIL approach to science education implies the teaching of both science content and foreign language in the same classroom and frequently by the same teacher. Multilingual science education contexts are very varied with multiple models and structures existing in different European education systems.

19.3 Challenges of CLIL Primary Science Teacher Education

When developing CLIL approaches in primary science classrooms in Catalonia, teachers need to manage the teaching of science and three languages at the same time: Spanish, Catalan, and English, the last one being a foreign language. The primary teachers responsible for CLIL approaches to science teaching are mostly English specialists who lack confidence in their ability to teach science and especially inquiry-based science (Espinet et al., 2017; Martin, 2008; Navés & Victori, 2010). This is consistent with a large body of literature that illustrates primary teachers' reluctance to teach science (Abell & Roth, 1992; Appleton, 1995; Davis, Petish, & Smithey, 2006) due to several factors such as teachers' lack of content knowledge, lack of confidence in teaching science, and a negative attitude toward science. One useful strategy to confront these challenges has been the development of co-teaching practices where both primary English and Science teachers teach in the same classroom (Valdés-Sánchez & Espinet, 2013a, b).

Efforts have been made to investigate the type of experiences offered to pre-service primary science teachers during the university coursework that might be useful to overcome some of the problems stated above. Avraamidou (2013) found that inquiry-based investigations and outdoor field studies among others provided experiences that positively influenced pre-service teachers' teaching orientations. Ramos and Espinet (2013) focused on classroom interactions and found that pre-service primary teachers expanded their agency when participating in laboratory activities enacted in multilingual contexts. Finally, Espinet et al. (2017) have recently developed a triadic partnership through the establishment of communities of practice that support primary science teachers, primary English teachers, pre-service primary teachers, educational administrators, and university professors from science education as well as language education to work collaboratively. The purpose of the triadic partnership was to create a learning environment that facilitates the collaborative design and implementation of innovative teaching units aiming at the integration of inquiry-based science and English in primary classrooms.

19.4 Experience Narratives in Pre-Service Teacher Education

This study is framed under a narrative approach to research in education (Cortazzi, 1993) and teacher education (Goodson, 2003) and recognizes the importance of teachers' narratives for their professional development. Some educational researchers following Lortie's pioneer work (Lortie, 1975) have found that schooling experiences and life experiences have influenced teacher candidates' beliefs about education and ultimately the type of teachers they became (Calderhead, 1989). Making sense of one's own experiences through the crafting of life stories has become a useful strategy to foster pre-service and in-service teacher development. Life story is considered a subset of the narrative genre (Avraamidou & Osborne, 2009) and consists of narratives of self that reconstruct the past, connect to the present, and anticipate the future.

Rivera (2011) has investigated how a particular type of narratives, the experience narratives, contribute to pre-service science teachers' life story. In her study, Rivera focuses on the importance of the language experience narratives in the development of pre-service science teachers' trajectories so that these experiences become visible and ready to be critically scrutinized. This study adds something new to the research work on primary science teachers' experiences since it highlights the importance of language experiences in addition to science experiences.

We take the concept of *Experience Narrative* from Rivera (2011) and consider it as a subset of life story, as a lens through which to look at one's own life story. We think that this narrow sense of narrative is more appropriate for the capturing of more specific pre-service primary teachers' experiences. The present research departs from the assumption that pre-service teachers' experiences are important factors influencing beliefs and practices on teaching science and languages in primary CLIL science classrooms. The aim of the present study is to characterize and compare pre-service primary teachers' science and language experiences enacted in their science and language experience narratives (SN and LN from now on).

19.5 Approaching the Analysis of Science and Language Experience Narratives

19.5.1 Pre-Service Science Teacher Education Context

The Universitat Autònoma de Barcelona (UAB) in Catalonia, Spain, started to offer an 80% English-mediated Primary Education Program in the academic year 2012–2013. The pre-service teachers participating in this study come from this particular program. During the first 3 years, all pre-service teachers take compulsory courses such as science and language. It is in the fourth and last year that they have the chance to become either specialized primary teachers (music, special needs, foreign languages, or physical education) or remain general primary teachers. Although all pre-service teachers in this study started their undergraduate studies with a similar background with regards to language (they had to pass tests of Catalan, Spanish, and English language competence), their science background was very diverse. In fact, most of them did not take science courses in high school while others studied scientific disciplines until their entrance to tertiary studies.

Thirty-nine pre-service primary teachers, 35 female and 4 male, participated in this study on a voluntarily basis (50% of the course enrollment). They were informed about the purpose of the research and the ways used to preserve their anonymity. Each pre-service teacher wrote one SN and one LN when attending two different undergraduate science education courses. First, they were encouraged to write the LN within the course *Teaching and Learning about the Environment in Primary Education* offered in the 2nd year (2013–2014). Second, they were encouraged to write an SN within the course *Didactics of Science* offered in the 3rd year (2014–2015). The guidelines shown in Table 19.1 were used to orient students' writing of their narratives. Students had about a month to individually write their narratives and submit them through the university virtual platform. The three authors of this chapter were responsible of teaching both subjects during the data collection period.

Table 19.1 Questions guiding pre-service teachers' writing of SN and LN

Language Experience Narrative (LN)	Science Experience Narrative (SN)
Your relationship with different languages: What languages are you familiar with? How competent in writing, speaking, listening, and reading you feel you are? What languages would you have liked to learn? Why? What type of experiences have you undergone through your life in which language was an important component?	*Your relationship with Science outside school*: How much do you like science? Why? How competent in writing, speaking, thinking, and doing science do you feel you are? What family models in science do you have? What issues or topics would you have liked to learn? Why? What type of experiences have you undergone through your life in which science was an important component? (Science museums, science centers, science summer camps, TV programs, scientific literature, etc.)
Your experience with the teaching and learning of languages: What type of language teaching and learning experiences have you undergone through your life? How do you evaluate them? What languages have been important for you in your educational and working experiences? What have your feelings been in relation to the teaching and learning of different languages? What are your expectations in life and in the university in relation to the use and learning of languages?	*Your experience with the teaching and learning of science*: What type of experiences in teaching and learning sciences have you undergone through your life? How do you evaluate them? What sciences, if any, have been important for you in your educational and working experiences? What have your feeling been in relation to the teaching and learning of science? What are your expectations in life and in the university in relation to the learning of science?

19.5.2 Field of Experience as a Construct for the Analysis of Science and Language Experience Narratives

A content analysis was performed on each SN and LN (Holsti, 1969; Krippendorff, 1980; Weber, 1990). Our analysis has been informed by the works of Clandinin and Connelly (2000) and Avraamidou (2013) on the use of narratives in teacher education. They developed and used an analytical tool called "Three dimensional space narrative structure" which conceptualizes experiences as a construct having three dimensions: interactions, continuity, and situations. Instead, we have developed an analytical tool that is bidimensional and built around the concept of Field of Experience.

Two main dimensions were inductively developed during the analysis: (a) the *Experience Context*, and (b) the *Experience Nature*. The *experience context* dimension characterizes the type of social organization where the pre-service teachers' experiences are enacted such as the family, the school, the community, work, and finally the person itself. The *experience nature* dimension has been conformed around three main categories which include the educational nature, the psychological nature, and the social nature of the expressed experiences. The social nature has been divided into three subcategories: environment, value, and utility. The *experience nature* construct is applied to each of the five *experience context* categories conforming a richer conceptualization of pre-service teachers' experiences. The crossing of these two dimensions has created 25 *Fields of Experience* which have been relevant for the characterization and comparison of pre-service primary teachers' language and science experience narratives. The five categories constituting the *experience context* dimension, and the five categories constituting the *experience nature* within the analytical tool *Fields of Experience* are described and exemplified in Tables 19.2 and 19.3.

Table 19.2 Experience context dimension of the fields of experience in SN and LN

Experience context dimension of fields of experience in SN and LN	
Personal context	Experiences narrated in first person affecting directly the individual pre-service teacher and without any explicit connection to other people have been associated to a personal context, that is: "English has always been a very important language for me"; "I like learning science."
Family context	Experiences involving family members such as parents, brothers, and grandparents were included in the category of family context, that is: "Since I was able to talk I was bilingual due to the fact that my mother spoke to me in Catalan and my father Spanish"; "As a kid I always wanted to be a biochemists, like my father."
School context	The experiences involving teachers and students during the school or university years and located in formal education institutions were considered to have a school context. The experiences related to non-formal or informal education contexts are excluded, that is: "We learned English when I was in school but not enough or not so good in my opinion"; "During the primary school I was truly interested in all science topics."

(continued)

Table 19.2 (continued)

Work context	Experiences narrating activities involving job mates, clients, or bosses are associated to work context. These experiences might include working as a teacher or an educator, that is: "Two years ago I taught English to four ten year old kids"; "I think that if I enjoyed science in my childhood (even though I didn't like maths) I will be able to make my class enjoy this subject too."
Community context	The experiences narrated in relation to community members, friends, and foreigners encountered in non-formal, informal, or leisure activities were typical of a community context, that is: "Some years later I went to an academy to improve the English I learnt in school"; "I have been in scientific museums such as Cosmo Caixa in Barcelona and "El museo del Hombre" in A Coruña."

Table 19.3 Experience nature dimension of the fields of experience in SN and LN

Experience nature dimension of fields of experience in SN and LN	
Educational nature	Experiences related to the teaching and learning of either science or languages (Catalan, Spanish English, and other foreign languages). They include the interest for learning, appraisal on teachers' effectiveness, classroom activities and dynamics, subject electives, educational experiences of all types, that is: "I concluded that this type of exams do not really test your level of English, you only need to know how to do this exam"; "It's not that I didn't like Biology; in fact, I think that it could have been one of my favorites if I had had another teacher or my performance had been better."
Psychological nature	Experiences expressing beliefs on pre-service teacher's self-efficacy, relating own personal characteristics (abilities, capacities, etc.) and social or professional demands (social expectations from oneself, professional profiles) that is: "I have always been very confident with Catalan and Spanish languages and had no problem using them"; "When it comes to speaking, writing and doing science, I feel very comfortable with it because it is like explaining a story."
Social nature (a) social environment	Experiences emphasizing the social components of pre-service teachers' activities. They can be related to the social environment that is: "Along the years newcomers kept arriving into the town and many times we had to use Spanish to be able to relate with them"; "Professionally, my closest family is not related with science."
Social nature (a) social value	Experiences emphasizing the social components of pre-service teachers' activities. The social value nature of the experiences involves making explicit the importance of learning or knowing science or languages as for instance: "science is very interesting and necessary to understand our world, but it must be explained in a practical way with many examples in order to relate the content with our real life."
Social nature (a) social utility	Experiences emphasizing the social components of pre-service teachers' activities. The social utility nature of the experiences refers to perception of the utility of knowing science and/or languages to reach specific goals, that is: "German and English are two strong languages that can be very useful for my future professional development"; "Nowadays, our society is developed thanks to science, which has become a very important tool."

Table 19.4 Global relevance of science and language experiences (number of sentences from the pull of all narratives used to build a particular field of experience; $n = 1867$)

Experience nature experience context	Educational		Psychological		Social		Total	
	SN	LN	SN	LN	SN	LN	SN	LN
Personal	104	160	16	95	52	81	172	336
Family	53	53	2	4	42	54	97	111
School	443	308	33	50	32	47	508	405
Work	59	41	10	10	20	42	89	93
Community	34	192	0	31	20	92	54	315
Total	693	754	61	190	166	316	2,180	

19.5.3 Strategies for the Representation and Analysis of Data

The narratives were analyzed considering the sentence as the unit of analysis. In fact, narratives as texts are built through sentences constituting their basic building blocks. Each sentence from the narratives was associated to one or more fields of experience and thus it was assigned one or more Experience Context and Experience Nature codes.

A descriptive and comparative analysis, both qualitative and quantitative, of the SN and LN was performed to investigate the relevance of the experiences built in the narratives (Stuckey, Hofstein, Mamlok-Naaman, & Eilks, 2013). A first type of analysis to characterize the relevance of experiences considered the absolute number of sentences associated to each field of experience emerging from the pool of all narratives (Table 19.4). This analysis provided a global view of each field of experience' relevance. A second type of analysis focused on the relative presence of the different fields of experience within each particular experience narrative represented in Table 19.5. This analysis is complementary to the previous one and focuses on the individual narrative and thus on the individual student. Finally, a third type of analysis includes the positive or negative appraisal given by pre-service primary teachers to their science and language experiences reconstructed in their LN and SN. We considered evidences of positive and negative appraisal the use of linguistic resources at the sentence level emphasizing positively or negatively a particular field of experience. The appraisal was represented in two graphs showing the frequency of positive appraisal (Fig. 19.1) and negative appraisal (Fig. 19.2).

19.5.4 Relevance of Science and Language Experiences Reconstructed Within SN and LN

The 78 narratives written by the 39 pre-service primary science teachers participating in this study provide data to support the idea that the science and language

Table 19.5 Individual relevance of science and language experiences (number and percentage of SN and LN including a particular field of experience; SN = 39; LN = 39)

Experience nature experience context	Educational		Psycological		Social					
					Environment		Value		Utility	
	SN	LN	SN	LN	SN	LN	SN	LN	SN	LN
Personal	30	36	10	29	9	15	16	22	4	14
	81%	97%	27%	78%	24%	41%	43%	59%	11%	38%
Family	20	18	1	3	19	26	1	7	1	4
	54%	49%	3%	8%	51%	70%	3%	19%	3%	11%
School	36	37	18	13	4	18	12	10	7	6
	97%	100%	49%	35%	11%	49%	32%	27%	19%	16%
Work	24	17	8	9	1	7	6	5	7	15
	65%	46%	22%	24%	3%	19%	16%	14%	19%	41%
Community	18	35	0	14	4	18	7	17	3	21
	49%	95%	0%	38%	11%	49%	19%	46%	8%	57%

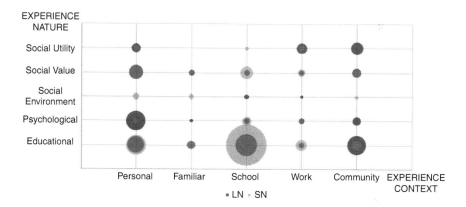

Fig. 19.1 Positive appraisal given to the experiences reconstructed through the SN and LN (percentage of sentences providing a positive appraisal on a particular field of experience: LN = 214; SN = 217)

experiences brought to the science teacher education programs are different. These narratives constitute a collection of 1,867 sentences developing different fields of experience. Whereas the LN add up to 1,050 sentences, the SN are shorter reaching 817 sentences as a whole. In our study, we consider the number of sentences as the bricks of the experience narratives used by pre-service primary teachers to reconstruct their science and language experiences. Whereas longer narrative texts such as LN might be associated to a more relevant experience, shorter ones such as SN might be an indicator of a less relevant experience.

Table 19.4 represents the total number of sentences that have been associated to each field of experience for both SN and LN. The most relevant fields of

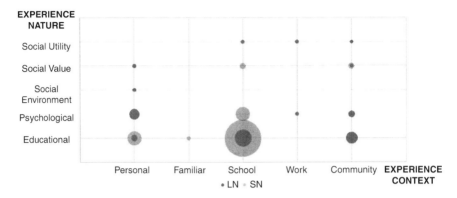

Fig. 19.2 Negative appraisal given to the experiences reconstructed through the SN and LN (percentage of sentences providing a negative appraisal on a particular field of experience: LN = 51; SN = 111)

experience are those having an educational nature and being contextualized at the personal and school levels. This is not surprising taking into account that the guidelines given to the pre-service teachers to write the narratives. However, we want to notice the differences found when comparing the experiences of SN and LN. Whereas the SN identified the school context as the most relevant when compared to LN, the LN privileged personal and community contexts when compared to SN. The family and work contexts provided experiences of a similar relevance for both the SN and LN although they are below the relevance of other contexts such community, school, and personal.

Differences can also be highlighted in relation to the nature of the experiences. Although both SN and LN are equally and highly relevant when reconstructing experiences of an educational nature, they are less relevant when this reconstruction refers to experiences of a psychological and social nature. In fact LN double the use of sentences in relation to SN when building experiences of a psychological and social nature indicating that language experiences of pre-service primary teachers are more relevant in the majority of the fields.

Figures 19.1 and 19.2 provide a graphical representation of the positive and negative appraisal given by pre-service teachers to the experiences associated to each field of experience. Pre-service teachers have been more positive than negative when appraising their science and language experiences as a whole.

On the positive side differences can be found when comparing the LN and SN. Whereas language experiences appear to be most positively valued when they are contextualized at the personal and community levels, the positive value of science experiences is concentrated mostly in school contexts (Fig. 19.1). On the negative side, Fig. 19.2 indicates that the SN accumulate most of the negative appraisals specially when valuing the experiences of an educational and psychological nature enacted within personal and school contexts.

Table 19.5 provides a quantitative account of the relevance that different fields of experience have in the reconstruction of science and language experiences through the writing of SN and LN at the individual level. A first glance at this table indicates that the percentage of LN developing each field of experience is higher than the SN.

The most relevant fields of experience reconstructed in both SN and LN are the educational experiences contextualized in school and at a personal level. Whereas almost all 39 students have reconstructed educational science and language experiences that are contextualized in school and at a personal level, around half of them have chosen family and work contexts.

What follows is a qualitative account of the most important differences between the science and language experiences reconstructed in the SN and LN. This account will be organized around the most relevant fields of experience such as those having a school, personal, or community contexts.

19.5.5 The School as the Most Relevant Context for Both Science and Language Experiences

Pre-service teachers' SN and LN provide extensive longitudinal accounts of their formal school experiences of an educational and psychological nature. The relevance of these fields of experience is supported first by the total number of sentences associated to them (Table 19.4), second by the number of individual narratives reconstructing such fields of experience (Table 19.5), and third by the concentration of most positive and negative appraisals given by these teachers in their SN and LN (Figs. 19.1 and 19.2). The reconstructed science experiences of an educational nature which have been developed in school point at two major aspects: the relationship with the specific school science subjects and their teaching methodologies.

Pre-service teachers provide accounts of science teaching methodologies that were both positively and negatively appraised. The negative experiences reconstructed in the SN describe science teaching methodologies as being theoretical and textbook driven such as the following statement illustrates: "I just can remember long classes doing exercises and reading theory from the book and how our teacher was constantly asking us to draw nice titles in our notebooks, instead of make science interesting and useful to us" (P30). Some SN however provide isolated positive accounts of motivating teachers and good science teaching methodologies emphasizing contextualization, and practical work in either primary and secondary education, as indicated by these two pre-service teachers: "Looking back, I can say that, for me, the science experiences that have been more meaningful in school are those in which learning occurred in a contextualized and practical way" (P14); "I remember loving these lessons because we were always doing experiments and applying what we have learnt to reality and see that what we learn is true" (P49).

The language experiences of an educational nature unanimously highlight the positive value of academic exchanges offered to pre-service teachers during their secondary education. One student stated: "During the fourth year in secondary education I lived through a very positive linguistic experience which I would recommend to everybody: I participated in an exchange program between my school and a German secondary school, and we needed to communicate in English" (P35). These exchanges almost always compensated the low quality of foreign language teaching during primary and secondary education. By low quality it was meant the use of foreign language teaching approaches that were oriented toward writing rather than speaking and listening.

The science experiences of a psychological nature enacted in school contexts are very homogeneous and focus mostly on a pivotal issue: the difficulty of learning mathematics and the influence this fact had in the learning of science especially in high school: "Later on, when I started High School, the difficulty of Science' subjects increased, and I realized I was not as good I was supposed to be at math, so I started disliking them… I was a little bit afraid because I was not good at math, neither chemistry nor physics, but I believed I would get over these subject-matters" (P25). The feeling of insecurity has been well documented throughout the research literature on primary science teachers' development and in this study we find the same results. In fact, this feeling gets well expressed throughout the sample of SN as exemplified in the following excerpt: "The fact that I had such low marks in math made me grow very insecure about myself and about my own abilities in science, as mathematics are an essential part of it" (P38). The LN instead, reflect a different picture in relation to the language experiences of a psychological nature enacted in school contexts. Pre-service teachers indicate feelings of comfort or easiness when reconstructing language learning experiences in schools. Although in some cases they describe difficulties in managing the three or four languages required in school, they recognize the importance of having the external support provided by English academies as exemplified in the following two excerpts: "After finishing the primary school and through the most of high school years I felt quite comfortable with Spanish and Catalan, not so much with English and I started speaking French" (P39); "In primary and secondary school I have always felt relaxed because those of us who went to an academy did better and for this reason I have never had problems with English" (P13).

19.5.6 The Personal Context of Science and Language Experiences

The most relevant experiences reconstructed within the SN and LN associated to a personal context are of an educational and psychological nature. The educational experiences include the likes and dislikes toward science and languages, and the psychological experiences deal with the perception of the personal ability to perform or succeed in learning science and languages without specifying a context such as school. Almost all SN and LN include experiences of an educational nature within a personal context. Pre-service teachers' passionate accounts of wanting to learn new languages which go beyond English, Catalan, or Spanish are an indication that they feel a strong curiosity for different languages and enjoy learning them: "I suppose that as it happens to many other people, I would like to have the opportunity to learn more languages since I like them; those I feel more curious about are German, Chinese and Italian" (P13). This is supported with the data shown in Fig. 19.1 indicating that teachers' narratives give positive appraisal to the fields of experience that involve personal experiences of an educational and psychological nature.

The likes and dislikes related to the personal context of science experiences are more ambiguous. Pre-service teachers are very explicit when pointing at the disciplines they do not like such as math or physics but acknowledge having of curiosity for knowing how the world functions in other fields such as health or nature: "For me, it is very interesting the curiosity of knowing and discovering things" (P23); "I like learning about science related with every day use since I consider that is a real, practical and useful knowledge" (P14). This is consistent with the negative appraisal data shown in Fig. 19.2. It is interesting to note, though, that two students have been able to experience a change of attitude toward science along their education indicating a general improvement: "Back when I started primary school, I never expected to become the avid science lover that I currently am" (P38).

Other interesting differences between LN and SN are related with the relevance of the personal experiences having a psychological nature. Whereas more than two-thirds of the LN reconstruct psychological experiences within the personal context, this percentage drops to less than one-third when it comes to the SN (Table 19.4). These experiences focus on the perception of the acquired level of competence. In fact, teachers show a sharp awareness of their communicative and linguistic competence level in the different languages as well as strategic approaches for their improvement: "In any case I think I need to improve my English speaking competence since although I consider I have good vocabulary and resources to speak correctly the insecurity about being able to speak correctly does not allow me to do so" (P13). This is not the case for personal experiences of a psychological nature reconstructed in the SN. Whereas pre-service teachers show a general awareness of science terminology and knowledge command, they rarely specify any domain nor a personal strategy for improvement: "In my opinion, I have a low level of scientific knowledge." (P49); "I cannot consider myself competent enough in the whole area of science" (P38).

19.5.7 *The Community as a Privileged Context for Language Experiences*

Pre-service primary teachers' language experiences contextualized in the community have shown to be more relevant than the science experiences contextualized in the same context. No matter what nature of experience is considered, around half of the teachers' LN consistently considered the community context as being important (Table 19.4). This stands in contrast with the SN where students hardly reconstructed their science experiences in the community with the exception of the educational experiences.

The educational nature of the community experiences is considerably different when comparing SN and LN. Half of the pre-service primary teachers indicate in their SN that visiting science museums, undertaking field trips in natural areas, watching scientific documentaries on TV, or participating in scout's activities have been important contexts for informal science learning. However, almost all students have written in their LN that they systematically attended an out-of-school English academy to improve the command of the foreign language, mostly being English language, and to get official certificates. The almost universal acceptance that the educational system was unsuccessful in relation to foreign language learning generated a strong pressure for young students to be involved in alternative non-formal language learning activities outside the school as stated by the following student: "As many of us can see, at least those belonging to the 90's generation, the basic level of mastering a foreign language is rather low, and when I was seven years old I enjoyed the privilege of attending an English academy to improve this language (P24)."

The psychological and social nature of the community experiences have been extensively reconstructed in the LN and very seldom in the SN. Pre-service teachers reconstruct in their LN a diversity of experiences enacted in all sort of multilingual social situations which include Spanish, Catalan, English, and many other European and non-European languages. They all feel they have been offered many opportunities to use different languages and are very explicit with their personal ability to communicate functionally. Those multilingual encounters in which they have participated outside school are related in one way or another to traveling, emigration, or visiting other locations and countries. Whereas the value of science is only associated to an external social realm being expressed in a detached way through sentences such as the following: "Nowadays, our society is developed thanks to science, which has become a very important tool" (P23), the value and the command of different languages is considered to be a very important tool for personal, social, and professional development. The following statement summarizes the centrality of languages in the pre-service teacher's life: "Summarizing and going over my life I realize that I have always been linked to languages, and I continue learning and practicing them, since for me it is more a life style and a way to understand the relationships between people of different places and cultures (P41).

19.6 Conclusion

This is a study on a specific group of pre-service primary teachers who are enrolled in the first English-mediated primary education degree ever offered by a public university in Catalonia, Spain. They will become general primary teachers who most likely will implement CLIL approaches integrating the teaching and learning of science and English. The analysis presented in this chapter supports the idea that this particular group of pre-service primary teachers construes its past experiences with science and languages in a very different way. Their science experiences appear to be located mostly in school contexts, oriented toward its teaching in school only, and associated to learning difficulties. This is in strong contrasts with language experiences which appear to be anchored in a wide variety of contexts including school, for purposes that go beyond school, and associated to progressive and positive learning trajectories. These results point at the issue of the isolation of science within the social and personal life of our students. Although non-formal and informal science education initiatives are increasing all over the world, they have not impacted our pre-service teachers' lives in the way language education initiatives have.

The question for science teacher education lies in how to use the language experiences of this particular group of pre-service primary teachers as contexts for the anchoring of their past, present, and future science experiences. Avraamidou (2013) found that university experiences could positively influence the development of pre-service primary science teachers. We suggest three possible ways to improve pre-service primary teacher education: increasing non-formal science education experiences, providing a values-oriented science education, and introducing a narrative approach within science education courses. We suggest on the one hand to promote university activities oriented toward the community as a way to expand the context of science and science education experiences such as staying in learning camps where they meet in-service teachers teaching science and language, among other subjects in less formal education scenarios. On the other hand, we suggest increasing the experiences of a social nature emphasizing the presence, value and utility of science, and science education through the promotion of context-based science teaching. Finally, following a narrative approach to teacher education, we suggest to undertake a systematic work with pre-service primary teachers based on the establishment of fruitful dialogues between their SN and LN. In doing so, pre-service primary teachers would be able to reconstruct their life stories and introduce new and valuable experiences collected through their university-based teacher education programs.

Acknowledgements This work has been partially funded by grant 2014SGR1492 from the AGAUR of the Catalan Government, and by grant EDU2015-66643-C2-1-P from the Spanish Ministry of Economy and Competitivity.

References

Abell, S.K., & Roth, M. (1992). Constraints to teaching elementary science: A case study on a science enthusiast student. *Science Education, 76*(6), 581–595.
Appleton, K. (1995). Student teachers' confidence to teach science: Is more science knowledge necessary to improve self-confidence? *International Journal of Science Education, 17*(3), 357–369.

Avraamidou, L. (2013). Prospective elementary teachers' science teaching orientations and experiences that impacted their development. *International Journal of Science Education, 35*(10), 1698–1724.

Avraamidou, L., & Osborne, J. (2009). The role of narrative in communicating science. *International Journal of Science Education, 31*(12), 1683–1707.

Calderhead, J. (1989). Reflective teaching and teacher education. *Teaching and Teacher Education, 5*, 43–51.

Clandinin, D.J., & Connelly, F.M. (2000). *Narrative inquiry: Experience in story in qualitative research*. San Francisco: Jossey Bass.

Cortazzi, M. (1993). *Narrative analysis*. London: The Falmer Press.

Coyle, D., Hood, P., & Marsh, D. (2010). *CLIL. Content and Language Integrated Learning*. Cambridge: Cambridge University Press.

Cummins, J. (2008). Teaching for transfer: Challenging the two solitudes assumption in bilingual education. In J. Cummins & H. Hornberger (Eds.), *Encyclopedia of language and education: Vol. 5. Bilingual education* (2nd ed., pp. 65–75). Boston: Springer Science.

Dalton-Puffer, C., Nikula, T., & Smit, U. (2010). *Language use and language learning in CLIL classrooms*. Amsterdam: John Benjamins https://doi.org/10.1080/13670050.2012.666117

Davis, E.A., Petish, D., & Smithey, J. (2006). Challenges new science teachers face. *Review of Educational Research, 76*(4), 607–651.

Escobar Urmeneta, C., Evnitskaya, N., Moore, E. & Patiño, A. (2011). *AICLE-CLIL-EMILE: Educació plurilingüe. Experiencias, research & polítiques*. Bellaterra: Servei de Publicacions Universitat Autònoma de Barcelona.

Espinet, M., Valdés-Sánchez, L., Carrillo, N., Farro, L., Martínez, R., López, N., et al. (2017). Promoting the integration of inquiry based science and English learning in primary education through triadic partnerships. In A. Oliveira & M. Weinburgh (Eds.), *Science teacher preparation in content-based second language acquisition*. New York: Springer.

Eurydice. (2006). *Content and Language Integrated Learning (CLIL) at School in Europe*. Brussels: Eurodyce European Unit. Retrieved from http://ec.europa.eu/languages/documents/studies/clil-at-school-in-europe_en.pdf

European Commission. (2007). High level group on multilingualism. Final report. Belgium: Official Publications of the European Community.

Goodson, I.F. (2003). Hacia un desarrollo de las historias personales y profesionales de los docentes. *Revista mejicana de investigación educativa, 8*(19), 733–758.

Guasch, O., & Nussbaum, L. (Eds.). (2007). *Aproximacions a la competència multilingüe*. Bellaterra: Universitat Autònoma de Barcelona.

Holsti, O.R. (1969). *Content analysis for the social sciences and humanities*. Reading: Addison-Wesley.

Krippendorff, K. (1980). *Content analysis: An introduction to its methodology*. Newbury Park: Sage.

Lortie, D. (1975). *Schoolteacher: A sociological study*. London: University of Chicago Press.

Martín, M. J. F. (2008). CLIL implementation in Spain: An approach to different models. In C.M. Coonan (Ed.), *CLIL e l'apprendimento delle lingue. Le sfide del nuovo ambiente di apprendimento* (pp. 221–232). Venezia: Libreria Editrice Cafoscarina.

Navés, T., & Victori, M. (2010). CLIL in Catalonia: An overview of research studies. In Y. Ruiz de Zarobe & D. Lasagabaster (Eds.), *CLIL in Spain: Implementation, results and teacher training* (pp. 30–54). Newcastle upon Tyne: Cambridge Scholars.

Ramos, S.L., & Espinet, M. (2013). Expanding the agency in multilingual science teacher education classrooms. In N. Mansour & R. Wegerif (Eds.), *Science education for diversity* (pp. 251–275). Dordrecht: Springer.

Rivera, M. (2011). Language experience narratives and the role of autobiographical reasoning in becoming an urban science teacher. *Cultural Studies of Science Education, 6*, 413–434.

Stuckey, M., Hofstein, A., Mamlok-Naaman, R., & Eilks, I. (2013). The meaning of "relevance" in science education and its implications for the science curriculum. *Studies in Science Education, 49*(1), 1–34.

Valdés-Sánchez, L., & Espinet, M. (2013a). La evolución de la co-enseñanza de las ciencias y del inglés en educación primaria a partir del análisis de las preguntas de las maestras. *Enseñanza de las ciencias. Nº Extra (2013)*, 3588–3594.

Valdés-Sánchez, L., & Espinet, M. (2013b). Ensenyar ciències i anglès a través de la docència compartida. *Ciències. Revista del professorat de Ciències d'Infantil, Primària i Secundària, 25*, 26–34.

Weber, R. P. (1990). *Basic content analysis* (2nd ed.). Thousand Oaks, California: Sage.

Chapter 20
Examining Teachers' Shifting Epistemic Orientations in Improving Students' Scientific Literacy Through Adoption of the Science Writing Heuristic Approach

Brian Hand, Soonhye Park and Jee Kyung Suh

Abstract The role of language is critical both in how science is done and through the products of scientific practices. Importantly, language is viewed as an epistemic tool that enables learners to engage with construction and critique in the practices of science. The Science Writing Heuristic (SWH) approach places particular importance on engaging in the critical language practices of science while building the conceptual understandings of science topics through immersion in an argument-based inquiry approach to learning. This chapter focuses on a study with 28 middle school science teachers who were taking part in a 3-year research project centered on the implementation of the SWH approach for the teaching of science. Teachers were involved in a professional development program where they were introduced into four critical areas for implementing the approach: Learning, Language, Scientific Practice, and Pedagogy. Importance was placed on encouraging teachers to engage in the critical written and oral discourse practices of science that underpin the SWH approach and are essential features of scientific literacy. To track teacher change over the study, an Epistemic Orientation Survey (EOS) was developed that enabled us to examine the alignment between Language as an Epistemic Tool, Science as an Epistemic Practice, and Learning as an Epistemic Act. Building on previous work on the SWH approach, we have examined students' critical thinking growth rates to the change in teachers' epistemic orientations as a way to gauge how well such a language-based approach can provide cognitive resources for future growth in developing understanding of science concepts.

B. Hand (✉)
University of Iowa, Iowa City, Iowa, USA
e-mail: brian-hand@uiowa.edu

S. Park
North Carolina State University, Raleigh, North Carolina, USA
e-mail: spark26@ncsu.edu

J.K. Suh
University of Alabama, Tuscaloosa, Alabama, USA
e-mail: jksuh@ua.edu

Keywords Epistemic orientation · Science Writing Heuristic (SWH) · scientific literacy · construction and critique · language use in science

20.1 Literacy in Science Education

The foundational work of Norris and Phillips (2003) provided a framework for the concept of literacy in science which shifted away from issues related to concepts of the nature of science and content knowledge, to focus on the role of language. Their argument was that language is fundamental to science – you cannot do science without language. They introduced the terms fundamental and derived senses of science into the debate of the role of language. This was important in helping the science education community begin to understand that being literate in science was not simply a function of being able to replicate the language of science, but needed to incorporate the concept that understanding any scientific endeavor requires language. As such this argument began a process of shifting the role and function of language in science classrooms from being able to read the science textbook and replicate the language of the textbook, to having to think about what learning environments are required for students' understanding of language and how it promotes learning of science.

Much of the work in terms of examining the teaching and learning of science within school classrooms was framed around the work of Halliday and Martin (1993) who emphasized a focus on genre. As such there was much emphasis placed on ensuring that students learnt the genres of science in order for them to be able to engage in science. A very clear example of this perspective is the work by Osborne and his colleagues (2001) on the use of argument approaches to science. Cavagnetto (2010) labelled such approaches as structural approaches because students were to be taught the structure of argument prior to being able to use argument – a key epistemic practice of science.

The earlier work of Halliday emphasized the need to engage with language as a "lived" process. He suggests that students learn about language as they learn through language as they live the language. Such a perspective highlights the basis of what Cavagnetto (2010) calls immersion approaches to argument-based inquiry. Immersion approaches require that students become engaged with learning the language of science and the language of argument as they live the language of both. This notion is important because it shifts the conversation about language as one in which the product is the focus as occurs with traditional or structured approaches to science to one in which the language serves as an epistemic tool. That is, as students engage in the process of writing or talking about the science they learn about the science concepts because they have to use the language, not as a means to reproduce what the teacher gave them, but as a means to build their own constructions of the science concepts.

As Hand (2008) has highlighted, the shift in the ways in which teachers need to operate within classrooms is not easy and requires investment of time and energy for them to be able to fully utilize the epistemic nature of language.

Critically, teachers need to begin to engage with a number of essential ideas – in particular teachers need to be challenged about their ideas of learning, how knowledge is constructed, and the role of language as underpinning all scientific endeavors. Teachers' understandings of these critical elements in essence require a paradigmatic shift from the teacher determining the language experiences in classrooms, to recognizing that students are constantly negotiating with language and their current understandings as a means of trying to come to terms with the science concepts being engaged with. Norton-Meier (2008) suggests that this will require teachers to provide much more space within the learning environment of the classroom to role of dialogue, both oral and written, if they are going to be successful in creating discourse spaces where language is viewed as an epistemic tool.

In building on this early work of Norton-Meier (2008), Ardasheva, Norton-Meier, and Hand (2015) have suggested that there is a need for teachers to understand the nature of nonthreatening learning environments as critical places where students can have power to have a voice in the various forms of dialogue occurring within the science classroom. In recent work, Chen, Park, and Hand (2016) have shown that when provided with a nonthreatening learning environment – as proposed by Watts and Bentley (1987) – students begin to use both talk and writing simultaneously to both construct and critique arguments as they live the language of argument in learning the science concepts.

In summary, shifting perspectives on science literacy from replication of language to language as an epistemic tool is necessary to promote students' deep understanding of science concepts through construction and critique of arguments. However, in order for this new perspective to be successfully adopted, one needs to consider what type of learning environment is required, if a shift in teacher understanding is required, and how immersing students in argument-based inquiry promotes this perspective of science literacy.

20.2 The SWH Approach and the Role of Language

The Science Writing Heuristic (SWH) approach places particular importance on engaging in the critical language practices of science while building the conceptual understandings of science topics through immersion in an argument-based inquiry approach to learning. The approach is thus closely aligned with the following definition of argument-based inquiry:

> Argument-based Inquiry is inquiry that is intended to build students' grasp of scientific practices while simultaneously generating an understanding of disciplinary big ideas. Construction and critique of knowledge, both publically and privately, are centrally located through an emphasis on the epistemological frame of argument by engaging them in posing questions, gathering data, and generating claims supported by evidence. (Hand, Nam, Cavagnetto, & Norton-Meier, 2013, p. 1)

The core notion of the SWH approach is that students are provided with opportunities to pose questions, gather data, make claims based on evidence, and check their assertions against current norms. Throughout the learning process, students are required to negotiate their understandings of the "big ideas" across multiple representations, such as pictures, text, graphs, or equations, as well as negotiate across multiple situations. In small groups, students work together to pose questions, debate their data to generate evidence, and prepare material for public whole class negotiation. At the whole class level, students are required to publicly present and defend/debate their claims and evidence. It is through these multiple discourse opportunities that students are engaged in that a conceptual understanding of the topic can be generated. Students are expected to participate in the multiple oral and written negotiations that form the basis of the epistemic practices both of science and language.

In terms of types of writing used within the SWH approach students are required to use both argumentative writing and summary writing. Argumentative writing is done at the end of each inquiry activity, after students have negotiated both in small group and whole class settings about the relative merits of their claims and evidence. Major components of the argumentative writing include questions, claims, and evidence. Summary writing is completed at the end of the unit and considered as a consolidation task that helps students understand relationships among conceptual ideas dealt with in the unit. The students are generally asked to write to an audience of their peers, parents, or younger students, about the big ideas of the unit. The intention of this task is for the students to explain the big ideas of the unit as they understand in a way that becomes understandable for a particular audience such as peers or younger students. As McDermott and Hand (2010) have shown, students find this task very cognitively demanding and they believe that it helps them understand science concepts in much greater depth. These two different writing types provide students with multiple opportunities for them to engage with, and understand, the epistemic nature of language.

20.3 Structure and Focus of an SWH Professional Development (PD) Program

The SWH approach places particular importance on engaging in the critical language practices of science while building the conceptual understandings of science topics through immersion in an argument-based inquiry approach to learning. The approach is grounded in Halliday's (1975) notion that the best way to learn about language is through using the language by being immersed within it. When applied to the context of science learning, it implies that only when students use language as an epistemic tool with which they are able to construct their own understanding in science (Kelly, Chen, & Prothero, 2000), does learning the disciplinary knowledge of science as well as learning the language of science occur.

In this regard, SWH PD programs are designed to help teachers understand the essential role of language as an epistemic tool through engaging them in four critical areas for implementing the approach: Learning Theory, Language Practices,

Scientific Practices, and Pedagogy. To effectively incorporate argumentation into science instruction, it is suggested that teachers need to shift their epistemic orientations to be more aligned with key ideas of argument-based inquiry (Hand, Cavagnetto, Chen, & Park, 2016). In particular, for successful implementation of SWH, appropriate epistemic orientations as to learning theory, language use, and the nature of science are necessary (Hand, 2008). Given that, the SWH PD program primarily focuses on challenging teachers' existing orientations on how students learn, what science is, and use of language in science to shift toward being aligned with fundamental notions of the approach. Along with challenging epistemic orientations, the SWH PD program purport to support the teachers' adoption of pedagogical approaches aligned with the shifted orientations. Table 20.1 summarizes learning targets for each of the four focus areas of the SWH PD program.

To achieve the learning targets, the PD program is designed to consist of an intensive summer workshop and ongoing on-site professional support throughout the academic year. Goals of the summer workshop are:

1. *Immersing the teachers in argument-based inquiry activities using the SWH approach* within selected science units (e.g., Forces and Motion, Properties of Matter, or Ecosystem Dynamics). Teachers experience the process by themselves as learners to develop familiarity with generating questions, gathering data, seeking information, and making claims and evidence. Within each unit, inquiry activities center on big ideas drawn from the Next Generation Science Standards (NGSS Lead States, 2013). Based on their immersion experience, teachers are asked to create unit plans including assessment plans to be implemented in their classrooms compatible with their own curriculum applying what they learned from the summer workshop on the SWH approach.
2. Constructing a strong *understanding of learning theory* to inform and guide the development of the necessary pedagogical practices.
3. Developing an understanding of the *critical role of language* in science from the construction and critique of knowledge to the representational demands of science.
4. Building the required *pedagogical approaches* necessary for successful implementation of argument-based inquiry, including (1) creating dialogical interactions, (2) diagnosing students' prior knowledge, (3) focusing on big ideas in science, (4) facilitating group work, and (5) developing questioning skills.

The ongoing on-site professional support is to scaffold teachers' unit planning and enactment facilitating necessary adjustments for the successful changes in implementing the approach. Support through individual or focus group meetings with the SWH PD team throughout the academic year aims to assist teachers to build an understanding of the theoretical and practical requirements of creating classrooms that engage students in posing questions, gathering data, seeking information, making claims and evidence, and negotiating this both publically and privately.

One thing that we need to emphasize here is that even though we describe the four key areas of the SWH approach separately, most critical to teachers' move toward higher level implementation of the approach within their classrooms is building a strong alignment among them. In other words, the shift from traditional

Table 20.1 Four focus areas of the SWH PD program

	Learning targets focus areas	What teachers should understand	What teachers should be able to do
Epistemic orientations	Nature of science	• "Big ideas" in science. • Construction and critique of ideas as core practices of science. • Scientific ideas are represented, communicated, and validated by the community of scientists through argumentation. • Science moves forward through argumentation process.	• Identify and design lessons on "big ideas" in science. • Align content standards (i.e., NGSS) with big ideas. • Immerse students in practices of construction, critique, and conceptual growth. • Emphasize that science is a process of negotiation based on test against nature.
	Learning theory	• Learning involves conceptual changes. • Knowledge is constructed through social interactions that build on individuals' prior knowledge. • Meaning is constructed through public and private negotiation. • Students are capable and responsible for their own meaning making.	• Focus on cognition not behavior when facilitate student learning. • Focus on student prior knowledge. • Provide consistent opportunities for social/individual construction and public/private negotiation.
	Language use	• Language of argument (e.g., *Questions*, *Claims*, and *Evidence* structure and difference between data and evidence) • Nature of dialogue in argument-based inquiry. • Language is learned through living the language and by using the language. • Structural supports for promoting language use in science.	• Consistently use and require language of argument. • Consistently promote and move between public and private languages, and between individual/group languages. • Provide opportunities for immersion into scientific discourse, knowledge construction/critique, and community of practice. • Consistently use multimodal representations of big ideas.
Pedagogical approaches		• Pedagogical approaches aligned with desirable epistemic orientations as to science, learning theory, and language central to the SWH approach.	• Create discourse space and dialogical interactions. • Diagnose and value student prior knowledge and ideas. • Focus on big ideas in science. • Facilitate group work. • Develop questioning skills.

approaches to the SWH approach requires changes across the four areas and more importantly coordinated alignment among those changes. The SWH is not a prescribed curriculum or a set of activities, but an *approach* to science curricula that requires teachers to align their orientations and practices with the key concepts and features of the SWH. Such alignment occurs through necessary adjustments across the areas of learning theory, language use, science practices, and pedagogy based on understanding of their students and teaching context. When teachers develop a strong alignment between epistemic orientations and pedagogy, the alignment can be applied to various teaching contexts or across topics to create opportunities for students to pose questions, gather data, make claims based on evidence, and critically evaluate the value of evidence and explanations of their own or those of their peers. Stated differently, the alignment serves as a framework that enables teachers to make professional decisions on how to adapt an activity or a curriculum in ways that are student-centered, argument-based, and inquiry-oriented given the context they are working within. This notion is at the heart of our conceptual model of teacher change for implementing the SWH approach as shown in Fig. 20.1. In this model, we conceptualize that a positive impact of the SWH approach on student achievement can be attributed to the critical alignment between teacher changes in epistemic orientations and pedagogical practices that are promoted through the SWH PD program including intensive summer workshop and ongoing academic year support for teachers.

20.4 Impact of the SWH PD on Teachers

Based on the conceptual model of teacher change (Fig. 20.1), we designed a 3-year grant project that aims to aid secondary science teachers shift epistemic orientations and pedagogical practices to effectively implement the SWH approach, and further positively impact student learning outcomes. Taking part in the research project was completely voluntary. All of the participating teachers and students were asked to sign an informed consent form and an assent form, respectively, if they would like to participate. For confidentiality, all teachers were given and addressed by pseudonyms while all students were given and addressed by ID numbers. We are currently in the second year of the project and we want to discuss what we have learned throughout the first year especially in terms of teacher changes in this chapter.

In Year 1, 25 secondary science teachers (20 females and 5 male) from a Midwest State of the USA participated in our professional development (PD) program and research project. Their teaching experiences ranged from 1 to 31 years; and only five had less than 5 years. The teachers were introduced to the four key areas shown in Table 20.1 through an intensive 10-day summer workshop and on-site academic year support during the first year. To trace changes in teachers' epistemic orientations and pedagogical approaches, we collected teacher data at different points in time as summarized in Fig. 20.2. To facilitate the identification

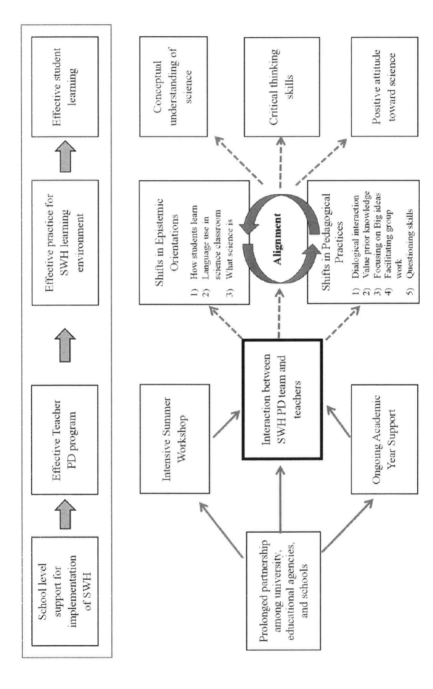

Fig. 20.1 Conceptual model of teacher changes for implementing the SWH approach

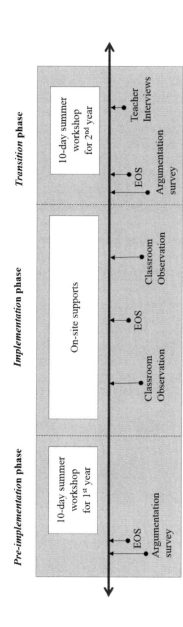

Fig. 20.2 Summary of data collection for each phase

of the teacher changes over time, we divided the first year into three phases and collected a set of data in each phase: (1) Phase 1, *preimplementation* phase (before and during the 1st summer workshop); (2) Phase 2, *implementation* phase (after the 1st year summer workshop, during the semesters with on-site academic year support); and (3) Phase 3, *Transition* phase (at the end of the 1st year, transition into 2nd year).

Teacher data sources included two teacher surveys (Epistemic Orientation Survey (EOS) and Argumentation Survey), classroom observations, and individual and focus group interviews with the teachers. The EOS (Suh & Park, 2016) is a Likert-scale survey consisting of 30 items that ask teachers to indicate the degree to which they agree with statements on epistemological ideas of learning science using a five-point scale from strongly disagree to strongly agree. The reliability estimates of this measure ranged from 0.80 to 0.93. To measure changes in epistemic orientations over time, the EOS was administered once per phase, three times per year in total. Meanwhile, the Argumentation Survey was distributed once before and after the implementation phase, respectively. This survey comprises three open-ended questions designed to gauge the teachers' views on the use of argument and argumentation in learning and teaching science.

In addition, each teacher was observed once per semester by PD facilitators, to determine her/his implementation level. This assessment focused on seven pedagogical practices that stem from the five essential pedagogical approaches for successful implementation of the SWH approach described in Table 20.1:

(1) Determining big ideas for a topic,
(2) Planning/enacting learning activities around the big idea,
(3) Creating classroom environments for active dialogical interactions,
(4) Diagnosing students' prior knowledge,
(5) Facilitating group work that promotes student conceptual understanding of big ideas,
(6) Asking questions that demand high-order cognitive thinking,
(7) Using different modes of language (e.g., writing, talking, multi-modal representation).

Using a scoring rubric as shown in Table 20.2, individual teachers' implementation levels were decided using a ten-level scale: None (NN), Low (L1, L2, and L3), Medium (M1, M2, and M3), and High (H1, H2, and H3) implementation. Each teacher's implementation level was first determined by how many essential practices listed above were presented in their lessons. Then, the quality of those practices was evaluated by three levels: Little, Somewhat, or A great deal.

20.4.1 Teacher Change in Epistemic Orientations

Analysis of the teacher responses to the EOS reveals that half of the teachers made substantial changes in epistemic orientation during the first year. The result is not surprising since previous studies suggest teachers need at least 18 months to

Table 20.2 A scoring rubric for teacher implementation level on the SWH

Level	Description	Little (1)	Somewhat (2)	A great deal (3)
None	No evidence of the SWH approach;		NN	
Low	Lessons demonstrate 1–3 essential practices of the SWH approach.	L1	L2	L3
Medium	Lessons demonstrate 4–6 essential practices of the SWH approach.	M1	M2	M3
High	Lessons demonstrate all seven essential practices of the SWH approach.	H1	H2	H3

Table 20.3 Results of the EOS

	No. of teachers	Gender	Yrs. of teaching science	Avg. EOS score [range]		
				Preimplementation	Postimplementation	Growth
Group 1	4	3 females 1 male	2–22 yrs.	107.0 pts. [97–112]	127.0 pts. [123–131]	20.0 pts. [15–26].
Group 2	8	7 females 1 male	2–31 yrs.	112.8 pts. [104–118]	124.5 pts. [116–134]	11.8 pts. [9–14]
Group 3	12	9 females 3 males	8–31 yrs.	114.3 pts. [109–121]	116.3 pts. [109–125]	2.2 pts. [−5–7]
Total	24[a]	19 females 5 males	2–31 yrs.	112.5 pts. [97–121]	120.8 pts. [109–132]	8.3 pts. [−5–26]

[a]1 missing data.

fully adopt the SWH approach (Martin & Hand, 2009). Based on their scores on the EOS over time, we categorized the teachers into three groups: Group 1 (Fast change), Group 2 (Gradual change), and Group 3 (No change). The Table 20.3 presents the general information and the EOS scores of the teachers for each group. There was one teacher who did not provide her responses to the 2nd EOS, and thus she was excluded from the data set for this analysis.

Among 25 teachers, only four teachers were categorized into Group 1 that showed the most growth in epistemic orientation. Although they began with lower scores compared to the other two groups, their scores in epistemic orientation increased fast with PD supports throughout the year (see Fig. 20.3). Interestingly, three Group 1 teachers were fairly new to teaching science (less than 3 years of experience) and were willing to learn a new approach to improve their science instruction. For example, Ms. Moore has taught 6th grade biology for 2 years. Although she had 9 years of teaching experience in other subject areas, she was relatively new to teaching science. Hence, she was open to a new teaching approach and willing to change her ideas about teaching and learning science. Her interviews helped us understand that her changes in epistemic orientation mainly

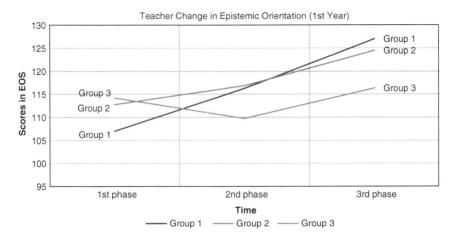

Fig. 20.3 Each group's change in epistemic orientation over the 1st year of the study

stemmed from her shift toward a student-centered view away from a teacher-centered view, as indicated in the following quote.

> Before doing the SWH, I was concerned a lot about how much I need to cover the curriculum instead of being concerned about my students' learning. Now I know my teaching should be more student-centered. I feel much more comfortable with teaching science than before because I do not need to cover everything and my students do not have to memorize a list of scientific terms anymore in my class. (Interview during Transition Phase with Ms. Moore)

Ms. Moore's ideas about learning fundamentally shifted from rote memorization to negotiation of meaning, generating rapid growth in her EOS score (27 points increase) throughout the year. She was very excited to see that her students were engaged in their own learning processes, not memorizing scientific facts and terms.

Group 2 includes eight teachers who showed gradual changes. They all reported positive values of the SWH approach and expressed willingness to use the approach continuously. Although this group of teachers did not show dramatic change in their epistemic orientations, they seemed to steadily develop their own ideas about learning and how to implement the new approach. In-depth analysis of the interviews revealed that most of the teachers began to recognize positive learning outcomes in their classrooms, such as high engagement level, growth in test scores, and students' positive attitude as they developed understanding of the approach. Unlike Group 1 and 2, Group 3 teachers did not show substantial change in their epistemic orientations. Many teachers in this group were still struggling with the underlying ideas of the approach and did not report any positive changes in both learning and teaching. However, there were some teachers who began to see positive aspects of the approach and wanted to better implement the SWH.

To measure change in orientations toward the use of language in science classrooms, we analyzed the teachers' responses to the Argumentation Survey using the constant comparative method (Strauss & Corbin, 1990). This analysis showed

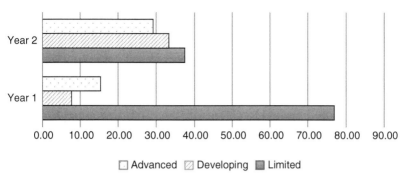

Fig. 20.4 Analysis of the teachers' view of language of argument

that the teachers began to see science argument and language as a tool that helps students develop conceptual understanding. The first Argumentation Survey showed that more than 75% of the teachers had a limited understanding of argument and argumentation, viewing argument as a mere structure and argument-based inquiry as a sequential data-collection process (Fig. 20.4). For example, most teachers described scientific argumentation as a process where students share their data collected from an investigation; they did not make a deep connection between scientific argument and learning science. In addition, most teachers described argument-based inquiry as an investigation of questions and collection of data.

The results of the second Argumentation survey, however, demonstrated that about 60% of the teachers began to develop or had developed more advanced view of language of argument. They came to use the terms, Question, Claim, and Evidence more frequently and described argument-based inquiry as an integrated learning process. They also began to regard argument and argumentation as a critical element in learning science. More importantly, they began to develop the idea that language is an epistemic tool that helps students construct their own understanding of science.

20.4.2 Teacher Change in Pedagogical Practices and Its Effects on Student Learning

In Year 1, most teachers incorporated some of the key features of the SWH approach into their instruction. Although about 40% of the teachers (10 teachers out of 25 teachers) reached a medium level of implementation by the end of the year, many teachers appeared to still be struggling with implementing the approach (see Fig. 20.5). Interviews with the teachers indicated that they found the following pedagogical approaches most challenging to implement: (1) facilitating group work to promote conceptual understanding of big ideas, and (2) creating dialogical interactions. In addition, there were some constraints that the teachers

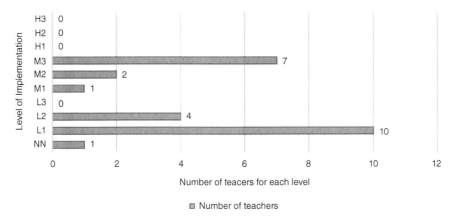

Fig. 20.5 Teachers' SWH implementation levels in Year 1

Table 20.4 Teacher change and student learning

Implementation level		Number of teachers	Avg. EOS scores (3rd phase)	Student Growth in CCTT
Low	L1	10	119.9	0.59
	L2	4		
	L3	0		
Medium	M1	1	124.0	2.58
	M2	2		
	M3	7		
Total		24	121.2	1.3

experienced in implementing the approach. The most prevalent were limited resources and unsupportive school contexts. The teachers frequently faced time constraints and had a hard time designing lessons with this new approach mostly due to a lack of resources. School contexts where the teachers were asked to cover school curriculum or teach multiple grades and subjects at a time were also challenging to them. Despite the challenges, we expect this group of teachers to improve significantly in their SWH implementation during the 2nd year through interactions with and support from our project team, as they develop a concrete understanding of the approach and align their epistemic orientations accordingly.

To understand whether their implementation levels were related to their epistemic orientations, we compared the Low and Medium implementation groups' EOS scores. As indicated in Table 20.4, Medium group's EOS score ($n=10$, $m=124.0$) was higher than Low group's score ($n=14$, $m=119.9$), suggesting teacher implementation level and epistemic orientation are seemingly related. However, further investigations with statistical tests are needed to understand the relationship between these two. We also collected students' Cornell Critical

Thinking Test (CCTT) (Level X) scores before and after the implementation of the SWH approach to examine how the teachers' changes in practices affected student learning. Although it is difficult to make concrete conclusions with 1 year data, our preliminary data analysis suggests that as a teacher's implementation level is higher, his or her students are more likely to improve critical thinking skills (see Table 20.4). That is, as teachers were shifting in their epistemic orientations toward being more aligned with tentativeness of scientific knowledge and empirical nature of scientific explanations, they appeared to be creating learning environments where students were being required to justify their claims based on evidence they generated from their inquiries. As such, this was requiring the students to adopt a much stronger critical thinking and reasoning approach which was in turn reflected in the increase growth rate in critical thinking test scores. However, this relationship should be further examined with Year 2 and Year 3 data.

20.5 Closing Remarks

The research presented in this chapter represents some new areas of work that we have begun to examine in terms of how we help teachers better understand the critical role language plays in learning science. It is important for teachers to understand that language is not only fundamental to the learning of science, but in its derived form, it also brings into focus the nature of writing and talk as an epistemic tool. Understanding that language is an epistemic tool is critical for the development of science literacy. Science cannot be done without language – science cannot advance without language. However, for teachers to be successful in using language in this manner there is a critical need for them to have an epistemic orientation toward the science classroom that recognizes the role of language. This shift means that teachers begin to highlight and use dialogical interactions as a critical element of their science classrooms. Requiring students to negotiate through, and with, the language of science means that students begin to live the language of science as they learn about science. That is, they can develop their science literacy as a tool for building understanding of the concepts of science and as a vehicle for building more sophisticated uses of the language of science. Science language becomes a dynamic tool for all students to use a future resource for learning. Not as a language to be replicated and given back to the teacher. However, this change is difficult and takes time. Teachers need to create environments where science is undertaken as a lived experience – where language is recognized as being critical for engagement in science and in helping students build rich understandings of the topic. Science literacy is not about replication of knowledge or being able to read science text divorced from how scientists actually use the language to advance the discipline. The initial efforts of this 3-year study have shown that time is required for teachers to undertake this change. However, in making the shift, there are student benefits to be gained.

The results from this study do raise several interesting questions. For example, how many of the 25 teachers will shift? Will we see consistent benefit when they do make the shift? Will their epistemic orientation remain consistent once they have shifted? We believe that these are important questions that need to be engaged with, particularly as we begin to better understand the use of immersive type approaches to argument-based inquiry. Shifting engagement for students to "live" the language of science as they learn science is something we believe is critical for learning. We can no longer ignore the critical role that language plays in learning – especially for science.

20.6 Dedication

We have not had the opportunity to publically express the incredibly valuable contribution that Steve Norris made to the field of science education. While it has been some time since his passing, we wish to dedicate this chapter to his efforts. His framing of the work in writing pushed us all to have to think much more deeply and carefully about the work that we engage in. We lost not only a valuable scholar but also a friend. We are hoping that we can in some small way continue to build on his efforts.

References

Ardasheva, Y., L. Norton-Meier, & B. Hand. (2015). Negotiation, embeddedness, and nonthreatening learning environments as themes of science and language convergence for English language learners. *Studies in Science Education, 51*(2), 201–249.

Cavagnetto, A. R. (2010). Argument to foster scientific literacy: A review of argument interventions in K–12 science contexts. *Review of Educational Research, 80*(3), 336–371.

Chen, Y. C., Park, S., & Hand, B. M. (2016). Examining the use of talk and writing for students' development of scientific conceptual knowledge through constructing and critiquing arguments. *Cognition and Instruction 34*(2), 100–147. DOI 10.1080/07370008.2016.1145120

Ennis, R. H., & Millman, J. (2005). *Cornell critical thinking test, level X* (5th ed.) Seaside: The Critical thinking Company.

Halliday, M. A. K. (1975). *Learning how to mean: Explorations in the development of language*. London: Edward Arnold.

Halliday, M. A. K., & Martin, J. R. (1993). *Writing science: Literacy and discursive power*. Pittsburgh: University of Pittsburgh Press.

Hand, B. (2008). Introducing the science writing heuristic approach. In B. Hand (Ed.), *Science inquiry, argument and language: A case for the science writing heuristic*. Rotterdam: Sense.

Hand, B., Cavagnetto, A., Chen, Y-C., & Park, S. (2016). Moving past curricula and strategies: Language and the development of adaptive pedagogy for immersive learning environments. *Research in Science Education, 46*, 223–241.

Hand, B., Nam, C., Cavagnetto, A. R., & Norton-Meier, L. (2013). The Science Writing Heuristic (SWH) approach as an argument-based inquiry. Roundtable discussion at 1st International Conference on Immersion approaches to Argument-based Inquiry (ABI) for Science Classrooms.

Kelly, G. J., Chen, C., & Prothero, W. (2000). The epistemological framing of a discipline: Writing science in university oceanography. *Journal of Research in Science Teaching, 37*, 691–718.

Martin, A. M., & Hand, B. (2009). Factors affecting the implementation of argument in the elementary science classroom. A longitudinal case study. *Research in Science Education, 39*(1), 17–38.

McDermott, M., & Hand, B. (2010). A secondary reanalysis of student perceptions of non-traditional writing tasks over a ten-year period. *Journal of Research in Science Teaching, 47*, 518–539.

NGSS Lead States. (2013). Next Generation Science Standards: For states, by states. Washington, D.C.: The National Academies Press.

Norris, S. P., & Phillips, L. M. (2003). How literacy in its fundamental sense is central to scientific literacy. *Science Education, 87*(2), 224–240.

Norton-Meier, L. (2008). Creating border convergence between science and language: A case for the Science Writing Heuristic. In B. Hand (Ed.). Science Inquiry, Argument and Language: A case for the Science Writing Heuristic. Rotterdam: Sense Publishers.

Osborne, J., Erduran, S., Simon, S., & Monk, M. (2001). Enhancing the quality of argument in school science. *School Science Review, 82*(301), 63–70.

Simon, S., Erduran, S., & Osborne, J. (2006). Learning to teach argumentation: Research and development in the science classroom. *International Journal of Science Education, 28*(2–3), 235–260.

Strauss, A. L., & Corbin, J. M. (1990). *Basics of qualitative research: Grounded theory procedures and techniques*. Newbury Park: Sage.

Suh, J., & Park, S. (2016, January). *Epistemic orientation toward teaching science: Toward better conceptualization and measurement*. Paper presented at the International meeting of Association for Science Teacher Education (ASTE), Reno, Nevada, USA.

Watts, M., & Bentley, D. (1987). Constructivism in the classroom: Enabling conceptual change by words and deeds. *British Educational Research Journal, 13*(2), 121–135.

Chapter 21
Developing Students' Disciplinary Literacy? The Case of University Physics

John Airey and Johanna Larsson

Abstract In this chapter we use the concept of *disciplinary literacy* (Airey, 2011a, 2013) to analyze the goals of university physics lecturers. Disciplinary literacy refers to a particular mix of disciplinary-specific communicative practices developed for three specific sites: *the academy*, *the workplace* and *society*. It has been suggested that the development of disciplinary literacy may be seen as one of the primary goals of university studies (Airey, 2011a).

The main data set used in this chapter comes from a comparative study of physics lecturers in Sweden and South Africa (Airey, 2012, 2013; Linder, Airey, Mayaba, & Webb, 2014). Semi-structured interviews were carried out using a disciplinary literacy discussion matrix (Airey, 2011b), which enabled us to probe the lecturers' disciplinary literacy goals in the various semiotic resource systems used in undergraduate physics (i.e. graphs, diagrams, mathematics, language).

The findings suggest that whilst physics lecturers have strikingly similar disciplinary literacy goals for their students, regardless of setting, they have very different ideas about whether they themselves should teach students to handle these disciplinary-specific semiotic resources. It is suggested that the similarity in physics lecturers' disciplinary literacy goals across highly disparate settings may be related to the *hierarchical, singular* nature of the discipline of physics (Bernstein, 1999, 2000).

In the final section of the chapter some preliminary evidence about the disciplinary literacy goals of those involved in physics teacher training is presented. Using Bernstein's constructs, a potential conflict between the hierarchical singular of physics and the horizontal region of teacher training is noticeable.

J. Airey (✉)
Department of Physics and Astronomy, Uppsala University, Uppsala, Sweden

Department of Mathematics and Science Education, Stockholm University, Stockholm, Sweden

Department of Languages, Linnaeus University, Kalmar/Växjö, Sweden
e-mail: john.airey@physics.uu.se

J. Larsson
Department of Physics and Astronomy, Uppsala University, Uppsala, Sweden
e-mail: johanna.larsson@physics.uu.se

Going forward it would be interesting to apply the concept of disciplinary literacy to the analysis of other disciplines—particularly those with different combinations of Bernstein's classifications of hierarchical/horizontal and singular/region.

Keywords Disciplinary literacy · undergraduate physics · knowledge structures · singulars versus regions · comparative education

21.1 Introduction

Traditionally, science—and in particular undergraduate physics—has been viewed as a difficult subject for students to master. From the early 1990s to the present day, there has been a great deal of concern internationally about falling enrolment, student dropout and the quality of education given to undergraduates (American Association of Physics Teachers, 1996; European Commission Expert Group, 2007; Forsman, 2015; Johannsen, 2013; Seymour & Hewitt, 1997). In an attempt to understand and address these concerns both the National Research Council in the USA and the Council on Higher Education and Institute of Physics in South Africa have recently undertaken reviews of the undergraduate physics education offered in their respective countries (CHE-SAIP, 2013; National Research Council, 2013). One of the themes that emerged in the US report is that 'Current practices in undergraduate physics education do not serve most students well' (National Research Council, 2013, p. 18). This theme is also echoed in the South African report where it was concluded that it is 'imperative that deep-seated changes regarding the length of the undergraduate programme and the way it is taught and monitored be introduced' (CHE-SAIP, 2013, p. 32).

In this chapter we use the notion of *disciplinary literacy* (Airey, 2011a, b, 2013) to both describe and problematize the goals of undergraduate physics lecturers in Sweden and South Africa. This is potentially important because the development of disciplinary literacy may be seen as one of the primary goals of university studies (Airey, 2011a) and high school physics.

The aspects of disciplinary literacy we have chosen to discuss are issues that have bearing on undergraduate physics and that we hope others may find interesting. Whilst our aim is to present a comparative analysis of the disciplinary literacy goals of physics lecturers in two countries, the thinking behind this chapter, as with the rest of this book, is that it can also function as a point of entry for the reader into aspects of the field of literacy in higher education. However, the description we present is selective and far from exhaustive.

21.2 Literacy

Traditionally, literacy has been framed in terms of the ability to read and write. Indeed the first truly international definition of literacy stated that, 'a literate person is one who can, with understanding, both read and write a short simple statement on his or her everyday life' (UNESCO, 1958). This definition was later broadened so that literacy was associated with 'the ability to identify, understand, interpret, create, communicate and compute, using printed and written materials associated with varying contexts' (UNESCO, 2004, p. 13).

An even broader definition of literacy was later put forward at a Nordic conference on the theme:

> ... communicative practices that shape the world we live in, determine how we read and write the world, how we see, understand and shape the relationship between ourselves, nature and our communal life. (Nordic Educational Research Association, 2009)

Here we can see a movement away from a strict focus on reading and writing towards a range of communicative practices, a theme to which we will return later.

21.3 Literacy in the Academy

In the academy the term literacy has mainly been used when referring to courses where students learn to read and write academic texts. Traditionally these courses have attempted to provide students with the necessary tools to complete their study program and are often given as a type of remedial help. Such courses frequently use genre analysis and corpus linguistics to provide students with general advice on the various forms of academic writing (Björk & Räisänen, 2003; Swales, 1990; Swales & Feak, 2004). In many of these courses academic writing is seen as a sort of academic *acculturation* or *socialization* where student writing gradually starts to mirror the broad differences in different disciplinary norms that have been identified (Becher & Trowler, 1989). A number of writers have questioned the learning outcomes of such courses since they rarely deal with questions of power, identity construction and the development of a personal academic voice in academic writing—see for example Ivanič (1998) and Lillis and Scott (2007). Here, Lea and Street (1998) adopt the term *academic literacies* (plural) in order to highlight the fact that there are competing voices in the academy (see Duff, 2010 for a presentation of this academic literacies discussion).

21.4 Language Choice in the Academy

In many countries, two or more languages are used in university education. This naturally has consequences for what is viewed as relevant to the term literacy. In the Nordic countries, for example, the term *parallel language use* has been proposed, where two (or more) languages are expected to be used in parallel in higher education (Josephson, 2005). As Phillipson (2006, p. 25) points out, although parallel language use is 'an intuitively appealing idea', it is also a 'somewhat fuzzy and probably unrealistic target'. Here, Airey and Linder (2008, 2011) have linked the idea of parallel language use to the notion of *bilingual scientific literacy*.

In their studies of Swedish undergraduate physics courses they found that very few physics lecturers had concrete goals for the parallel development of disciplinary language skills. Airey and Linder conclude with a recommendation that each course syllabus should specify not only disciplinary learning outcomes, but also in which language(s) those outcomes will be. This proposal has received some international attention, with at least one university designing its language policy around the idea (Fortanet-Gomez, 2013). An overview of the discussion of language choice in the academy can be found in Airey, Lauridsen, Raisanen, Salö, and Schwach (2017).

21.5 Multimodality

Thus far the description of literacy in the academy presented here has been limited to written language—that is literacy as the ability to read and write in the academy (cf. Norris and Phillips' (2003) notion of fundamental scientific literacy in the following section). This traditional focus on written text has been problematized by Lemke (1998) who points out that scientists integrate resources over a range of semiotic systems in order to handle problems that would otherwise be impossible to solve. Other semiotic resource systems used in science are, for example, graphs, diagrams, sketches, gesture, mathematics, spoken language, tables, apparatus and simulations. Following Gibson (1979), Kress, Jewitt, Ogborn, and Tsatsarelis (2001) wonder whether different semiotic resource systems have different *affordance*s for knowledge representation, that is, whether say, spoken language is better for certain tasks and diagrams are better for other tasks. Based on this work, Fredlund, Airey, and Linder (2012) suggested the term *disciplinary affordance*. Airey (2015) defines disciplinary affordance as '*the agreed meaning making functions that a semiotic resource fulfils for a particular disciplinary community*'. In this respect Airey (2009) has argued that there is a critical constellation of semiotic resources that students need to become fluent in as a necessary (but not sufficient) condition for appropriate construction of disciplinary knowledge. When teachers understand the disciplinary affordances of the range of semiotic resources available with respect to a given concept they are better placed to design learning tasks that activate this critical constellation (see e.g. Fredlund, Airey, & Linder, 2015). Often, the disciplinary affordances of semiotic resources are not immediately apparent to students. In such cases, the semiotic resources will need to be 'unpacked' (Fredlund, Linder, Airey, & Linder, 2014). Disciplinary-specific resources that have been unpacked in this way lose much of their disciplinary power in the process, but their pedagogical affordance is greatly increased (Airey, 2015). For teachers, then, striking a balance between the disciplinary affordance and pedagogical affordance of the semiotic resources they use is crucial for effective teaching and learning (see for example Airey & Linder 2017). Finally, Airey and Linder (2008) point out that each individual semiotic resource requires two types of control, interpretive (equivalent to reading a written text) and generative (equivalent to writing a text). Thus literacy in the academy entails not only a question of which semiotic resources students need to learn to control, but also which type of control is needed, interpretive or generative.

21.6 Scientific Literacy

There are many meanings ascribed to the word literacy. For example a simple internet search easily identifies: biological, historical, engineering, musical, medical, economic, computer, psychological and cultural literacies. The list could probably become quite long. When literacy is used in this way it often signals a metaphorical relationship—readers are expected to take their own associations to literacy and apply them to a new situation. The same can be said for scientific literacy. Here, Norris and Phillips (2003) have characterized scientific literacy in terms of two aspects, fundamental and derived. Fundamental scientific literacy has a direct link to the original definition of literacy and refers to the ability to read and write in the natural sciences. Derived scientific literacy, however, refers to a range of competencies such as knowledgeability about science, the ability to think scientifically, the ability to distinguish science from nonscience, an understanding of the nature of science, feeling comfortable discussing science topics and the ability to critically appraise science. By extension, fundamental scientific literacy can be seen to apply to all the semiotic resource systems mentioned in the previous paragraph.

A further division in scientific literacy has been observed by Roberts (2007) who identifies two visions of scientific literacy, where vision I refers to scientific literacy for use in the academy, whilst vision II refers to scientific literacy for use in society. We will return to these two visions of scientific literacy in our discussion of disciplinary literacy in the next section.

Laugksch (2000) enumerates some of the ways in which scientific literacy has been used. He points out that ever since its introduction by Hurd (1958) its meaning has been undefined and difficult to pin down. Laugksch concludes by suggesting that when researchers use the term scientific literacy they should be very clear about presenting what they mean. In the next section we address Laugksch's critique of scientific literacy by turning to a new term, *disciplinary literacy*, the development of which has previously been suggested as one of the overarching goals of undergraduate studies (Airey, 2011a).

21.7 Introducing Disciplinary Literacy

The main criticism of the term scientific literacy, then, is that it does not have a clear, unambiguous definition—that is the term means different things to different people. In this section we briefly describe the thinking behind our use of the term disciplinary literacy and offer a definition. In comparison with scientific literacy, to date there have only been a small number of people who have used the term disciplinary literacy (McConachie et al., 2006; Shanahan & Shanahan, 2012; Tang, Ho, & Putra, 2016). An extensive overview can be found in Moje (2007). Here one can already see the different directions that the term is starting to take. Following Laugksch's advice we will now explain what *we mean* when we use the term disciplinary literacy.

Gee (1991) sees language as divided into one primary and many secondary discourses. Primary discourse is the oral language learned as a child, whereas secondary discourses are other specialized communicative practices that we learn for use in other sites in society outside the home. Gee defines *literacy* as the control of these secondary discourses. Building on Gee's definition and broadening it to include semiotic resource systems other than language, Airey (2011a) claims disciplinary literacy involves learning to control a range of disciplinary communicative practices. He defines disciplinary literacy as follows:

> The ability to appropriately participate in the communicative practices of a discipline. (Airey, 2011a)

One remaining question is that of the site in society that disciplinary literacy refers to. Clearly, disciplinary literacy refers to communicative practices for the academy; however, Airey (2011a) argues that disciplinary literacy, like scientific literacy, can also involve developing communicative practices about the discipline for use in society in general. Similarly, one further potential site is the world of work, since there are, of course, a number of vocational disciplines where the majority of the communicative practices developed relate to future requirements on the job market. Thus, we argue that all disciplines potentially develop disciplinary literacy for three specific sites: society, workplace and academy (Airey, 2011a, 2013). This can be represented visually by the disciplinary literacy triangle shown in Fig. 21.1 (Airey, 2011a).

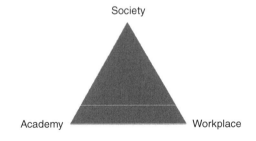

Fig. 21.1 The disciplinary literacy triangle. Disciplinary literacy involves developing communicative competence for three specific settings: Society, Workplace and Academy. The positioning of a given discipline within the triangle is dependent on the relative emphasis placed on developing communicative competence for each of the three settings Airey (2011a)

21.8 The Disciplinary Literacy Triangle

The disciplinary literacy triangle is a diagram that can be used to represent in broad terms the disciplinary literacy goals envisaged for a course or degree programme. Using the definition of disciplinary literacy, the relative emphasis placed on developing communicative practices for the three sites: society, workplace and academy can be visualized by placing a cross somewhere within the triangle.

Clearly different disciplines will have different priorities. So whilst history lecturers might place their emphasis on developing communicative practices for the academy say (i.e. less of a workplace emphasis), we might expect lecturers in vocational programmes such as nursing, to place more emphasis on communicative practices for the workplace and society (i.e. less of an academy focus). Moreover, we can also expect different specialists within the same discipline to have different priorities.

21.9 Knowledge Structures

Bernstein (1999) attributes the differences between disciplines to differences in knowledge structures. He describes two quite different knowledge structures within the academy: *hierarchical* and *horizontal*. In this division, disciplines such as the natural sciences have predominantly hierarchical knowledge structures. These disciplines develop through integration of new knowledge with knowledge that has previously been developed. In this way, disciplines with hierarchical knowledge structures manage to '[…] create very general propositions and theories, which integrate knowledge at lower levels' (Bernstein, 1999, p. 162). On the other hand, disciplines with predominantly horizontal knowledge structures (such as the humanities) develop by introducing new ways to describe the world. Crucially, these new descriptive 'languages' need not be compatible with each other. For example, a postcolonial approach to literature need not be coherent with a feminist reading of the same text, rather it is the new aspects that are brought into focus in these two approaches that are of interest. Martin (2011) compares development in hierarchical knowledge structures to a growing triangle whilst he compares development in horizontal knowledge structures to the development of new languages of description (Fig. 21.2).

Fig. 21.2 Progression in hierarchical and horizontal knowledge structures (adapted from Martin, 2011, p. 42)

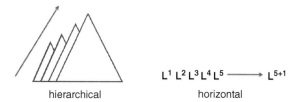

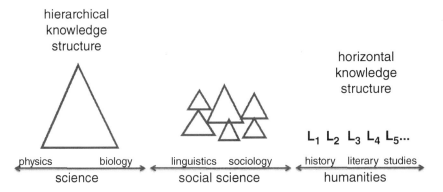

Fig. 21.3 Bernstein's knowledge structures across disciplines. Physics has a hierarchical knowledge structure, whilst disciplines such as education have more horizontal knowledge structures (adapted from Martin, 2011)

Bernstein (1999) pointed to physics as the discipline with the most hierarchical knowledge structure (Fig. 21.3).

21.10 Singulars and Regions

Bernstein (2000) also introduced the analytical categories singulars and regions. A singular is a discipline with strong boundaries such as physics, history and economics. Singulars generate strong inner commitments centred around their perceived intrinsic value. Regions are disciplines in which a number of singulars are brought together in an integrating framework. While singulars face inwards, regions face both inwards and outwards recontextualizing singulars for use in everyday life. In Bernstein's terms, education is a horizontal region, whereas physics is a hierarchical singular.

21.11 Disciplinary Literacy: A Summary

Figure 21.4 shows schematically the four main areas we have discussed thus far in this chapter. These are (proceeding anticlockwise around Fig. 21.4 from the right hand side) the parallels between disciplinary literacy and scientific literacy, the widened focus on a range of semiotic resources (rather than just written language), the definition of disciplinary literacy with the focus on three different sites, and, following Bernstein (1999, 2000), the role of the type of discipline on attitudes to disciplinary literacy.

21 Developing Students' Disciplinary Literacy? The Case of University Physics

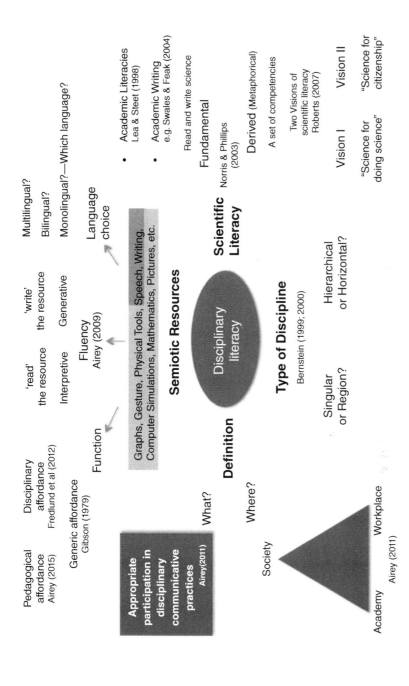

Fig. 21.4 Disciplinary literacy. Schematic diagram highlighting some of the aspects related to disciplinary literacy (adapted from Airey, 2013)

21.12 Using Disciplinary Literacy: An Empirical Study in University Physics

21.12.1 Data

The main data set used for this chapter is taken from a larger comparative research project where 30 university physics lecturers from a total of 9 universities in Sweden (4) and South Africa (5) described the disciplinary literacy goals they have for their students (Airey, 2012, 2013; Linder, Airey, Mayaba, & Webb, 2014). A disciplinary literacy discussion matrix (Airey, 2011b) was used as the basis for in-depth, semi-structured interviews (see Appendix A). All interviews were conducted in English and lasted approximately 60 minutes each. In the interviews, the lecturers were encouraged to talk about their disciplinary literacy goals for their undergraduate students, the site(s) in society that this disciplinary literacy is developed for, the semiotic resources they believe students need to learn to control and the type of control (interpretive or generative) that students need to develop. Ethical clearance for the study was not required for Sweden, but was applied for and granted in South Africa. In both countries participation in the study was voluntary with lecturers being selected on the basis of their involvement in some capacity with undergraduate physics. The anonymity of the lecturers was guaranteed and transcripts were only seen by the research team.

21.12.2 Methodology

The analysis drew on ideas from the phenomenographic research tradition by treating the interview transcripts as a single data set or 'pool of meaning' (Marton & Booth, 1997, p. 133). The aim was to understand the expressed disciplinary literacy goals of the physics lecturers interviewed.

Following the approach to qualitative data analysis advocated by Bogdan and Biklen (1992), iterative cycles were made through the data looking for patterns and key events. Each cycle resulted in loosely labeled categories that were often split up, renamed, or amalgamated in the next iteration. More background and details of the approach used can be found in Airey (2012).

21.12.3 Findings

Analysis of the 30 transcripts resulted in the identification of five themes with respect to the lecturers' disciplinary literacy goals.

1. Teaching physics is not the same thing as developing students' disciplinary literacy.
2. Disciplinary literacy in a wide range of semiotic resources is a necessary condition for learning physics.

3. Developing the necessary disciplinary literacy is not really the job of a physics teacher.
4. Some teachers were prepared to take responsibility for the development of certain aspects of students' disciplinary literacy.
5. The type of disciplinary literacy developed is focussed on the academy.

We will now present each of the themes illustrating them with quotes from the interviews.

1. Teaching physics is not the same thing as developing students' disciplinary literacy.
All the lecturers expressed a strong commitment that physics is independent of the semiotic resources used to construct it. For them, developing disciplinary literacy and teaching physics were quite separate things.

> These are tools, physics is something else. Physics is more than the sum of these tools it's the way physicists think about things—a shared reference of how to analyse things around you.
>
> Interviewer: Do you see yourself as a teacher of disciplinary Swedish for Physics?
>
> Lecturer: No, only in a very broad sense. Physics is a way of looking at nature and understanding things in simplified models. These other things are for presenting this way of thinking.
>
> Interviewer: So is your focus on scientific writing?
>
> Lecturer: No, you don't have time for that, there is content that you need to sort out.

This theme challenges contemporary thinking in education and linguistics. Halliday and Martin (1993, p. 9) for example insist that communicative practices are not some sort of passive reflection of *a priori* disciplinary knowledge, but rather are actively engaged in bringing knowledge into being. In science education, an even more radical stance has been taken by Wickman and Östman (2002), who insist that learning itself should be viewed as a form of discourse change.

Why, then, do lecturers view physics knowledge as separate from its representation? Here, we suggest that the hierarchical, integrated nature of physics knowledge leads to a belief that it will remain unaltered through processes of transduction between different semiotic resource systems. Thus, whilst Kuteeva and Airey (2014) have shown how there are strong preferences in physics for the use of one language—English—in the discipline, this is not a commitment to English *per se*, but rather a rational choice born out of a push towards standardization and the belief that physics is the same (and separate from) whichever language is used (see also discussion in Airey, 2012).

2. Disciplinary literacy in a wide range of semiotic resources is a necessary condition for learning physics.
All the lecturers in the study felt it was desirable that students develop disciplinary literacy in a range of semiotic resources in order to cope with their studies. In many ways this finding is unremarkable, with a number of researchers having commented on the wide range of semiotic systems needed for appropriate knowledge

construction and communication in physics (Airey & Linder, 2009; Lemke, 1998; McDermott, 1990; Parodi, 2012). The following quote sums up this point:

> If you want to come out with an undergraduate degree in physics you will need to be able to interpret and use graphs, tables, diagrams and mathematics for an undergraduate degree in physics and then there is also the communication part of it which is the language. We work in English and so all the communication is in English. You need to be able to read the question and understand problems you know from reading. You need to write, to be able to communicate your answers. You need to be able to listen to the lecturers, you need to be able to speak in order to verbalise what problems there are with your answers.

The lecturers essentially reported that they would prefer students to develop disciplinary literacy in *all* the semiotic resources mentioned in the disciplinary literacy discussion matrix (see Appendix A).

3. Developing the necessary disciplinary literacy is not really the job of a physics teacher.

All physics lecturers expressed frustration at the level of disciplinary literacy of their students, feeling that they really should not have to work with the development of these skills:

> As a physicist I'm not there to solve the problem of the maths. They must be able to understand the maths sufficiently at that level and know why … I'm not there to teach maths, they must go to the maths department if they need to learn it.

> I cannot say that I test them or train them in English. Of course they can always come and ask me, but I don't think that I take responsibility for training them in English. I don't correct their work in English.

Northedge (2002) holds that the role of a university lecturer should be one of a discourse guide leading 'excursions' into disciplinary discourse. However, the physics lecturers interviewed in this study did not agree with this position.

4. Some teachers were prepared to take responsibility for the development of certain aspects of students' disciplinary literacy.

Nonetheless, some physics lecturers did say that the development of students' disciplinary literacy would be something that they would work with. In these cases, lecturers (grudgingly) took on Northedge's (2002) role of a discourse guide. This position was most common for mathematics which was seen as essential for an understanding of physics (See Airey, 2012, p. 75, for further discussion of this theme).

> Interviewer: Do you then have to spend time going back over the maths?

> Lecturer: Yes, what we do most of the time maybe he needs background on these topics. Differentials—you don't take it as granted that they know. Because of time constraints I invite them in their free time, then I brush up on the maths.

> So we would explain to them how to plot a graph, heading, labels—I mean our students don't know this! […] They don't know about scales so you see we would spend a lot of time explaining to them look why do we need a graph, why do we do this, so we would explain why we want them to do it in a particular way and then it takes a lot of practice and exercise to get better.

Fig. 21.5 The disciplinary literacy triangle for physics. The lecturers in the study situated their disciplinary literacy goals firmly within the academy

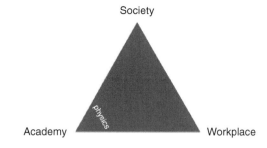

> The thing is that to be able to express it in a precise enough way you need mathematics. Language is more limited than mathematics in this case. So they need to use mathematics to see physics rather than language in this case.

5. The type of disciplinary literacy developed is focussed on the academy

The lecturers all report that they direct their teaching towards the academy, placing physics disciplinary literacy in the bottom left-hand corner of the disciplinary literacy triangle (Fig. 21.5).

> Interviewer: We are interested in how you decide on the learning goals for your students ….
>
> Lecturer: Physics has been around for a long time, you know it changes very slowly, anyway I would say that that is much given by the next level what you need to, to go on in physics
>
> What I teach, society doesn't really need to know—it would be nice if society knew and understood … but you don't have to know it.

21.13 Disciplinary Literacy Across Disciplines

The fundamental starting point for the conceptualization of disciplinary literacy presented in this chapter is that different disciplines emphasize the development of quite different communicative practices. Drawing on Bernstein (1999, 2000), Airey and Larsson (2014) suggest that it is differences in disciplinary knowledge structures that lead to these strikingly different disciplinary literacy goals. This could cause problems for inter-disciplinary work (see also argument in Airey, 2011b).

As we have argued, in physics, meaning is taken as agreed and unchanging across contexts (see discussion in Airey, 2012). This in turn leads to a commitment that physics itself is independent of the semiotic resources used to construct it. However, this argument is more difficult to uphold in the case of more horizontal knowledge structures where Bernstein (1999) suggests that development is actually driven by the creation of new 'languages' to describe the world around us. In the humanities meaning is contested and inextricably bound up with language. In extreme cases it has even been argued that meanings made in one language might be impossible to appropriately construe in another (Bennett, 2010).

In what follows, we present anecdotal evidence of this clash of disciplinary literacy goals using data from a study of physics teacher training (Larsson & Airey,

2014, 2015). First, the movement from the discipline of physics into teacher training entails a radical change in communicative practices:

> For me it was a shock to be thrown into an institution where you have to write essay-type exam questions. The students who had read History, Swedish, Social science, they passed these exams without any problems. For me the first time it was like ok, how do I do this? (School physics teacher)

Airey and Larsson (2014) show how different ideas about what counts as knowledge in the disciplines of physics and education have the potential to cause problems for trainee physics teachers. Students who are steeped in the epistemological commitments of a coherent, hierarchical, positivist, physics knowledge structure may experience the contingent nature of educational science as disjointed, incoherent and unscientific.

> These new values that they've included in the curriculum now—they don't seem so natural to me. There are competencies that I'm supposed to develop that can't be measured—it's silly! The whole thing falls like a house of cards because you just can't measure these things. (School physics teacher)

Here we see how a commitment to coherence and measurability (values of hierarchical physics) leads to the rejection of other forms of knowledge. This problem is compounded by the attitude of physics lecturers who insist that trainee physics teachers need the same experience as those reading for a physics degree:

> Interviewer: Do you have different goals for physicists and engineers?
>
> Lecturer: Yes, I suppose … but only slightly different.
>
> Interviewer: And for the teachers is it the same?
>
> Lecturer: Yes, I don't really distinguish between them. You need to understand physics to be able to teach it.

This quote also illustrates the inwardness of Bernstein's singulars (such as physics). This can be contrasted with the recontextualizing agenda of regions (such as education) where singulars are 'repackaged' for use in the society and workplace (Fig. 21.6).

In Fig. 21.7 we bring together Bernstein's two concepts of knowledge structure and disciplinary classification in one diagram. Singulars such as physics and history can have different knowledge structures, the same can be said of regions such as engineering and education. Here we see that that physics (hierarchical singular) and education (horizontal region) are diametrically opposed within Bernstein's two systems. This

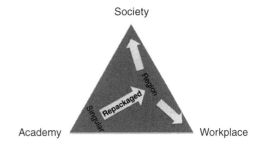

Fig. 21.6 Disciplinary categorization in the disciplinary literacy triangle. Singulars face inwards, developing disciplinary literacy for the academy, whilst regions face both inwards and outwards, recontextualizing singulars for use in society and the world of work

Fig. 21.7 Categorization of disciplines using Bernstein's (1999, 2000) constructs. Bernstein categorized physics as the singular with the most hierarchical knowledge structure of all. In the same diagram, education is categorized in a radically different manner—as a horizontal region (adapted from Airey & Larsson, 2014)

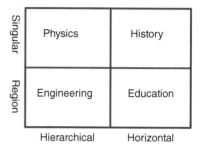

surely has consequences for the types of communicative practices that are valued in the two disciplines and the ease with which students can move between the two.

21.14 Conclusions

In this chapter we discussed the concept of disciplinary literacy and applied it to the goals of university physics lecturers. Lecturers reported that disciplinary literacy in a wide range of semiotic resources is a necessary condition for physics learning. However, the same lecturers do not view the development of this disciplinary literacy as their job. Although some lecturers were prepared to help students develop disciplinary literacy, all the lecturers interviewed believed that teaching physics is something that is separate from teaching disciplinary literacy. Here, Airey argues that:

> Until lecturers see their role as one of socialising students into the discourse of their discipline … [they] will continue to insist that they are not [teachers of disciplinary literacy] and that this should be a job for someone else. (Airey, 2011b, p. 50)

In the final section of this chapter we tentatively addressed the issue of disciplinary literacy across disciplines. Here, we suggest that the differences between the disciplinary literacy goals of physics and teacher training are an inevitable consequence of the differences between the two disciplines in terms of Bernstein's disciplinary classifications. Regions such as education always have to reformulate disciplinary literacy goals. Trainee physics teachers come from a singular with a strong disciplinary identity. This identity needs to be renegotiated into a teacher identity. As such we believe that teacher trainers should anticipate these issues and discuss them with their trainees. In particular, we suggest that some trainee teachers from disciplines with hierarchical knowledge structures may struggle to see the validity of other types of knowledge.

Our intention in this chapter has been to examine the disciplinary literacy goals of university physics lecturers. As such the concluding discussion of disciplinary literacy across disciplines has necessarily been tentative in nature. Going forward it would be interesting to apply the concept of disciplinary literacy to the analysis of other disciplines—particularly those with different combinations of Bernstein's classifications of hierarchical/horizontal and singular/region.

Appendix A

Disciplinary Literacy Discussion Matrix

This matrix contains some of the many representations used in physics (down the left hand side). Given the overloaded nature of many physics degrees, please tick the boxes that you think should be emphasized for students to master **during an undergraduate physics degree course**.
Please do this with respect to **where** students need this physics representational skill (for physics, for the workplace or for participation in society in general).

		Where for?		
		Physics	Job	Society
Graphs	interpret			
	use			
Tables	interpret			
	use			
Diagrams	interpret			
	use			
Mathematics	interpret			
	use			
→ _____	interpret			
	use			
→ _____	interpret			
	use			
→ _____	interpret			
	use			
Language				
_____	Reading			
	Writing			
	Listening			
	Speaking			
_____	Reading			
	Writing			
	Listening			
	Speaking			
_____	Reading			
	Writing			
	Listening			
	Speaking			

References

Airey, J. (2009). Science, language and literacy. Case studies of learning in Swedish University Physics. Acta Universitatis Upsaliensis. Uppsala Dissertations from the Faculty of Science and Technology, 81, Uppsala. Retrieved from http://publications.uu.se/theses/abstract.xsql?dbid=9547Accessed 27 April 2017.

Airey, J. (2011a). The disciplinary literacy discussion matrix: A heuristic tool for initiating collaboration in higher education. *Across the disciplines, 8*(3) Unpaginated.

Airey, J. (2011b). Initiating collaboration in higher education: Disciplinary literacy and the scholarship of teaching and learning. In *Dynamic content and language collaboration in higher education: Theory, research, and reflections* (pp. 57–65). Cape Town: Cape Peninsula University of Technology.

Airey, J. (2012). "I don't teach language." The linguistic attitudes of physics lecturers in Sweden. *AILA Review, 25*, 64–79.

Airey, J. (2013). Disciplinary literacy. In E. Lundqvist, L. Östman, & R. Säljö (Eds.), Scientific literacy–teori och praktik (pp. 41–58). Malmo: Gleerups.

Airey, J. (2015). Social semiotics in higher education: Examples from teaching and learning in undergraduate physics. In: SACF Singapore-Sweden excellence seminars, Swedish Foundation for International Cooperation in Research in Higher Education (STINT), 2015 (p. 103). Retrieved from urn:nbn:se:uu:diva-266049.

Airey, J., & Larsson, J. (2014). What knowledge do trainee physics teachers need to learn? Differences in the views of training staff. International Science Education Conference ISEC 2014, National Institute of Education, Singapore, 25–27 November 2014.

Airey, J., Lauridsen, K., Raisanen, A., Salö, L., & Schwach, V. (2017). The expansion of English medium instruction in the Nordic Countries. Can top-down university language policies encourage bottom-up disciplinary literacy goals? *Higher Education. 73*(4), 561–576. https://doi.org/10.1007/s10734-015-9950-2.

Airey, J., & Linder, C. (2008). Bilingual scientific literacy? The use of English in Swedish university science programmes. *Nordic Journal of English Studies, 7*(3), 145–161.

Airey, J., & Linder, C. (2009). A disciplinary discourse perspective on university science learning: Achieving fluency in a critical constellation of modes. Journal of Research in Science Teaching, 46(1), 27–49.

Airey, J., & Linder, C. (2011). Bilingual scientific literacy. In C. Linder, L. Östman, D. Roberts, P.-O. Wickman, G. Ericksen & A. MacKinnon (Eds.), *Exploring the landscape of scientific literacy* (pp. 106–124). London: Routledge.

Airey, J. & Linder, C. (2017) Social semiotics in university physics education. In D. Treagust, R. Duit & H. Fischer (Eds.), *Multiple representations in physics education*. (pp.95–122). Cham: Springer.

American Association of Physics Teachers (1996). Physics at the crossroads. Retrieved from http://www.aapt.org/Events/crossroads.cfm

Becher, T., & Trowler, P. (1989). *Academic tribes and territories*. Milton Keynes: Open University Press.

Bennett, K. (2010). Academic discourse in Portugal: A whole different ballgame? *Journal of English for Academic Purposes, 9*(1), 21–32.

Bernstein, B. (1999). Vertical and horizontal discourse: An essay. *British Journal of Sociology Education, 20*(2), 157–173.

Bernstein, B. (2000). *Pedagogy, symbolic control and identity: Theory, research and critique*. Lanham: Rowman and Littlefield.

Björk, L., & Räisänen, C. A. (2003). *Academic writing: A university writing course* (3rd ed.). Lund: Studentlitteratur.

Bogdan, R. C., & Biklen, S. R. (1992). *Qualitative research for education: An introduction to theory and methods*. Boston: Allyn and Bacon, Inc.

CHE-SAIP. (2013). *Review of undergraduate physics education in public higher education institutions*. Retrieved from http://www.saip.org.za/images/stories/documents/documents/Undergrad_Physics_Report_Final.pdf.

Duff, P. (2010, March). Language socialization into academic discourse communities. *Annual Review of Applied Linguistics, 30*, 169–192.

European Commission Expert Group (2007). *Science education now: A renewed pedagogy for the future of Europe*. Brussels: European Commission.

Forsman, J. (2015). Complexity theory and physics education research: The case of student retention in physics and related degree programmes. Digital comprehensive summaries of Uppsala dissertations from the Faculty of Science and Technology. Uppsala: Acta Universitatis Upsaliensis. Retrieved from http://www.diva-portal.org/smash/record.jsf?pid=diva2%3A846064&dswid=-4668.

Fortanet-Gomez, I. (2013). *CLIL in higher education: Towards a multilingual language policy*. Bristol: Multilingual Matters.

Fredlund, T., Airey, J., & Linder, C. (2012). Exploring the role of physics representations: An illustrative example from students sharing knowledge about refraction. *European Journal of Physics, 33*, 657–666.

Fredlund, T., Airey, J., & Linder, C. (2015). Enhancing the possibilities for learning: Variation of disciplinary-relevant aspects in physics representations. *European Journal of Physics, 36*(5), 055001.

Fredlund, T., Linder, C., Airey, J., & Linder, A. (2014). Unpacking physics representations: Towards an appreciation of disciplinary affordance. *Physical Review Special Topics Physics Education Research, 10*, 020128.

Gee, J. P. (1991). What is literacy? In C. Mitchell & K. Weiler (Eds.), *Rewriting literacy: Culture and the discourse of the other* (pp. 3–11). New York: Bergin & Garvey.

Gibson, J. J. (1979). *The theory of affordances. The ecological approach to visual perception* (pp. 127–143). Boston: Houghton Miffin.

Halliday, M. A. K., & Martin, J. R. (1993). *Writing science: Literacy and discursive power*. London: The Falmer Press.

Hurd, P. D. H. (1958). Science literacy: Its meaning for American schools. *Educational Leadership, 16*, 13–16.

Ivanič, R. (1998). *Writing and identity: The discoursal construction of identity in academic writing*. Amsterdam: John Benjamins.

Johannsen, B. F. (2013). *Attrition and retention in university physics: A longitudinal qualitative study of the interaction between first year students and the study of physics*. Doctoral dissertation, University of Copenhagen, Faculty of Science, Department of Science Education.

Josephson, O. (2005). Parallellspråkighet [Parallel language use]. *Språkvård, 2005*(1), 3.

Kress, G., Jewitt, C., Ogborn, J., & Tsatsarelis, C. (2001). *Multimodal teaching and learning: The rhetorics of the science classroom*. London: Continuum.

Kuteeva, M., & Airey, J. (2014). Disciplinary differences in the use of English in higher education: Reflections on recent policy developments. *Higher Education 67*(5), 533–549.

Larsson, J., & Airey, J. (2014). Searching for stories: The training environment as a constituting factor in the professional identity work of future physics teachers. *British Educational Research Association Conference BERA 2014*, London, September 2014.

Larsson, J., & Airey, J. (2015). The "physics expert" discourse model – counterproductive for trainee physics teachers' professional identity building? Paper presented at the 11th Conference of the European Science Education Research Association (ESERA), Helsinki, August 31 to September 4, 2015.

Laugksch, R. C. (2000). Scientific literacy: A conceptual overview. *Science Education, 84*, 71–94.

Lea, M. R., & Street, B.V. (1998). Student writing in higher education: An academic literacies approach. *Studies in Higher Education, 23*(2), 157–172.

Lemke, J. L. (1998). Teaching all the languages of science: Words, symbols, images, and actions. Retrieved from http://academic.brooklyn.cuny.edu/education/jlemke/papers/barcelon.htm. Accessed 16 September 2005.

Lillis, T., & Scott, M. (2007). Defining academic literacies research: Issues of epistemology, ideology and strategy. *Journal of Applied Linguistics, 4*(4), 5–32.

Linder, A., Airey, J., Mayaba, N., & Webb, P. (2014). Fostering disciplinary literacy? South African physics lecturers' educational responses to their students' lack of representational competence. African Journal of Research in Mathematics, *Science and Technology Education, 18*(3), 242–252. https://doi.org/10.1080/10288457.2014.953294.

Martin, J. R. (2011). Bridging troubled waters: Interdisciplinarity and what makes it stick. In F. Christie & K. Maton (Eds.), *Disciplinarity* (pp. 35–61). London: Continuum International Publishing.

Marton, F., & Booth, S. (1997). *Learning and awareness*. Mahwah, NJ: Lawrence Erlbaum Associates.

Moje, E. B. (2007, March). Developing socially just subject-matter instruction: A review of the literature on disciplinary literacy teaching. *Review of Research in Education, 31*, 1–44.

McConachie, S., Hall, M., Resnick, L., Ravi, A. K., Bill, V. L., Bintz, J., & Taylor, J. A. (2006). Task, text, and talk: Literacy for all subjects. *Educational Leadership, 64*(2).

McDermott, L. (1990). A view from physics. In M. Gardner, J. G. Greeno, F. Reif, A. H. Schoenfeld, A. A. diSessa, & E. Stage (Eds.), *Toward a scientific practice of science education* (pp. 3–30). Hillsdale: Lawrence Erlbaum Associates.

National Research Council. (2013). *Adapting to a changing world — challenges and opportunities in undergraduate physics education*. Committee on undergraduate physics education research and implementation. Board on physics and astronomy division on engineering and physical sciences. Washington, DC: National Academies Press.

Nordic Educational Research Association. (2009). Literacy as worldmaking. *Congress of the Nordic Educational Research Association*. Retrieved from http://www.neracongress2009.com.

Norris, S. P., & Phillips, L. M. (2003). How literacy in its fundamental sense is central to scientific literacy. *Science Education, 87*(2), 224–240.

Northedge, A. (2002). Organizing excursions into specialist discourse communities: A sociocultural account of university teaching. In G. Wells & G. Claxton (Eds.), Learning for life in the 21st century. *Sociocultural perspectives on the future of education* (pp. 252–264). Oxford: Blackwell Publishers.

Parodi, G. (2012) University genres and multisemiotic features: Accessing specialized knowledge through disciplinarity. *Fórum Linguístico, 9*(4), 259–282.

Phillipson, R. (2006). English, a cuckoo in the European higher education nest of languages. *European Journal of English Studies, 10*(1), 13–32.

Roberts, D. (2007). Scientific literacy/science literacy: Threats and opportunities. In S. K. Abell & N. G. Lederman (Eds.), *Handbook of research on science education* (pp. 729–780). Mahwah: Lawrence Erlbaum Associates.

Seymour, E., & Hewitt, N. (1997). *Talking about leaving: Why undergraduates leave the sciences*. Boulder: Westview Press.

Shanahan, T., & Shanahan, C. (2012). What is disciplinary literacy and why does it matter? *Topics in Language Disorders, 32*(1), 7–18.

Swales, J. (1990). *Genre analysis: English in academic and research settings*. Cambridge: Cambridge University Press.

Swales, J., & Feak, C. (2004). *Academic writing for graduate students: Essential tasks and skills*. Ann Arbor: University of Michigan Press.

Tang, K. S. K., Ho, C., & Putra, G. B. S. (2016). Developing multimodal communication competencies: A case of disciplinary literacy focus in Singapore. In *Using multimodal representations to support learning in the science classroom* (pp. 135–158). Springer International Publishing.

UNESCO (2004). *The plurality of literacy and its implications for policies and programmes.* Paris: UNESCO.

UNESCO (1958). *Recommendations concerning the international standardization of educational statistics.* Paris, UNESCO.

Wickman, P.-O., & Östman, L. (2002). Learning as discourse change: A sociocultural mechanism. *Science Education, 86*(5), 601–623.

Part 7
Commentary

Chapter 22
Commentary on the Expanding Development of Literacy Research in Science Education

Larry D. Yore

Abstract This commentary situates, summarizes, and critiques the global attempts, as documented in this book, to address the complex language/literacy-science education problem space involving curriculum, integrated learning, classroom practices, challenges, instruction embedded in an inquiry-oriented context, and teacher education and development issues focused on the fundamental sense of science literacy. Few science education policies or curriculum documents recognize a contemporary view of learning science or specify that language is a critical component of science literacy and that instructional attention must be afforded science language, scientific metalanguage, and other fundamental abilities and strategies as part of inquiry-oriented programs. Many countries rely on the language arts or literacy curricula to justify disciplinary literacy in science education. The infusion of literacy goals into science programs requires the reallocation of effort and time, which many educators and teachers will view as impeding the content objectives emphasized in most science curricula, teaching, and assessment. This infusion will necessitate the development of a robust operational definition of science literacy amongst the language/literacy and science education communities that respects the epistemic and ontological nature of science, the development, verification, and implementation of innovative science literacy opportunities in argument-based, multiple information resources and technology-rich science instruction, and new professional learning approaches for language and science teachers in primary, middle, secondary, and postsecondary institutions. Furthermore, the language/literacy and science education research communities may wish to consider secondary analyses of existing research results in order to set the agenda and designs for future research.

Keywords 5-E inquiry cycle · argument-based inquiry · classroom literacy practices · epistemic · communicative and rhetorical functions of language · global perspectives · integrated · interactive and dynamic framework of science

L.D. Yore (✉)
Distinguished Professor Emeritus, University of Victoria, Victoria, BC, Canada
e-mail: lyore@uvic.ca

literacy—derived · fundamental and applied components · just-in-time LLSE instruction and tasks · Just-in-time professional development and ongoing support · LLSE communities/researchers · meta-analyses and metasyntheses · models of learning science and reading/writing · multimodal representations · networks of diverse new and experienced researchers · science and engineering practices · science language (L3) challenges · science literacy pedagogical content knowledge (PCK) · science writing heuristic · the 3-language problem · the three μετά- (metas): metatalk · metacognition and metalanguage · theory-practice gap

22.1 Introduction

This three-part commentary will provide a scan of the language, literacy, and science education (LLSE) landscape; a brief summary and critique of the parts of the book to highlight the results reported and relationships amongst science curriculum reforms, content and language integration, classroom literacy practices, disciplinary literacy and science inquiry, and teacher development; and foci and approaches for future research and development efforts in LLSE. This commentary is from the perspective of an experienced science educator; therefore, it will not capture all the nuances of the authors and may run contrary to perspectives and research preferences in the language and literacy education communities and some in science education. However, it will try to highlight ideas, concerns, and approaches for further considerations and to provoke deliberations.

22.2 Landscape of Language, Literacy, and Science Education

There have been a series of science education reforms since the 1960s that have emphasized various learning theories, goals, and outcomes, teaching approaches, and assessment techniques. These reforms were frequently based on influences, desires, and opinions (e.g., political, international competition, economic development, globalization) that originated outside of education communities, did not consider the pervasive challenges and implementation barriers within the societies, educational systems, and classrooms, and lacked compelling evidence of their achievability and effectiveness. Many of the current international reforms are also lacking informed input from the collective LLSE communities and evidentiary support for the advocated curriculum, teaching, and assessment recommendations. Inquiry-oriented teaching and learning—the goal of most international science education reforms—is still questioned by many classroom teachers, and the evidence for its effectiveness has only gradually been amassed with secondary analyses of research results (Minner, Levy, & Century, 2010). A similar situation has occurred for the current science literacy and the language and literacy in science efforts. Many publications have reported fragmented and isolated research results on the science literacy problem space with very few long-term research agendas that approximated the *gold standard* design of randomized large-scale control and experiment groups. These big agendas require big money and interdisciplinary teams enacting increasingly more rigorous and robust designs as evidence is amassed and the research agenda matures. Such projects are not available to many LLSE researchers. Therefore, these research communities must seek innovative solutions to inadequate funding using novel designs and analysis methods and developing international networks of interdisciplinary researchers focused on this problem space. This book illustrates the potential for developing such global perspectives and networks of diverse new and experienced researchers interested in LLSE.

How people learn science is the essential foundation for curriculum, teaching, and assessment reforms. A brief overview of the related literature reveals various models of learning science and reading/writing, isolated explicit reading and writing instruction independent of science learning experiences, and science textbooks that were encyclopedias of knowledge with readability generally higher than assigned grade use as well as little attention to coordination of print and visual adjuncts (Yore & Tippett, 2014). The pre-1960 *read first, do later* instructional practices reflected the stimulus-response-reinforcement approach of the behaviorist view of learning, where activities, if they occurred, became verifications of what was read, and reading strategies, if provided, were generally bottom-up (i.e., skill and drill) with little attention to top-down (i.e., prior knowledge and literacy of the reader) or interactive-constructive approaches involving concurrent experiences, prior knowledge, and information resources. The 1960s science education reforms were founded on the rejection of reading science textbooks, which was shortened to texts and then generalized to all text and language activities other than listening and speaking in favor of hands-on experiences by many science educators. These inquiry programs did not fully reflect how scientists actually use other information resources and language modes to construct, argue, and communicate their ideas.

Today, contemporary interactive-constructivist views of learning generally assume that learners, young or old, make meaning from concurrent experiences and information and stored knowledge and experiences in working memory within a sociocultural context using public negotiations and private processes. Language—especially written and other learner-generated forms—is an essential resource in learning science; furthermore, language—the placenta for a culture—reflects cultural beliefs, values, and traditions. Therefore, science literacy instruction focused on any population, especially minority and indigenous, needs to consider beliefs and values inherent in their language and their views of knowledge and wisdom about nature and naturally occurring events inherent in their culture.

Science literacy is an old construct, circa 1958, but it does not have a widely shared definition within the LLSE communities. There appear to be three isolated definitions in common use: knowledge about science, reading and writing in science, or participation in the public debate about science-related issues. Some researchers use an integrated, interactive, and dynamic framework of all three views involving a derived component (e.g., knowledge about the science, the nature of science, and scientific enterprise), a fundamental component (e.g., cognitive/metacognitive abilities, scientific dispositions/habits of mind, processes/practices, critical thinking, constructive-interpretative language arts—speaking/listening, writing/reading, representing/interpreting, and scientific metalanguage—enterprise language), and the application of these components in the literate citizens' public debate about science, technology, society, and environment (STSE) or socioscientific issues (SSI). This book and the authors focus mostly on the second view dealing with disciplinary literacy (i.e., fundamental sense) embedded in science learning environments and its manifestations in curriculum, classroom teaching and assessment practices, and teacher education/professional development.

22.3 Summary and Critique of the Contributions

The interesting contributions in this book represent a rekindling of international-interdisciplinary LLSE research and development (R&D) scholarship as many of the current issues are as much about technology and engineering (design/mission-driven innovations) as pure scientific research (curiosity-driven inquiry). Several contributions addressed the theory-practice gap, while others help redefine the problem space with contemporary considerations of second-generation science education reforms and contemporary school and social contexts globally. This book is a start on establishing a global collaborative network of researchers and setting a research agenda in LLSE.

22.3.1 *Part 1: Curriculum Issues*

Curricula, specific types of education policy, are products of political processes and policymakers (i.e., educators, scholars, public stakeholders, politicians) involved in complex negotiating and lobbying; such policies are influenced by a variety of inputs and persuasion from groups with different degrees of power and influence. Thereafter, educators and teachers spend much time interpreting and enacting curricula without fully understanding how such policies were developed, the sociopolitical factors that influenced their production, and the subsurface intentions. The new USA framework was the product of the National Research Council (NRC, 2012) composed of scientists, university faculty members, and members of society. Elizabeth Moje, P. David Pearson, and I were invited to address an NRC Steering Committee hearing. We lobbied for a clearer definition of science literacy that was composed of dynamic, interacting, fundamental, derived, and applied components that would embrace the communicative, epistemic, and rhetorical functions of language in doing and learning science. Unfortunately, our effort had less than the desired impact on the new framework. Although science literacy is not explicitly mentioned, some features advocated can be implied in the science and engineering practices: "#2—Developing and using models [representations], …

#7—Engaging in argument from evidence [rhetorical function], ... #8—Obtaining, evaluating, and communicating information [communicative function]" (NRC, 2012, p. 42).

Several other international science curricula have evolved foundational assumptions, expanded their goals, and recognized instructional approaches over the earlier views of learning and limitations of science teaching as solely inquiry-oriented. The USA's framework has assumed learning and teaching involves interactive-constructivist views, science is as much about argumentation as inquiry and engineering is about design not applied science, and broadens the goals—core ideas, science and engineering practices, and crosscutting concepts (NRC, 2012). The core ideas listed are idiosyncratic to the composition of the curriculum committee and development process and would likely vary if these factors were changed slightly. The interdisciplinary crosscutting concepts and the science and engineering practices provide a potential context for anchoring and justifying the fundamental and applied components of science literacy; that is, the cognitive, social, and physical activities that scientists and engineers do to investigate, evaluate (argue, critique, analyze), and develop explanations, solutions, and innovations. The specificity of the science and engineering practices and crosscutting concepts can be debated, but they are meant to stress the commonalities across the life, earth-space, physical, and engineering sciences and provide foundation for developing interdisciplinary programs like science, technology, engineering, and mathematics (STEM). Unfortunately, the science and engineering practices do not fully reflect the epistemic, communicative, and rhetorical functions of language in making, arguing, and reporting meaning and understanding. This limited view of language and literacy in science curricula requires advocates, as illustrated by several authors in this book, to rely on their prescribed language arts curricula to justify a fuller range of language and literacy strategies and to persuade teachers of science about their inclusion.

The curricula in Australia, Norway, and Singapore, which were influenced by the recognition of the ever-increasing diversity in schools and the value of disciplinary literacy, are similar to the emphases in other parts of the world. They recognize the importance of science literacy and how it differs from traditional definitions of literacy and the belief that science literacy is connected to improved achievement. A hierarchical framework for disciplinary literacy abilities (Shanahan & Shanahan, 2008) that progresses from the basic level (applicable to most disciplines) to intermediate (applicable to some disciplines) and advanced levels (applicable to specific disciplines) underpins these studies.

Knain and Ødegaard reported on the Budding Researchers project evolving from the general curriculum reform across the disciplines that infused literacy and opportunistic instruction (as needed) within science inquiries and projects. Norwegian classroom and lead teachers collaborated to develop and evaluate embedded writing, reading, and speaking in their science teaching that considered science processes, basic literacy skills, and scientific metalanguage. The authors believed that permanency of printed language and representations allow the reflections necessary to analyze data and to generate and check evidence-based claims, theory-based explanations, and cause-effect mechanisms. Davison and Ollerhead reported similar results in Australia where science teachers implemented English as a second language and addressed concerns about declining achievement of low-SES students. These students' underdeveloped language ability limited both their participation in inquiry-based settings and group discussions and their understanding of evidence-based reasoning and argumentation. The teachers believed that the overcrowded curriculum and their inability to develop and infuse authentic science literacy opportunities and practices in their classroom teaching were barriers to implementing literacy-rich SSI and problem-based learning. Ho, Rappa, and Tang reported on professional learning programs involving design-based research (lesson studies) situated in Singapore's attempt to implement the Whole School Approach to Effective Communication in English mandate. The science teachers benefited from just-in-time professional development and ongoing support from school, university, and ministry of education resource people. One exciting design study involving four teachers who planned and enacted science lessons using a premise—reasoning—outcome approach focused on teaching students how to construct scientific explanation (see Tang, 2016, for ontological attributes of scientific explanations).

These three chapters revealed the lack of a clear, concise, and shared definition of science literacy and confirmed the reluctance and limited awareness of teachers of science, especially science specialists, to provide embedded language and literacy support or explicit instruction within the context of science inquiry and projects by diverting instructional time and effort away from traditional content outcomes. The struggle to convince teachers of science would be much easier if the authorized science curricula specifically identified fundamental literacy and application components and the parts, abilities, and strategies in these components as prescribed learning outcomes.

The research designs used reflect the current R&D into the policy, curriculum implementation, and professional learning problem space—but they have limited generalization and strength of claims. LLSE researchers would be well served by ensuring that future case, participatory action, and lesson design studies have common data sources and interpretative frameworks to allow for meta-analyses (quantitative results) and meta-syntheses (qualitative results) from and across a number of small-scale studies. Furthermore, these contributions illustrate the need for policy and curriculum research to inform LLSE researchers about how they could more fully and effectively participate in these endeavors and become influential change agents in promoting science literacy in science curricula.

22.3.2 *Part 2: Content and Language Integrated Learning*

All students are science or other disciplinary language learners (L3); language conventions and traditions are essential parts of a discipline. The ever-increasing diversity in classrooms worldwide has highlighted globalization, political unrest, and dislocation of peoples and the related needs of students in schools where their home language/native tongue (L1) does not align with the language of instruction (L2) or the target disciplinary language (L3). Visits to schools in Vancouver, Minneapolis, Stockholm, Berlin, Melbourne, London, Capetown, and other urban areas will document students speaking numerous nondominant/official languages. Many multilingualism and multiculturalism students frequently demonstrate lower achievement than dominant-language-speaking students.

Clearly, these multilingual sociocultural contexts complicate the three-language problem, which only involves variations of nonstandard and standard forms of a single language. However, the diversity represents richness of values, beliefs, and experiences rather than a deficit in the construction of understanding. The difficulty is not the richness of resources but rather how to access, engage, and coordinate these memories and concurrent experiences on the cognitive workbench. The following contributions report on some content and language integration efforts that addressed science literacy in diverse settings.

Markic reported on participatory action research involving science and German-as-a-second-language teachers planning and developing materials. She found success in helping linguistically disadvantaged Grades 5–8 Turkish- and Arabic-speaking students in science lessons using small group (2–3 students) and individual methods that included intercultural understandings to engage the diverse resources that these students bring to the learning environment. The science teachers asserted that the second-language students needed less support in understanding and writing when using these materials, while the students reported they were more motivated and willing to share their writing products. Lo, Lin, and Cheung used lesson studies in Hong Kong to document the rationale and collaborative approach of science teachers, English language teachers, and university faculty to provide scaffolding and help reasonably proficient English-as-a-foreign language students (age 13–15). They developed, used, and evaluated integrated genre-based content and language lessons focused on writing sequential explanatory texts (e.g., science talk, terminology, representations, words, and phrases) that described and explained the target phenomena to bridge the three languages. Their results suggested that the materials and scaffolding enhanced students' science literacy. Msimanga and Erduran explored South Africa's diverse multilingual classrooms where the language and science problem is compounded by the facts that students learn English as their third or fourth language and their teachers are not proficient in English. However, it was informally reported elsewhere that many parents support the use of English during instruction as they see it as necessary for future work or education. Lesson transcripts illustrate how the participating teachers' metatalk focused on the conceptual content as a discursive tool but did not address the language of science and its demands and functions. Wu, Mensah, and Tang conducted case studies in New York and Singapore secondary schools focused on English language learners (ELL). These ELL populations have different socioeconomic and sociocultural backgrounds and motives; that is, immigrants from the Dominican Republic seeking to complete high school certificates and students in a private school seeking entrance to English-speaking universities. The New York case study revealed that L1 can be used for learning scientific content but is seen by some students as a hindrance to their acquisition of English. The Singapore study of teacher-directed dissemination of knowledge and procedures provided few opportunities for student–teacher or student–student interactions, but the students did use their L1 in laboratory and small-group settings. Analysis revealed that comfort with using the English language was a significant predictor of students' science achievement.

These small-scale studies focused on content and language integration where students are learning a majority language and language of instruction (L2) at the same time as they learn the language of science (L3). Such studies are needed to better understand the complexity and competing factors involved in learning and teaching environments with students of diverse linguistic and cultural backgrounds. Applying an interactive-constructive view of learning and teaching suggests the problem is not lack of cognitive resources required to construct meaningful understandings, but rather that they are stored in students' L1, which differs from the desired dominant language and the access and retrieval of these stored resources from long-term memory may require the use of students' home language.

Pragmatics as well as theoretical considerations need to be addressed regarding the theory-practice gap in literacy-science teaching focused on science literacy. English is often referred to as the lingua franca for international science, but findings from an English language context may not be fully applicable to other language and science literacy spaces (e.g., Mandarin, Swedish). Integration of language and literacy into science learning and teaching is a very challenging task for specialist teachers who lack insights into the complexities of the language system or the nature of science. Science literacy has received increased attention in recent years, but language and literacy educators appear to concentrate on the *fundamental* component while science educators appear to concentrate on the *derived* component. Furthermore, LLSE researchers do not always address the functions of language completely with many concentrating on communications and less so on meaning making (epistemic function) and argumentation (rhetorical function).

Fundamental science literacy instruction, the focus of this book, needs to be opportunistic by capitalizing on authentic science learning environments that require just-in-time instruction and tasks. The contextual fabric of the inquiry will avoid the so-called transfer problems encountered by much of the language instruction outside of science classrooms. Opportunistic literacy instruction requires convinced, confident, and proficient science teachers or teacher teams that can grasp available opportunities and provide metatalk (i.e., talking about the discourse being used or targeted rather than simply talking about the concepts being explored) about the language or literacy strategies. Integrated LLSE instruction needs to recognize that students are not deficient in background, but rather they bring a rich array of resources for making sense of the natural world—These ideas may be encoded and stored in long-term memory using native languages that are different from the language of instruction. Few teachers will be proficient in these native languages; therefore, instructional strategies will need to be developed to use the collective language abilities of the class and low-demand visual tasks (e.g., student drawings or other representations) to help individual students access, engage, and use these cognitive resources in their meaning making.

22.3.3 Part 3: Classroom Literacy Practices

Classroom literacy instruction can involve a variety of tasks, activities, and interventions focused on enhancing students' communicative, epistemic, and rhetorical strategies. Strategies are assumed to be clusters of commensurate operations, moves, and skills that can be substituted for one another within the cluster and used to accomplish the same function or outcome. Unlike skill development based on rote memorization and drill and practice, effective strategies instruction should involve mindful choice between alternatives or informed selection amongst options. An example of this perspective applied to science literacy, such as data interpretation to reveal empirical relationships (evidence-based claims), might involve critical thinking about, data manipulation, and representations of observations, data tables, numerical calculations, diagrams, graphs, flow charts, other data displays. Each of these options could partially illustrate potential patterns between the dependent and independent variables. The decision to use one, or a combination, of these options will depend on other factors—audience, type of data, presentation media, available technologies, and resources, etc. The following contributions illustrate literacy instruction and methodologies in various countries, sciences, and classroom settings.

Wilson and Jesson used case studies of New Zealand science teachers (2 each in Grades 7, 9, and 11) to document the enactment of the national curriculum on subject-specific literacy. Interpretations of classroom observations, teacher interviews, and measures of subject literacy pedagogical content knowledge (PCK) used to document the teachers' beliefs, knowledge, and practices indicated that the teachers were using traditional teacher-directed approaches with supplemental vocabulary definitions. The authors believed that there was a need to expand the learning and assessment beyond content outcomes to include reading, writing, and critical literacy. Cavalcanti Neto, Amaral, and Mortimer investigated the role of discursive interactions in three multilingual Brazilian Grades 6 and 7 classrooms. Results revealed differences in the teachers' use of language and literacy: one used an initiate-response-evaluate method, one used an interactive-dialogic to access and partially use students' ideas, and one used an interactive-dialogic method to engage environmental issues. The authors believed that science literacy is often limited to an authoritative reading and writing of scientific texts and could be made more dialogic by including student-generated language, texts, and representations, discussion, and argumentation. Jakobson, Danielsson, Axelsson, and Uddling investigated Swedish multilingual Grade 5 students' interactions and meaning making. Results revealed that the teacher and students engaged in meaning-making activities involving a variety of semiotic resources (e.g., representations, speech, gestures, writing) to develop science literacy. However, some classroom practices (e.g., stress of exactness, meticulousness, writing forms) appeared to hinder meaning making. He and Forey examined a Grade 9 science classroom's meaning making with various resources and their affordances (e.g., language, gestures, animation) as part of an Australian professional development project. Analysis of a video recording revealed that gestures and animation provided temporal and spatial meaning while language mediated knowledge and established conceptual links and organization.

These four interesting contributions generally lacked a shared working definition of science literacy; and the participating teachers focused on content knowledge with little consideration of learning the epistemic and rhetorical functions of language and the ontological requirements of science. However, these chapters provided foundations for defining and specifying contextual language in science demands/actions. The listing of teacher actions and strategies, the multimodal resources involved in meaning making, and the measurement of science literacy PCK were important contributions that could be useful to LLSE researchers.

Explicit instruction about strategies (i.e., clusters of commensurate operations, moves, and skills) should involve the three μετά- (metas): metatalk, metacognition, and metalanguage. Metatalk involves talking about the target concept, which is reasonably common in conceptual change teaching where learners need to be convinced to give up or modify their existing conception for a more compelling, robust alternative concept and to link the new concept to established ideas and practices. However, literacy instruction also needs to involve metatalk about the literacy strategies that considers the metacognitive awareness (declarative knowledge—what, procedural knowledge—how, and conditional knowledge—why and when) and metacognitive self-management or executive control (planning, monitoring, and regulating) of the specific strategy and other strategies in the commensurate cluster. Furthermore, literacy instruction needs to consider scientific metalanguage (enterprise terms) associated with the nature of science (evidence supports not proves as in mathematics, relationships amongst theory, model, hypothesis, prediction, inference, and observation, etc.).

22.3.4 Part 4: Disciplinary Literacy Challenges

Science language (L3) incorporates terms from everyday, academic, and other discipline-specific languages and enterprise terminologies and unique symbolic, visual, genre (form/function), and style features that makes it challenging for many producers and users of scientific oral and print texts. It is not uncommon that highly proficient academic English students find the move into comprehending and producing scientific English language and text problematic and variable across different science disciplines with their dense terminology and heavy reliance on symbolic representations and mathematical features. These problems are multiplied for ELL or other official language learners with nonstandard home and minority everyday native languages. Students without some prior informal or formal understanding of the target ideas and experiences with the oral or print science text are expected to face challenges of lack of prior conceptual and experiential resources. The following studies explored some of these challenges for proficient and nonproficient dominant language speakers as they navigate amongst their home, instruction, and scientific languages and texts.

Liu examined the language and symbols in introductory secondary school chemistry textbooks used in Singapore. The functional analysis of selected textbooks illustrated that chemical formulas and equations involve several numerical and symbolic conventions to denote chemical structures/compositions and the mechanics of reactants and products in chemical reactions. These well-established conventions challenge many novice and nonexpert readers of chemistry. Danielsson, Löfgren, and Pettersson examined the use of metaphors in a Swedish and two Finnish-Swedish secondary chemistry classrooms. Analysis of video recordings of classroom interactions indicated that the teachers used a variety of scientific, everyday, and anthropomorphic metaphors as foundations for the properties of the atom. However, the native language (Swedish rather than English) made a difference in whether a concept label might be metaphorical in nature. Ge, Unsworth, Wang, and Chang explored the design of visual adjuncts on reading comprehension and understanding of print-visual texts in Taiwan. This clever two-group quasi-experimental study examined the effects of image design on reading comprehension and meaning making involving visual and verbal text using a five-phase interview (i.e., image only, addition of caption, addition of text, text with synonymous image, selection, rationale of most appropriate image) to partition the reading comprehension of 12 Grade 7 students in different textual conditions; a comparison group read the text with textbook images, and a treatment group read the same texts but with a tree-structure image. Results suggested that the textbook image did not activate as many themes as the tree-structure representation, but surprisingly the influence of prior knowledge was negligible.

The results from this part illustrate the need to consider language's sociocultural context, the visual and print resources involved, and linguistic features to be considered. Much LLSE research has been done in English-language settings. However, one needs to be cautious about generalizing these results to other languages and settings because of linguistic differences. These studies reveal that sociocultural beliefs/values, traditions, and conventions are embedded in the language. The systemic functional linguistics and social semiotics interpretative frameworks used in these contributions provide a sound basis for considering other sciences and topics as well as language modes or resources (Liu's explanation of semiotics appears to be useful in physics as well as chemistry topics).

22.3.5 Part 5: Disciplinary Literacy and Science Inquiry

The collaboration and integration of the language and literacy community (disciplinary literacy) and the science education community (science as inquiry, engineering as design, evidence-based argument, and science and engineering practices) is the central focus of the next three studies. These lesson studies implicitly assumed that literacy in the science classroom should reflect what scientists do, support students in learning the concepts and practices of science, and enhance their application to the public debate about STSE or SSI problems leading to sustainable evidence-based solutions. These assumptions closely approximate a contemporary definition of science literacy composed of fundamental, derived, and applied components; they appear to use a constructive-interpretive view of the language arts (speaking-listening, writing-reading, representing-interpreting) where students generate oral, print, and visual texts as epistemic, rhetorical, and communicative tools in learning about, persuading others, and applying science.

Ødegaard explored how six elementary teachers implemented the Budding Scientist program as part of the Norwegian emphasis on disciplinary literacy. This program embedded students' use of multiple sources of evidence (primary hands-on experiences and secondary experiences: text-based inquiries, external information sources, representational tasks, etc.) to construct understanding in argument-based inquiry. Analysis of classroom video recordings, observations, and interviews revealed multiple learning modalities (read-it, write-it, do-it, talk-it adapted from the *Seeds of Science/Roots of Reading* program) distributed across different phases of inquiry (preparation, data, discussion, communication). The teachers' greatest challenge was to find the time and courage for consolidating conceptual learning in the discussion and communication phases. Students expressed concerns that literacy and the role of text in science were not clear. Tang and Putra explored the implementation of Singapore's subject-specific literacy mandate using design studies where four secondary school chemistry and physics teachers developed, enacted, and tested integrated literacy and science lessons. The instruction-infused literacy strategies were designed to support students in constructing scientific explanations using the 5E Inquiry Cycle (engage, explore, explain, elaborate, evaluate). Interpretation of classroom activities and interactions illustrated the literacy activities and support of scientific practices and suggested specific ways of reading science, translating information amongst or between various formats, and writing and evaluating explanations. Tytler, Prain, and Hubber explored students' construction and use of multimodal representations in Australia. They engaged urban junior secondary school teachers and students in collaborative lesson studies about the rock cycle. Analyses of lesson plans, classroom videos, instructional artefacts, and teacher–student interactions revealed partially how to address the theory-practice gap and challenges within authentic/meaningful science inquiry. The locus of control during the professional learning project was transferred to teachers as they gained self-confidence and took increasing leadership in planning and enacting the guided-inquiry approach (student-generated representation, experimental or alternative sources of evidence, discussion and evaluation of representation, assessment of learning).

These contributions implicitly endorse an interactive, dynamic relationship amongst the three senses (fundamental, derived, and applied) of science literacy; their design and results demonstrate how enhancement of fundamental literacy strategies helped improve content understandings and promote participation in the public debate about science-related issues. The opportunistic infusion of science literacy strategies into authentic inquiry learning and use of multiple information sources place increased demands on teachers and an expanded need for science literacy PCK not available to many preservice and practicing science teachers from their previous professional education. The studies started to outline the demands as well as the planning and classroom practices needed to address integrated science literacy and science learning. Participating teachers developed their science literacy PCK *in situ* as they planned, enacted, and evaluated science lessons on a variety of topics and grade levels with multiple resources and language modes using the collaborative support and mentoring of peers and experts. Each instructional approach avoided transfer problems by infusing the literacy strategies into actual science learning environments. These studies may lack generalizability, but they indicate the need for teacher education and professional development involving ongoing support, mentoring, and cascading leadership that is not common in many programs. Furthermore, they indicate the need for science curricula to explicitly identify fundamental and applied components of science literacy along with the commonly identified derived understanding component. Without this endorsement in the authorized science curricula, it is much more difficult to convince science teachers of their fundamental and applied science literacy responsibilities.

22.3.6 Part 6: Teacher Development

This part naturally flows from earlier parts of this book by exploring issues and tensions faced in preparing science teachers to integrate disciplinary literacy into their teaching and the ongoing difficulty in teacher education related to changing the effects of teachers' previous experiences in school and university science classes. Many university students selecting science education as a teaching area have been successful in their prior science courses, which were frequently taught with teacher- or professor-directed lectures, verificational laboratory work, and knowledge-focused assessments. Students see little need to change such personally effective methods (the *It's not broke; why fix it?* perspective); therefore, they adopt these well-engrained instructional methods. Language in these approaches assumes a communication function used to disseminate knowledge, evaluate understanding, and manage behavior. Contemporary language- and literacy-oriented science instruction is different because it assumes epistemic and rhetorical functions for language as well as the communicative function. The three contributions outline efforts to expose, convince, and empower preservice and practicing teachers and university lecturers of these functions and related tasks and strategies.

Espinet, Valdés-Sanchez, and Hernández illustrated how the three-language problem can become more complicated in places like Catalonia, Spain, where there are at least three common public languages as part of belonging to the European Union and regional aspirations for nation status. This context makes learning the language of science even more complex than in many countries and likely places it at a lower priority than where English is spoken at home and is the basis for learning scientific English in school. They examined 39 primary school preservice teachers' beliefs and expectations about the Content and Language Integrated Learning approach. Analysis of the participants' science and language narratives revealed that their science experiences were more related to negative school contexts, whereas their language experiences were connected to a variety of positive out-of-school contexts. The implications for teacher education are related to how to connect these formal and informal experiences and to establish the value and utility of language, science, and science education. Hand, Park, and Suh tracked changes in 28 middle school teachers' epistemic orientations and pedagogical practices as they experienced and implemented the Science Writing Heuristic (SWH) approach during a three-year immersion argument-based inquiry professional development project in the USA. Analysis of teachers' epistemic orientation and students' critical thinking revealed that teachers started to view science as argument and language as an epistemic tool and that improved implementation of the SWH approach led to enhanced critical thinking by the students. The authors suggested that professional development is not a quick fix, teachers need to be aware that language is essential for learning of science, and science cannot be done without language, especially written language. Airey and Larsson explored the disciplinary literacy goals related to university, workplace, and society of 30 undergraduate physics lecturers in Sweden and South Africa. These differences pose significant challenges for preservice physics teachers who have to navigate across the disciplines of physics (hierarchy structure) and education (horizontal structure). Analysis of semi-structured interviews revealed the lecturers had similar disciplinary literacy goals for their students and very different ideas about their responsibility to teach literacy and the use of the semiotic resources. Results indicated that the lecturers moved toward a broadened view of science literacy that includes using cognitive resources and various information sources in different contexts, but was still limited to the communicative function, neglecting the rhetoric and epistemic functions.

The professional education and learning of science teachers to incorporate science literacy into their beliefs and values, instructional goals, and PCK cannot be achieved by lecture or increased time in traditional coursework. It requires coordinated efforts across university departments and the teaching profession with authentic learning experiences involving planning, classroom engagement, and reflection-on/reflection-in action. Contemporary views of science literacy are a major departure from the traditional expectations and experiences of preservice and practicing teachers of science. Many science teacher education programs involve several departments in the science faculty and the general education, language and literacy, and science education departments of the education faculty—these two faculties' views about science literacy are frequently not aligned. Many science courses stress and reward content mastery, while general, language/literacy, and science education courses do not provide consistent views about goals, teaching, and assessment across the integrated components of science literacy for citizenship. Therefore, many teachers leave their initial education with rather poorly organized and justified traditional beliefs, values, and practices about effective science teaching and assessment. This claim can be verified by visits to early-career and experienced science teachers' classrooms where teacher-directed lectures are the most common teaching approach to be found. Professional development takes time. A long-term conceptual change approach to teacher education and professional learning with ongoing clinical experiences and mentoring is needed to achieve the goal of teachers facilitating student-directed learning with a variety of experiences and resources.

22.4 Closing Remarks

It has been both pleasant and educational reading—consolidating my reactions and commenting on this book that gives a global perspective and overview to the complexity of the language/literacy and science problems space and that considered curriculum, content and language integration, classroom practices, disciplinary literacy within science inquiry, and professional learning components. My closing remarks recognize the pragmatics of an edited book endeavoring to achieve these worthwhile goals and different preferences about research design within and across the LLSE research communities. These remarks are not intended to be viewed as negative or to reignite the *paradigm wars*. Rather, they are provided to reinforce a few ideas and to move the concerned communities toward shared deliberations, insights, and consensus about their commonalities and differences, and relationships amongst science literacy, science understanding, and participatory citizenship. These collaborative efforts should provide a basis on which to (a) develop more useful operational definitions, compelling arguments, and empirical claims, (b) explicitly recognize the limitations of hastily drawn global assertions/claims and recommendations, and (c) outline potential actions and research addressing integrated language, literacy and science curricula, learning, teaching, and assessment.

22.4.1 Science Literacy

A consensus operational definition of science literacy is lacking in LLSE literature. Science literacy was originally defined as knowledgeable in science (derived sense) and later revised to include a language component (fundamental sense) and recently evolved to include an application component (citizen participation sense). The derived sense, which reflects authorized curricula, can include knowledge about the nature of science, big ideas such as core ideas and crosscutting concepts, and science, technology, and social interactions. The fundamental sense can include cognitive/metacognitive abilities, emotional dispositions/habits of mind, attitudes, science and engineering practices/processes, critical thinking, and scientific language (speaking-listening, writing-reading, representing-interpreting, enterprise terms). The application sense can involve the fundamental and derived senses required of an informed, active citizen in the consideration of public science, technology, and environment-related issues to make informed decisions and produce sustainable solutions.

The definition of science literacy continues to evolve toward the interacting perspective, as no component should or can standalone. The NRC (2016) report has provided an expanded view of science literacy that goes beyond the individual to include the community/society; it "identified four additional aspects of science literacy that, while less common, provide some insight into how the term has been used: foundational literacy, epistemic knowledge, identifying and judging scientific expertise, and dispositions and habits of mind" (p. 5). Science literacy and its subsumed components should consider the criticality needed in our rich, diverse, and un-reviewed information communication technology age that can be elaborated and repositioned to address an elite version focused on STEM careers and expertise by increasing the specificity and proficiency levels.

The expressed intention of this book was to focus on the fundamental sense (disciplinary literacy), but just about every contribution considered the fundamental sense in conjunction with the derived or applied senses. However, several authors do not consistently recognize the functions of language in doing and learning science—communication, construction, and argumentation of/about knowledge—nor the nature of science involving unique epistemic features and ontological requirements.

22.4.2 Explicit Views of Science Learning

Researchers and research reports about science literacy need to specify their assumed view of learning, which will influence beliefs, values, and practices about teaching for science literacy and the interpretation of data and results. Taking a behavioral view would lead to assumptions that science literacy is a collection of language skills applied to science that could be achieved by a drill-and-practice approach. Taking an interactive-constructive view involves learners making meaning from a combination of prior knowledge, experiences, and beliefs and concurrent sensory experiences and information sources within a sociocultural context and defined content area with public and private negotiations (see Hand et al., in this book). The interactive-constructivist view moves science literacy instruction toward strategic clusters and interacting abilities, the three μετά (metatalk, metacognition, metalanguage), and group and individual negotiations using multiple modes of language in constructing and representing understanding.

22.4.3 Science Education Policy and Curricula

Policy and curriculum development do not always consider the realities of schools, classrooms, students, and teachers fully. Sometimes the most powerful members of a development group can unknowingly move the policy and curriculum toward unachievable ends. The 1960s process versus product dilemma and inquiry-oriented teaching are illustrations of such ends brought about by well-meaning scientists and philosophers.

The current science reforms and curricula continue to emphasize science as inquiry but have added engineering as design, science and engineering practices, and implicitly recognized the importance and some functions of language (communications and argumentation) as epistemic tools in doing and learning science. However, they stress approaches without fully considering the problematic features and challenges for teachers. Many generalist and specialist teachers with limited science knowledge, PCK, and experience working in challenging linguistically diverse environments are unable to successfully implement the outcomes and teaching methods. The barriers—lack of background, support, equipment, resources, preparation, and instructional time; large class sizes; overcrowded curricula—overwhelm these teachers.

22.4.4 Theory-Practice Gap in Science Literacy

Collectively, the chapters in this book have provided partial evidence for several literacy strategies and identify the need to address the complex and potential interactions amongst educational policy, curriculum, science literacy instruction, and teacher education and profession development within argument-based inquiry environments. This is important as analysis of teacher magazine articles on classroom practice involving language and literacy activities embedded or associated with the science education program revealed that most of the recommended practices, regardless of their efficacy, do not have sufficient research foundation and, therefore, do not qualify as evidence-based practices (Jagger & Yore, 2012).

22.4.5 Teacher Education and Professional Learning

Teacher education and professional learning must address the difficulty of changing teachers' established beliefs and practices—many of which go back to their experiences as an elementary, middle, or secondary school student or their postsecondary science courses. Clearly, initial teacher education and continuing professional development cannot be viewed as quick fixes. One contribution in this book used a PCK measure for science literacy that holds promise for further efforts. Furthermore, policy scholars need to explore the internal politics within curriculum development and teacher education programs to determine the factors influencing program, recruitment, and enrolment management efforts. Based on my experience, science and disciplinary literacy educators hold minority positions with little power to influence these decisions.

22.4.6 Building More Compelling Research Claims

This edited collection has illustrated the potential influences and differences among native/home languages, cultures, societal and environmental contexts on the use and interpretation of language in doing and learning science. Any generalization to other non-English languages, schools, and societies based on English language settings and results must be questioned based on the different linguistic structures of these languages and the classroom settings, cultural values, and beliefs associated with Anglo communities. The results of some chapters have demonstrated how sociocultural and linguistic contexts change the classification and interpretation of data—verb-based compared to object-based iconic languages like Mandarin, nucleus as metaphor in Swedish, etc.

More literacy-science research using all types of designs is needed, but it may be time to encourage convergence of existing results before striking out on divergent R&D agendas. How can we naturally integrate language, literacy, and science into argument-based inquiry, design, and science and engineering practices? Case, participatory action, lesson, and design studies and quasi-experimental studies have been useful in surveying the problem space, defining driving questions, and illustrating unique and potentially powerful teaching/learning approaches, data collection, and data analysis techniques. But the need for (a) inclusive definitions of science literacy, (b) understanding the distinctive nature of science, and (c) models of science learning that respects the ontological requirements, epistemic practices, and metalanguage of science overrides doing more of what has been done without these definitions.

The integrated language, literacy, psychology, measurement, and science communities must form multidisciplinary, cognitive science, and multi-methodological research networks to achieve fiscal efficiencies and address the more complex issues in the language-science learning, teaching, and assessment problem space because of the multiple information sources and communication technologies available. The international nature of the author teams and research environments in this book illustrates the potential influence of home/native language on the demands and requirements of doing and learning science in different cultural, social, and environmental contexts. A first step would be to conduct meta-syntheses and meta-analyses of existing interpretative and quantitative results to establish a firmer foundation and landscape of the language-science problem space and compelling evidence-based practices (Rossman & Yore, 2009). The history (1999–) of the SWH approach based on the authors' opinions, numerous related qualitative and quantitative studies, and the meta-analysis and meta-synthesis of these results illustrated how the theory-practice gap was addressed and how an evidence-based science literacy practice was established (Jagger & Yore, 2012). Finally, there are multiple needs for policy research and action—participatory action research that clarifies the basis for curriculum decisions and teacher education program revisions involving science literacy education.

References

Jagger, S., & Yore, L. D. (2012). Mind the gap: Looking for evidence-based practice in science literacy for all in science teaching journals. *Journal of Science Teacher Education, 23*(6), 559–577.

Minner, D. D., Levy, A. J., & Century, J. (2010). Inquiry-base science instruction—what is it and does it matter? Results from a research synthesis years 1984 to 2002. *Journal of Research in Science Teaching, 47*(4), 474–496.

National Research Council. (2012). *A framework for K-12 science education: Practices, cross-cutting concepts, and core ideas*. Washington: National Academy of Sciences.

National Research Council. (2016). *Science literacy: Concepts, contexts, and consequences*. Washington: National Academy of Sciences.

Rossman, G. B., & Yore, L. D. (2009). Stitching the pieces together to reveal the generalized patterns: Systematic research reviews, secondary reanalyses, case-to-case comparison, and metasyntheses of qualitative research studies. In M. C. Shelley II, L. D. Yore, & B. Hand (Eds.), *Quality research in literacy and science education: International perspectives and gold standards* (pp. 575–601). Dordrecht: Springer.

Shanahan, T., & Shanahan, C. (2008). Teaching disciplinary literacy to adolescents: Rethinking content-area literacy. *Harvard Educational Review, 78*(1), 40–59.

Tang, K-S. (2016). Constructing scientific explanations through premise–reasoning–outcome (PRO): An exploratory study to scaffold students in structuring written explanations. *International Journal of Science Education, 38*(9), 1415–1440. https://doi.org/10.1080/09500693.2016.1192309

Yore, L. D., & Tippett, C. D. (2014). Reading and learning science. In R. Gunstone (Ed.), *Encyclopaedia of science education* (pp. 821–828). Dordrecht: Springer. https://doi.org/10.1007/978-94-007-6165-0_130-2

Index

A
Academic literacies in science, 90, 91
Adolescent literacy, 135
Affordances, 180, 184, 188, 197, 201, 232, 233, 304, 360, 387
Analogies, 220, 221, 226, 227, 230, 232, 233
Argument-based inquiry, 340, 341, 342, 343, 344, 351, 354, 390, 392, 395, 396

B
Basic skills, 15–27, 262

C
Chemistry teaching, 222
Classroom discourse, 3, 7, 99, 100, 135, 153, 165, 269, 293
Classroom literacy practices, 5, 380, 387–388
Classroom practices, 54, 220, 272, 310, 314, 387, 390, 393
Communicative and rhetorical functions of language, 383
Comparative education, 358
Concepts, 3, 47, 48, 87, 92, 103, 105, 106, 107, 109, 145, 150, 190, 220, 221, 276, 292, 294, 304, 340, 353, 389
Conceptual understanding, 20, 30, 39, 92, 109, 116, 120, 140, 206, 269, 271, 272, 273, 283, 284, 287, 341, 342, 348, 351
Constructing explanations, 189, 283, 287, 292–294
Construction and critique, 341, 343
Content and language integrated learning (CLIL), 5, 6, 68, 80, 322–323, 384–386, 391

Content area literacy, 2, 3
Content teaching and learning, 45
Content-based language instruction, 115
Content-language tension, 113–128
Continuity, 55, 107, 169, 177, 180, 326
Curriculum, 19, 27, 34, 45, 52, 93, 262, 381, 382, 383, 384, 387, 394, 395
Curriculum reform, 5, 16–19, 26, 27, 281, 380, 383

D
Disciplinary literacy, 4, 5, 8, 9, 29–40, 45–58, 134, 135, 143, 144, 189, 221, 284, 287, 302, 310–314, 357–372, 383, 388–391, 392

E
English language, 48, 84, 98, 100, 118, 325, 334, 386, 388, 389, 395
English language learners (ELLs), 6, 31, 33, 81, 82, 97–109, 113–128, 385
English learners, 114
English medium education (EMI), 83, 84
English second language learners (ESLs), 98, 101, 108
Environmental issues, 150, 151, 152, 158, 159, 160, 161, 162, 163, 165, 387
Epistemic, 4, 292, 302, 382, 383, 387, 388, 390, 391
Epistemic orientation, 339–354, 392

F
5E inquiry cycle, 275, 390

G

Genre-based pedagogy, 80, 82–83, 85, 89, 91, 92, 93, 94
German language, 66, 68, 71, 72, 75
Global perspectives, 381
Grammar, 34, 100, 208, 240

I

Image design, 237–256, 389
Inquiry, 8, 20, 23, 261, 262, 267, 268, 273, 281, 282, 287, 297, 383, 389
Inquiry-based science education, 267
Inquiry-based science teaching (IBST), 23, 26, 273
Integrated curriculum, 94
Integrated genre-based content and language lessons, 385
Integrated inquiry and literacy approach, 269, 282, 298
Interactive and dynamic framework of science, 381
Intercultural understanding, 63–75, 385

K

Knowledge structures, 363–364, 369, 370, 371

L

Language, 5, 55, 64, 80, 81, 97–109, 127, 128, 176, 186, 189, 190, 191, 193–197, 198, 214, 220, 321–335, 380–382, 387, 392
Language across the curriculum, 84, 93, 94
Language experience narratives, 8, 321–335
Language use in science, 6, 344
Literacy, 20, 32, 97–109, 144, 262–264, 272, 285, 287, 361, 380–382, 384, 392, 393, 396
Literacy challenges, 7, 30, 205–216
Literacy practices, 2, 7, 33, 46, 54, 264, 272, 273, 274, 276
Lower secondary school, 25

M

Meaning, 206, 208, 211, 215, 216
Meaning-making, 3, 32, 33, 98, 99, 103, 145, 167–180, 183–201, 206, 209, 215, 238, 240, 249, 254, 256, 303, 387
Meta talk, 6, 97–109
Metaphors, 7, 219–234, 389
Multilingual students, 66, 168, 179
Multilingualism, 66, 322, 385
Multimodal, 4, 172, 184, 206–207, 216, 284, 311, 388, 390
Multimodal discourse analysis, 183–201
Multimodality, 3, 27, 85, 167, 168, 216, 360–361
Multiple representations, 239, 240, 304, 342

N

National curriculum, 2, 5, 16, 33, 150, 262, 314, 387
New Zealand, 2, 7, 133–146, 387

P

Participatory action research, 6, 68–70, 385
Pedagogy, 18, 51, 54, 83, 99, 102, 201, 216, 345
Preservice teacher education, 6, 9, 30, 33, 34, 35, 40

R

Reading comprehension, 7, 137, 238, 239, 241, 254, 256, 389
Representation, 140, 207, 220, 240, 255, 256, 269, 301–315, 328, 367, 390

S

Science, 30, 31, 35, 46, 65, 98, 134, 149–165, 168, 183, 196, 221, 225, 262, 264, 265, 272, 330, 332, 383, 384
Science communication, 101, 185, 188, 394
Science education, 1–9, 18, 27, 30, 31, 32, 54, 64, 65, 69, 101, 102, 114, 150, 151, 168, 184, 185, 187, 197, 216, 221, 222, 262, 265, 267, 282, 303, 323, 335, 340–341, 354, 367, 379–396
Science experience narratives, 326
Science inquiry, 8, 25, 49, 262, 263, 264, 269, 276, 282, 283–284, 302, 380, 384, 389–391, 393
Science Writing Heuristic (SWH), 8, 339–354, 392
Scientific explanation, 24, 49, 56, 196, 282, 284–286, 292, 293, 313, 384
Scientific literacy, 4, 5, 7, 15–27, 30, 31, 38, 48, 49, 64, 73, 114, 128, 134, 150, 151, 152, 163, 165, 168, 179, 185, 205, 233, 262–264, 304, 339–354, 360, 361, 362
Secondary science, 113–128, 141, 183–201, 222, 345

Semiotic affordances, 184, 185, 188, 189, 190, 197, 201
Sheltered instruction (SI), 115
Singulars versus regions, 364
Symbolic formulas, 205–216
Systemic functional linguistics (SFL), 3, 32, 169, 170, 186, 219, 224, 285, 389

T
Teacher development, 5, 324, 380, 391–392
Teacher education, 6, 9, 30, 33–35, 38, 101, 109, 323, 326, 329, 335, 391, 392, 395
Teacher observations, 137
Teaching strategies, 98, 106, 145, 149–165, 266, 287
Translanguaging, 114–115, 124
Tree structure, 7, 237–256, 389

U
Undergraduate physics, 9, 358, 360, 366, 392

V
Video study, 267, 268

W
Writing in the disciplines, 18, 27